Scalable VoIP Mobility

Scalable VoIP Mobility

Integration and Deployment

Joseph Epstein

AMSTERDAM • BOSTON • HEIDELBERG • LONDON
NEW YORK • OXFORD • PARIS • SAN DIEGO
SAN FRANCISCO • SINGAPORE • SYDNEY • TOKYO
Newnes is an imprint of Elsevier

Newnes is an imprint of Elsevier
30 Corporate Drive, Suite 400, Burlington, MA 01803, USA
Linacre House, Jordan Hill, Oxford OX2 8DP, UK

Library of Congress Cataloging-in-Publication Data
Application Submitted

British Library Cataloguing-in-Publication Data
A catalogue record for this book is available from the British Library.

ISBN: 978-1-85617-508-1

For information on all Newnes publications visit our Web site at www.elsevierdirect.com

Transferred to Digital Printing, 2011

Printed in The United States of America

To my little e.b., Noah Vinh, and his mother, my loving wife Regina

Contents

CHAPTER 1

Introduction to Voice Mobility

1.1 Why Voice Mobility?

The term *voice mobility* can mean a number of different things to different people. Two words that can be quite trendy by themselves, but stuck together as if forgotten at a bus station long past the last ride of the night, the phrase rings a number of different, and at times discordant, bells. Mobility is something that can be quite useful, but it is not always easy to justify spending large sums of money just for a network that can provide this feature. The lack of interest in *mobile IP* is just one example. Desktops don't seem to need mobility, and laptops can use whatever IP address they get. And although voice—especially voice over IP—has become the cornerstone for a number of large enterprise voice networks, as the days of analog or telephone-specific digital phone lines has past, it is not used for every application, at every time. Sometimes typing is just more efficient than talking. There seems to be a bit of *Web 2.0* in the concept, perhaps the setting up of lofty expectations that are impossible to reach, based on recurring themes that were once passé but have now returned in a slightly different, more confusing form.

However, one thing is for certain: voice mobility does have the ability to blend together the most obvious applications of voice and place the result directly into the hands of the enterprise. The driving force behind voice mobility is that voice *immobility* never quite made sense, with wires and heavy, strapped-in handsets that were necessary only because voice had not been easily sent without the copper cabling. The cellphone proved this true in a massive way, and landline phones are quickly withering on the vine, being replaced by these multipurpose, flashy, and quite useful devices that seem to do so much more now than a telephone ever did. This way of thinking, seeded into every person's mind—that voice belongs in your pocket and not on your desk or in some small closet—demands that the enterprise must provide voice mobility as the primary, if not the only, means for communicating with the spoken word.

doi:10.1016/B978-1-85617-508-1.00001-3.

Everyone has a cellphone, and so many want to expense it to the company, spending many hours a week doing nothing but work on the devices. This can be costly for organizations that need to avoid wasteful spending, and oftentimes the mobility need not extend quite as far as the other side of the world—perhaps the other side of the building will do. Or perhaps not, and the mobile workforce does really need to access their corporate voicemail, and email, and maybe even paper mail, at 3:00 a.m., and the people who need this access are not going to tolerate any plan that requires them to learn fancy software or new VPNs or magic remote solutions. That phone in their pocket, the one with a dozen or so buttons, or maybe a little keypad, might be the only thing they will use.

What finally made voice mobility a subject that could save real dollars was the maturation of the unlicensed enterprise mobile network, based on Wi-Fi. Now, every enterprise has or may shortly have a complete radio network within the office, providing access to everything that was once wired. The enormity of Wi-Fi helped in another way, causing phone manufacturers to place a Wi-Fi radio in their more modern, enterprise-oriented devices, so that people can use the one device at home or at hot-spots. It did not take long before people figured out how to combine the two functions—the voice network for the mobile operator and the data network for the enterprise—and allow one radio or both to do all of the tasks.

This whole concept is given the label *convergence*, quite rightly, as it describes how different market and technology trends have converged onto the same point, and into the same device. Convergence cannot solve every problem, but it can and has made the phone the most important device that many people carry. Even President Obama had to experience the shock of being asked, at first, to let go of his voice mobility device, his trusted BlackBerry that led him to victory over a worthy but less technologically savvy opponent. (Thankfully, his BlackBerry was replaced with one that had more advanced security.)

This book tells the story of voice mobility in a modern setting.

1.2 Audience and Expected Background

My hope is that this book can appeal to all technology-savvy audiences. If you have an interest in voice mobility, and are willing to explore it from a number of different angles, then this book is written for you. Admittedly, this book is a bit thin on case studies and examples of motivating implementations, mostly because of the difficulty in exposing the details of the voice mobility networks for organizations who have made the jump, as they consider their network—and the competitive advantages they gain—to be a company secret. However, the intuition behind every concept this book explains should be readily apparent. In fact, this intuition is the most important aspect of this book. There are times where concepts may be explored in excruciating detail, where packet-by-packet examples are shown and terms that make more sense to the people who built the technology behind the network than those who built the network itself appear. However, this is done only when the concepts within that example may be needed at a critical moment in running a real voice mobility network. For example, the entire login and connection process for a mobile device on a secure Wi-Fi network is shown, step by step, to give the appreciation for how many different parts are involved, and for why the process itself makes voice quality a challenge and network management a complicated thing. If it also lets you diagnose problems over the air, by comparing the steps with what you are seeing, then that is even better.

At other times, the paragraphs will become chatty, staying at a high level and exposing the reason behind why the network does what it does, and the consequences—both positive and negative—for each decision. This approach may not immediately be useful when setting up a voice mobility network, but it can give just the right "a-ha" moment for understanding how voice mobility comes together.

The details on specific implementations are not provided, and vendor names are mentioned only when necessary for recognition. This book is not a how-to guide, or a configuration or application manual, and I highly recommend the books and papers provided by the vendors of interest for learning the details, when those vendors have been chosen and the time for implementation arrives. This book is a concept book, a study in the whys and wherefores of voice mobility networking. After reading this book, you should have a good foundation to tuck into those instruction manuals, with this book providing the concepts that the manuals provide only slowly or among the instructions that compose the majority of the words between their covers.

1.3 How to Read This Book (Chapter Layout)

The first four chapters make up one unit, and establish how voice flows, what influences voice quality, and how voice is carried over IP-based networks. The next two chapters make up a second unit, and discuss how voice runs over the unlicensed, enterprise-owned Wi-Fi network. These chapters make up the majority of the book, as the subject of voice over Wi-Fi is where the most complexities that an organization will see and have the most influence in and responsibilities for occur. The remaining chapters make up the last unit, covering cellular voice mobility, security for networking in general, and a look to the future.

Throughout the book, the names of standards or specifications will be provided, be they from the ITU, IETF, IEEE, or some other initialism-based organization (I'll explain these abbreviations within the text). Feel free to explore the details of these standards. They are available on line, from *http://itu.int*, *http://ietf.org*, and *http://standards.ieee.org/getieee802*, respectively. (The final one has a longer URL because the IEEE is an academic and professional society first, and happens to include standards-writing as a part of their activities.) Digging through these references can be a chore, however, and is seldom necessary for understanding the technology. For this reason, along with the fact that the information contained within it is not truly verified, Wikipedia is also not always a good source.

CHAPTER 2

Voice Mobility Technologies

2.0 Introduction

This chapter dives into the technology behind voice mobility. We will look into what makes the phone work, peeling back the network and the features within it, and looking at how voice calls are made and ended, and how voice is carried over the network.

A voice call will be described as being made of not only the *bearer* channel, or the part of the voice call that actually carries the sounds, but also the *signaling* channel, which sets up the call by carrying the phone number, describes whether the other end is ringing, busy, or not found, and figures out when either side of the call hangs up. After that introduction, the voice-specific architecture is explored, followed by the various protocols that make voice work inside the enterprise.

2.1 The Anatomy of a Voice Call

Placing a call may seem very simple to us modern users of voice mobility. Remove the phone from your pocket, purse, automobile cup holder, or wherever you may keep your phone, find the name of the person you wish to call, and press the Send or Yes button to dial the caller. However, underneath that simple experience lies a vast wealth of technology, with electronic gears turning to produce a sound that not only sounds like a human, but is actually recognizable as the person on the other end of the phone, even if you are driving at 65 miles an hour down the interstate.

The voice call is made of a number of moving parts. Figure 2.1 illustrates the basic example.

Each handset contains all of the technology necessary to place a phone call. The phone must have a microphone to capture the voice call and a speaker to play out the audio from the other party. But beyond that, the phone must have a way of converting the analog voice information into encoded digital signals, using audio *codecs* with possible compression to ensure that high-fidelity voice quality can be carried over lower-bandwidth links. The phone must also have one or more wireless radios, complete with technology stacks and engines that allow the phone to connect to each of the networks, or even to hand off between the networks. These stacks must fully understand and adequately implement the necessary

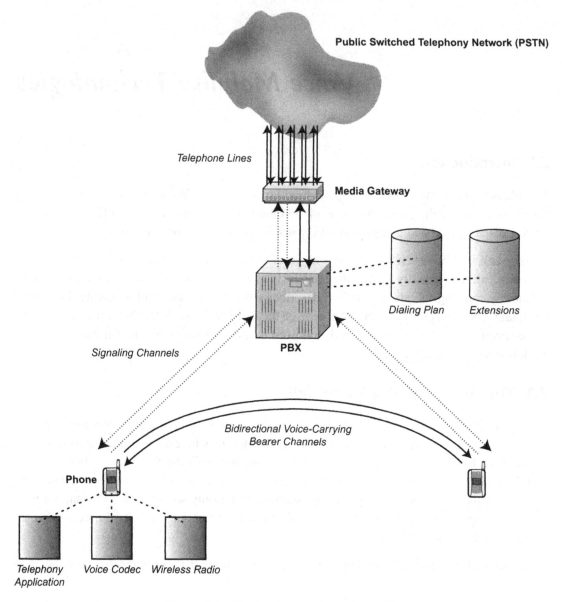

Figure 2.1: The Anatomy of a Voice Call

protocols to provide high voice quality over the network, something easy for voice-oriented technologies but much harder for data-oriented packet-switched networks. Finally, the phones must also support a rich telephony application. This application needs to look like, act like, and have the features of a common telephone, and yet, for voice mobility, must now often store the addresses and phone numbers of crucial contact information, provide access into the enterprise directories, allow for dialing as if in the office, and be manageable by network administrators. Many of these applications are not necessarily native to the phone itself, but created by third-party vendors to tie multiple telephone systems together to appear as one. The telephony applications must support all of the necessary protocols to create a phone call, set up the audio channels, and deal with advanced features such as voicemail and three-way calling.

The call itself is composed of two separate flows of information. The more obvious flow is what is known as the *bearer channel*. The bearer channel, a term borrowed from *Integrated Services Digital Network* (ISDN) technology, carries the voice of the callers. Looking back to analog telephones, there is only one channel, an analog audio channel that provides both the callers' voices and the necessary tones to tell the caller what to do. Dial tones, busy signals, and other beeps and clicks are used to communicate the number being dialed and the state of the call. In voice mobility networks, however, the audio is kept separate from the communication that is used to set up the call and keep it running. This is known as *out-of-band signaling*. Voice can be encoded in a variety of ways, and the bearer channel is more likely to be carried in a number of digital formats. The communication about what the phone call is for, how it is being set up, and what state it is in is encoded in the *signaling channel*. The signaling protocol is used to dial out to set up the call, provide feedback as to whether the other phone is ringing or busy, and to finally set up the bearer channel when the call goes through.

The signaling information needs to go somewhere, and that somewhere is to a *telephone switch*. For private telephone networks, where the handsets are maintained by the enterprise or organization who owns the network, these switches are also owned by the enterprise and are called *private branch exchanges* (PBX). The term *exchange* and *switch* are interchangeable, and these devices are the electronic equivalent of the old manual switchboard, creating connections between two calls. The PBX (or PBXs) for analog systems are proper switches, with internal analog phone lines that run from the handset directly into the PBX. But for Internet Protocol (IP)-based telephone networks, the PBX is more of the central call manager, providing for the list of extensions—users, basically, as well as the telephone numbers each user has—and the dialing plan—the notion of how to route calls based on the phone numbers dialed. PBXs also implement all of the advanced features of voice mobility calling, including voicemail, call forwarding, and multiple ring.

The PBXs define the internal voice network, but calls need to be able to be routed back onto the real, public telephone network, to allow calls in and out from the outside world. IP PBXs rely on a module or a separate device, known as a *media gateway*, which has the responsibility of converting the signaling and bearer traffic into a format and onto a wire that is provided by the telephone company. Media gateways thus bridge the inside world to the outside.

Let's get a better look at each component.

2.1.1 The People and Their Devices: Phones

Phones come in a number of shapes and sizes. Some have the latest features for consumers, such as music playing, video recording and camera functions, global positioning, and touch screens with tactile feedback. Some are designed for enterprise users, and have large screens, with strong email access and integration, spreadsheet and document editing capabilities, and large storage for use as a computer away from the laptop. Others are simple and rugged, meant for use in physically demanding environments where the phones need to withstand a beating. Some have nearly no buttons at all, and are even designed for nearly hands-free operation.

All in all, each phone may seem wildly different from the next. But, underneath, they are made of the same stuff: a microphone to pick up voice; a speaker to play it back; maybe another speaker for speakerphone operation or to play ringtones, so as not to deafen the user who happens to have accidentally hung up on a conversation and is being called back; some way to dial; some way to see or hear who is calling; a battery for mobility; and one or more radios to connect back to the network.

Within those components, the description becomes even more common. There is a digital sampler and a codec engine, to convert voice into digital data and back. There is a CPU somewhere, orchestrating everything, along with memory and some nonvolatile storage. The radios have antennas, all folded neatly into the small device. Voice operates the same on every one of these devices, and users will become just as irritated by poor audio quality as they will be pleased by good quality. The best voice mobility network is the one that users forget is even there, or is anything unique or special. They get out of the way, so to speak, and let the voice mobility user do her work.

2.1.2 The Separate Channels: Signaling and Bearer

In the analog telephone days, there was only one line per extension. This analog line has to do everything. It carries the voice, but it also has to ring the phone, send busy or dial tones, and handle the beeps corresponding to each button being pressed.

With digital phone calling, the signaling and bearer channels are separated. All the beeping, humming, and chirping that is meant to tell the caller what is going on with the call is removed from the audio stream and sent separately in the *signaling* channel. The *bearer* channel holds the human voice, and nothing else. The advantage of this having been done for IP-based voice mobility networks is that it allows the call setup part of the network to operate differently from the voice encoding and decoding part. IP PBXs may be configured to never carry a single voice packet, because their job is simply to figure out how to route calls—much like how Internet DNS servers are so critical in figuring out what "www.google.com" refers to without carrying a byte of Google's web traffic. Media gateways can be created that specialize in conversion of media formats, and then only need to implement the basic signaling protocols, and do not need to be concerned with advanced PBX features.

2.1.3 Dialing Plans and Digits: The Difference Between Five- and Ten-Digit Dialing

For all the advancements in the digital age, with email accounts, instant-messaging handles, avatars, and what not, phones still work with the concept of dialing a series of numbers. But not all numbers are created alike. In the telephone network, someone needed to determine what all of the digits mean. This meaning is known as the *dialing plan*.

Think of the dialing plan as a series of simple rules that tell the phone system when you are done dialing and where the numbers are to go. In the United States, the dialing plan for our public telephone lines specifies that every phone number is seven digits long. Type an extra digit, and the phone ignores it. However, some calls are not in the same area code. This area code concept is a part of the dialing plan. To get to other blocks of phone numbers, outside of the block of numbers that you can dial the most conveniently, you need to dial a "1", followed by the area code, followed by the seven-digit number. Other calls require even more digits. An international call requires dialing "011" before the country code, and then whatever digits are necessary to place a call in that country. And the first "0" must be followed by the first "1" quickly enough to prevent the phone from thinking you are done, and connecting you to the operator. Finally, some calls, like "411", require only three digits. In the office, things can be a bit more complicated. Many people may have four-digit extensions. Only those four need to be dialed. Some companies may use longer extensions, however, with access codes in front of them. Finally, to dial out to the public network, you may need to dial a "9". But not just any "9" will do! The "9" must be followed by a pause, to let the system present a new, outside-world dial tone, where the rest of the digits can be placed.

The dialing plan defines all of this behavior. Every PBX system provides an incredible amount of depth into how these dialing plans can be created, and whether some of the digits are just part of the extension number and others are meant to shift the call over to another PBX somewhere else (like the "9" did to dial outside, but even the "1" for long distance does the same thing) to figure out the meaning.

A lot is mentioned about having four- or five-digit dialing within voice mobility networks. There is an added convenience, and it is true that users of a PBX may not remember the outside number corresponding to an extension, especially if the rest of the number is different for different extensions. (Picture a system in which the 6xxx extensions are reached from outside the office by dialing 487-6xxx, but the 7xxx extensions are reached by dialing 935-7xxx.)

2.1.4 Why PBXs: PBX Features

PBXs serve as a lot more than just the anchor or administrative server of the phone network. They also provide a long list of features that people have come to expect from enterprise phone lines—features that they probably do not have at home, even with today's rich cellphone feature sets.

PBX vendors compete with each other by making the feature set as useful and fancy as possible. There are a number of important PBX features. Some are listed here:

- *Dial-by-name directory*: A computer voice system that allows callers to find out an extension and dial it by interactively pressing a few buttons, usually the first portion of the name. This directory is driven by the autoattendant feature.

- *Autoattendant*: The automated telephone operator, represented by a series of recorded prompts. Autoattendants allow users to access and even manage their account on the PBX simply by calling in. Autoattendants are also the anchor for the interactive voice response systems that outside callers might get into a call center line, whose PBX is advanced enough to guide callers through the menu of options.

- *Call forwarding*: The user can set the line up to forward to another extension, or an outside line, rather than ring the phone. This is useful for when the user is out of the office. Call forwarding is also done automatically when the user does not answer the phone after a certain number of rings.

- *Find-me/Follow-me/Hunting*: These three names for broadly the same feature allow the user to have a number of different alternative phone numbers. When the user does not answer his or her primary line after a certain number of rings, the system hunts down the list, forwarding the call to the next number until it gives up.

- *Simultaneous ring*: Instead of hunting through a series of numbers, the PBX can call out to each of them at once. The first one to answer gets the incoming call. This is useful when the user has a desk phone and a mobile phone, or multiple other phones, and might be at any of them.

- *Call transferring*: Allows the user to send the answered phone call to another phone.

- *Call park*: Allows the call that is already in place to be placed on hold and transferred to another extension, where the user can remove the call from hold. Unlike call transferring, which would ring the other phone and cause the user to have to run until voicemail picks up, call parking allows the user to take more time.

- *Call pickup*: Allows a user to answer another user's phone when it is ringing by entering their extension number. It can also be used in the same sense as simultaneous ringing can, in that an incoming call to a department might ring multiple extensions, and the first to pick it up wins.

- *Do-not-disturb*: Rejects the call before it rings the phone, usually sending it to voicemail or back to whomever transferred the call. Similarly, a user can often use this feature manually on an incoming call by pressing a button on the phone to terminate the incoming call and bounce it back.

- *Voicemail*: Answers the phone and records a message.

- *Hold music*: PBXs provide a series of options and selections for the caller to be subjected to while on hold. For some unknown reason, even advanced PBXs often play a short, few-second-long segment of supposedly relaxing music in an endless loop. Administrators can, however, often replace the hold music with a prerecorded selection. This is most useful for queuing of calls in call centers, where the hold music might be interspersed with the autoattendant informing the caller of the expected wait time.

- *Time-based policies*: PBXs can change their configuration based on the time of day, routing calls to the autoattendant instead of the corporate operator, for example, after hours.

- *Conference calling*: PBXs can join together a limited number of lines for ad hoc conferences, such as three-way calling, for which multiple parties are needed to be on at once.

As you can see, PBXs are designed to have a broad series of functions. Thankfully, PBX features are generally independent of voice mobility networking, in the sense that every PBX has a good number of features, and these features will generally work on IP PBXs, no matter what IP-based protocol the user is using. On the other hand, *fixed-mobile convergence* (FMC) solutions and PBXs do interact, and we will discuss that later, in Chapter 7.

2.2 Signaling Protocols in Detail

Signaling protocols, and the architectures on which they run, are responsible for carrying out the process of setting up a phone call. Their systems determine how to find out the network location of the other party being called, whether the other party can be reached or is out of the network, and help establish the flow of voice traffic.

The concepts from signaling protocols are roughly the same across signaling protocols. There is the notion of a *registrar*, where the phone number is registered for an extension and is mapped to the current network address or location of the handset. This allows for the system to maintain a list of active or available phones. There is a *gateway*, which is responsible for bridging the signaling and possibly the bearer protocols for a call between different formats and networks. Within every system, there needs to be a way to discover the location of users and their availability. Furthermore, there must be provided methods to initiate a phone call, indicate that the other side is ringing, and once the call is connected, to manage the session, allowing parts of it to be renegotiated (such as a change in the bearer channel), additional callers to be brought into the call, and other services to be invoked.

We'll go through many of the major protocols, but will spend more time on the first, the *Session Initiation Protocol* (SIP), as that is the common protocol on packet-based legs of voice mobility networks.

2.2.1 The Session Initiation Protocol (SIP)

The Session Initiation Protocol is, by far, the most common implementation for Wi-Fi-based phones. SIP runs over the *User Datagram Protocol* (UDP), and so can be run over any IP-based network. Transmission Control Protocol (TCP) is also an option, though it is not commonly used for plain SIP, given the shortness of a standard SIP message. SIP was created by the *Internet Engineering Task Force* (IETF), the group that standardizes basic protocols such as TCP, Hypertext Transfer Protocol (HTTP), and Transport Layer Security (TLS), among many others. The definition for SIP is in IETF RFC 3261.

SIP is loosely based on the concepts of another popular Internet protocol, HTTP, used by web browsers and servers to access web pages. This means that SIP is constructed around a request and response model, where one side sends a request for an action for a particular resource, and the other side reports with a response, complete with response code. Every SIP device has the ability to operate as a requester and as a responder, depending on which device is initiating the specific request/response exchange. Furthermore, every SIP message is in text, and so is theoretically human readable. (When you see some of the text that is used in a SIP message, you may beg to disagree!)

The goal of SIP is to provide a simpler method, compared to the prior H.323 protocol (in Section 2.2.2) and others, for performing the basic tasks of call signaling. The introduction of SIP opened up the development of *softphones*, or applications that run on computers and devices not originally designed for telephone usage, to interact with calling services and act like real phones. Even the Microsoft Messenger got into the act and used SIP for instant messaging and chat. The most interesting part of SIP is that a world of open-source or low-cost applications came into the industry, spurred on by its simpler, easier-to-use interface, free from significant intellectual property encumbrances. Now, major digital PBX vendors have gotten into the act, offering SIP services on their systems, eager to allow nontraditional devices onto networks created by their equipment.

2.2.1.1 SIP Architecture

The SIP name for a handset is a *user agent*. A user agent is an endpoint in the SIP communication, applying to both handsets and servers, and is capable of dialing out or receiving phone calls. User agents have IP addresses, and also have users. The users are identified by a SIP *Uniform Resource Identifier* (URI). These look like web URLs, and are based on the same concept, but apply to the domain of telephone calls, rather than web servers.

A URI for a typical caller might look like the following:

```
sip:5300@corp.com
```

This looks like an email address, but it is preceded by the "sip:" marker (in the same way as web pages are proceeded by the "http:" marker). The 5300 marks the phone number, and the @ sign and everything following it represents some notion of the system that the phone number lives on. More often than not, users can ignore the @ sign and the remainder of the string, and concentrate on the phone number before it, just as email users can on a corporate email network. The fact that the URI looks a lot like an email address lets you know that SIP can also use text "phone numbers," such as

```
sip:bob@corp.com
```

which requires the phone user to be able to type in letters rather than numbers, but performs the same way.

SIP phones register their presence with a SIP *registrar.* Registrars perform one of the major functions of the PBX, which is keeping track of phones, the users, their capabilities, and locations. The registrar is how one phone knows that another phone exists. When a phone is first turned on, or it changes IP addresses or networks, it registers itself with the registrar. Before doing that, phone calls placed to that number will be rejected, or possibly sent to voicemail. After registration, however, a phone call to the number will be sent to the registered phone.

This raises the question of whom a phone sends requests for phone calls to. The registrar needs to get its registrations out into the network, so that a placed phone call can find its way to the right party. This is where the second concept, the SIP *proxy*, comes in. The SIP proxy's job is to take requests for phone calls, look up the location of the called party in some database—the one created by the registrar would be ideal, but not required—and forward the call signals appropriately. In this sense, the SIP proxy is the switch, or PBX, for the signaling protocol. Registrars, in fact, are generally integrated into the SIP proxy, making for one device that performs the functions expected of a PBX, including endpoint, or extension management, permissions-checking, logging, and so forth.

The SIP proxy is called a proxy, however, because it does not exist transparently in the process. Rather, its job is to act as a server for the calling party and a client for the called party, responding to the caller's requests by creating nearly identical ones of its own and sending them to the called party. This looks a lot like a web proxy, which is intentional. We will get to the mechanics of SIP signaling shortly.

SIP does not get involved with the actual carrying of voice. In fact, it is not voice-specific, and works just as well for video calls. We will look at how the SIP signaling protocol specifies the different bearer protocol used for the voice (or video) call. One other thing that SIP was not designed to do is phone conference management. SIP is fundamentally call-based, and so is great for phones setting up a call into a conference server. However, the conference server is expected to have some other intelligence built on top that lets it tie the calls together into a conference and knows which users manage the conference and which do not.

Figure 2.2 shows the architecture diagram for SIP, mapped to the standard PBX model.

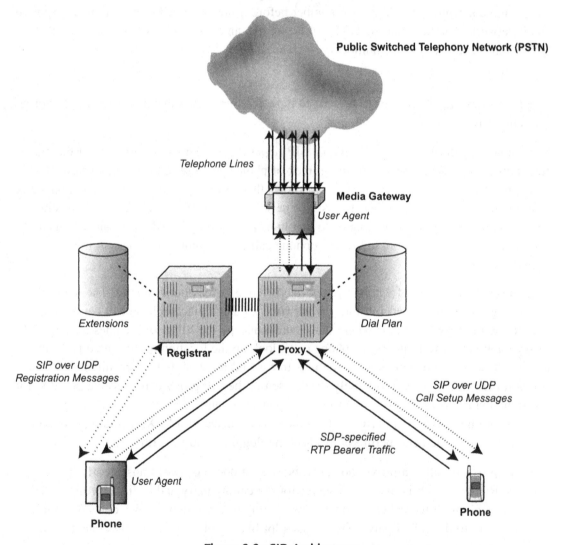

Figure 2.2: SIP Architecture

SIP is based on the concept of a caller inviting the other caller to join the call. Once the invitation goes out to the proxy, who knows where the other party is located, the endpoints and the proxy exchange messages until the call is established. Each invitation, and its successful response, both carry information that is used by other, non-SIP parts of the phone, to establish the bearer channels of the call. Invites are not just for new calls. A phone is allowed to send a new invite to a party while it is in the middle of a call to that device. This would be done when the caller wants to renegotiate the bearer channel, or perhaps to tear it down, such as when a call is placed on a silent (no music) hold.

SIP is heavily oriented toward the notion of the proxy. The proxy, being the switch or PBX, can take care of complex routing decisions that phones should not be bothered with. One wrinkle to this is what two phones do once they find out about the other one's addresses. Some SIP proxies will allow the contact information (which IP address an extension is currently at) to pass through the call, from one party to the other. This allows the two endpoints to take over after the call is set up, and exchange messages exclusively with each other. In this description, however, we will focus on proxies that intentionally hide the addresses of one side from the other. Doing so ensures that the PBX is always a party to every call, making network design simpler and enabling the PBX to support a larger number of features than if the clients communicated peer-to-peer.

Media gateways appear, in SIP, just as ordinary endpoints. The difference lies in how the registrar and proxies treat them. The proxy will know to forward all phone numbers in the dialing plan that must go to the next network (such as outside calls) to the media gateway, as if the gateway had registered for that number. Incoming calls from the other network operate in the same way as outgoing calls do from a phone: the call is routed to the proxy. In this way, the same protocol can work for bundles of lines or general routes as easily as it can for simple devices.

SIP includes provisions to allow for user authentication, and for encryption of parts of the packets.

2.2.1.2 SIP Registration

As mentioned before, the SIP registrar knows about the existence of a phone by the process of registration. When the phone is turned on, or when it changes its network address, or when its old registration has expired and it needs to refresh it, the phone sets up a SIP request to the registrar. This means that the SIP phone must know which IP address the registrar is at, as that registrar becomes the constant point of contact for the network.

Because registration is so important, we will use the SIP registration process as our way of understanding the format of SIP messages. For the examples in the section on SIP, we will use the following:

SIP Registrar and Proxy:	Name: corp.com.	Address: 10.0.0.10
Phone 1:	Number 7010.	Address: 192.168.0.10
Phone 2:	Number 7020.	Address: 192.168.0.20

Let's look at our first SIP message, then. SIP is sent in UDP packets to port 5060, and so the contents in Table 2.1 show the payload of the UDP packet, sent from Phone 1's IP address at 192.168.0.10, port 5060, to the registrar's IP address at 10.0.0.10, port 5060.

Table 2.1: SIP REGISTER request

```
REGISTER sip:corp.com SIP/2.0
Via: SIP/2.0/UDP 192.168.0.10:5060;branch=z9hG4bK1072017640
From: "7010"⟨sip:7010@corp.com⟩;tag=915317945
To: "7010"⟨sip:7010@corp.com⟩
Call-ID: 1422523958@192.168.0.10
CSeq: 1 REGISTER
Contact: ⟨sip:7010@192.168.0.10⟩;expires=3600
Max-Forwards: 70
Content-Length: 0
```

This is all text, with newlines given by a carriage return and linefeed, just as with HTTP. It is structured the same way, as well.

The first line begins with the action, in this case, to "REGISTER." The URI for the registration is that of the registrar, which is "sip:10.0.0.10". Finally, the version is "SIP/2.0", meaning, understandably, SIP 2.0. This message is a request to register with the registrar.

The rest of the lines are presented as SIP (HTTP) headers. That is, there is the text string naming the header, followed by a colon. The first header is the Via header, identifying the most recent sender of this message. Remember that all messages could potentially be proxied in the protocol, and the Via header allows the receiver to understand why the IP sender of the message is involved in sending it, especially if the From line doesn't match. In this case, the Via header just specifies the phone who sent the REGISTER message, as no one proxied it. The line can be broken down as follows. "SIP/2.0/UDP" just repeats that the phone sends UDP. "192.168.0.10:5060" is the IP address and UDP port of the phone. With this information, the recipient—the registrar—knows that the response has to go to 192.168.0.10:5060 using SIP 2.0 on UDP. The registrar has to use this, and not the IP and UDP sender (which is identical, of course), as this allows messages to be routed in stranger ways. Think of the Via as a "Reply-to" header from email. The last piece, the "branch" part, specifies a unique identifier for this request/response transaction. (The semicolon sets the branch and other pieces that might follow aside from what came before it, and the equal sign sets the value of the branch, until the end of the line or another semicolon.)

Because UDP has no real concept of a connection, this branch parameter is used to establish that concept.

The next line is the From line, which specifies the identity of the user agent making the transaction. This line looks like a From email header, for good reason. The quoted "7010" is the user-displayable phone number. Just as with email addresses, in which the account name may be "bob@corp.com" but the person's name would be "Bob Baker," a user might have a different name that the callers see than that of the SIP account he uses. The "⟨sip:7010@corp.com⟩" is the URI for the account, set aside in angled brackets. Finally, the "tag" serves the purpose of identifying the overall call sequence for this series of requests. Whereas the branch strictly identifies the request/response pair, the tag identifies the entire sequence of requests and responses that make up one action between callers.

The To line is similar to the From line. Here, there is no tag yet, because the "called party"—because this is a REGISTER, that party is just the registrar, and there is no real call from a user's point of view—is required to pick its own tag.

The Call-ID is unique for the particular call from that caller, and is given in the similar email-address format, with the IP address of the caller defining the part after the @ sign.

The CSeq field defines where we are in the back-and-forth of the particular action. The value of "1 REGISTER" tells us that this is message one of the handshake, and this is a REGISTER message. These are useful for human debugging of call problems, as it tells you where you are in the process, even if the earlier parts of the process are missed.

All of this previous stuff is just mechanics. The important part of the REGISTER message comes now. The Contact field tells the registrar that this is a registration for "⟨sip:7010@192.168.0.10⟩", meaning that the phone number is at 192.168.0.10, and goes by the name "7010". It is actually possible for one user agent to have multiple phone numbers, and this registration is for the one and only one phone number here. The "expires" tag states that the registration expires 3600 seconds, or one hour, from now.

The Max-Forwards header just states that any intervening proxy can proxy this message, for a total of 70 times, after which, the message is dropped. This protects the network from times when a proxy might be misconfigured to forward a message back along the path from where it came.

The Content-Length states that there is no SIP message body. Message bodies are used in INVITEs, which we will see later.

Now that the registrar has received the request, it will send a response. The response lets the client know that the registration went well, or had an error. Table 2.2 has the response.

The first line has the response. "SIP/2.0" is the version, but more importantly, "200 OK" means that the response was a success. Registrars can fail with different codes, such as

Table 2.2: SIP REGISTER response

```
SIP/2.0 200 OK
Via: SIP/2.0/UDP ⇒
 192.68.0.10:5060;branch=z9hG4bK1072017640;received=192.168.0.10
From: "7010"(sip:7010@corp.com);tag=915317945
To: "7010"(sip:7010@corp.com);tag=as7374d984
Call-ID: 1422523958@192.168.0.10
CSeq: 1 REGISTER
User-Agent: My PBX
Allow: INVITE, ACK, CANCEL, OPTIONS, BYE, REFER, SUBSCRIBE, NOTIFY
Supported: replaces
Expires: 3600
Contact: (sip:7010@192.168.0.10);expires=3600
Date: Tue, 27 Jan 2008 00:25:14 GMT
Content-Length: 0
```

when the extension is not known by the registrar. That would happen when the phone really does not belong to this particular telephone network.

The Via, From, and To fields serve the same purpose as before. Note that the To field now has a tag. The From field is still from the original caller, even though this is a response, and the To field is still from the called party. The Via line, and the indented one following it, are all one line in the packet: the description will use ⇒ to mark that the following line continues the current one.

The CSeq field has not changed, as this is still the REGISTER request/response pair.

The User-Agent is the vendor name of the registrar. In this case, it is "My PBX." Because you cannot go out and purchase "My PBX," you may have to settle for someone else's.

The Allow header states what types of actions a caller can request from the registrar. We will look at the more important ones—INVITE and BYE—shortly.

The Expires header states that the registrar agreed with the client, and is going to let the registration live for 3600 seconds. The registrar can shorten this amount, if it so chooses and the client will lose its registration if it does not come in before then. You will notice that the Expires information repeats as a separate field in the Contact header. There are, unfortunately, still multiple ways of encoding exactly the same information in SIP, and there will be a lot of redundancy. The problem, of course, with redundancy, is that no one really knows how a complex system will work if the redundant information actually changes.

Finally, the date of the response came in. Now, the registrar knows the phone exists, and the phone can make and receive calls. One good thing about registration is that the phone number and IP address bindings can be looked up in the PBX by the administrator. SIP-aware intervening networks, such as some Wi-Fi systems, also track this state and show which phones are wireless and where they are.

2.2.1.3 Placing a SIP Call

Our user, 7010, wants to place a call to his already registered coworker, 7020. In SIP, this process belongs to a request-response series kicked off by an INVITE. The idea behind the invite is that the caller *invites* the called party into the call.

Phone calls always start off with the user dialing the called party. Once that is done, the call should ring until the called party answers, or until the system forwards the call to voicemail or delivers a busy signal. Because there are phases of the call setup process—ringing, then a forward to voicemail—we will see how SIP deals with multiple responses for the same request. Figure 2.3 shows the messages that are exchanged in setting up the call, in order from 1 to 9. The top and bottom half of the diagram are of the same equipment, but are

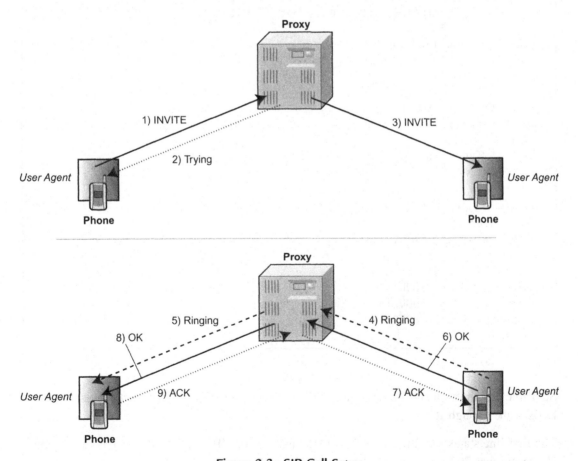

Figure 2.3: SIP Call Setup

separated to make the arrows clearer. Each arrow represents a message being sent from one part to the other. The black lines are for messages that represent the main setup messages, representing the call being placed or answered. The dashed lines are for messages that inform the caller of the progress. The dotted lines represent messages that are necessary for the protocol to work, if a message gets lost, but do not carry meaningful information about the call itself.

Table 2.3 shows the outgoing request for a phone call. This is delivered to the proxy, as the calling phone has no idea where the called party is. The proxy will ask the registrar for details, and will then forward the message appropriately.

Table 2.3: SIP INVITE Request from Caller to Proxy

```
INVITE sip:7020@corp.com SIP/2.0
Via: SIP/2.0/UDP 192.168.0.10:5060;branch=z9hG4bK922648023
From: "7010"(sip:7010@corp.com);tag=1687298419
To: <sip:7020@corp.com>
Supported: replaces,100rel,timer
Call-ID: 2114455679@192.168.0.10
CSeq: 20 INVITE
Session-Expires: 1800
Contact: <sip:7010@192.168.0.10>
Max-Forwards: 70
Expires: 180
Content-Type: application/sdp
Content-Length: 217
v=0
o=7010 1352822030 1434897705 IN IP4 192.168.0.10
s=A_conversation
c=IN IP4 192.168.0.10
t=0 0
m=audio 9000 RTP/AVP 0 8 18
a=rtpmap:0 PCMU/8000/1
a=rtpmap:8 PCMA/8000/1
a=rtpmap:18 G729/8000/1
a=ptime:20
```

So Table 2.3 shows a substantially more involved message than we saw with registration. Let's walk through it.

The first line states that this is an "INVITE", asking for URI "sip:7020@corp.com", or the phone number 7020. So now we know who the call is for.

The To field repeats that the message is really destined for URI "sip:7020@corp.com".

The CSeq field is "20 INVITE". This INVITE started at 20. There may be multiple INVITEs coming for this call, depending on what happens as the call progresses. So 20 is a starting point.

The Session-Expires header means that the entire session to start a phone call, including future changes, shouldn't live for more than 1800 seconds, or a half-hour. This has no influence on the endpoints, but the proxy may be tracking the call, and this expiration lets the proxy know when the call can be flushed from the system, if, for some reason, one side or another disconnects from the network without informing the proxy.

The Expires header, on the other hand, states how long the calling phone is willing to wait until it gets a yes-or-no answer from the other party. After 180 seconds, or three minutes, the caller will give up and cancel the invitation.

The Content-Type field states that there is a body to this message, and it is in the *Session Description Protocol* (SDP) format. As mentioned earlier, SIP has no understanding of voice traffic itself. It is charged only with setting up calls. The SDP, on the other hand, describes how to set up the voice traffic. We'll talk more about SDP in the section on bearer protocols, but for now, it's enough to note that the rest of the packet is for the client asking for an Real-time Transport Protocol (RTP) session with voice.

Table 2.4: SIP INVITE Trying Responses from Proxy to Caller

```
SIP/2.0 100 Trying
Via: SIP/2.0/UDP ⇒
 192.168.0.10:5060;branch=z9hG4bK922648023;received=192.168.0.10
From: "7010"〈sip:7010@corp.com);tag=1687298419
To: 〈sip:7020@corp.com)
Call-ID: 2114455679@192.168.0.10
CSeq: 20 INVITE
User-Agent: My PBX
Allow: INVITE, ACK, CANCEL, OPTIONS, BYE, REFER, SUBSCRIBE, NOTIFY
Supported: replaces
Contact: 〈sip:7010@192.168.0.10〉
Content-Length: 0
```

Table 2.4 describes the response from the proxy back to the caller. The code is "100 Trying", meaning that the proxy has more work to do, and a better result will be provided shortly. In the meantime, the calling phone may show some state reflecting that the call is being attempted, but no ringing should be heard by the caller.

The proxy will also proceed to forward the INVITE request to the called party.

Table 2.5: SIP INVITE Request from Proxy to Called Party

```
INVITE sip:7020@192.168.0.20 SIP/2.0
Via: SIP/2.0/UDP 10.0.0.10:5060;branch=z9hG4bK51108ed9;rport
From: "7010"(sip:7010@corp.com);tag=as13a69dc0
To: ⟨sip:7020@192.168.0.20⟩
Contact: ⟨sip:7010@10.0.0.10⟩
Call-ID: 524f6a41059ad64d43dd4cce4b001b69@10.0.0.10
CSeq: 102 INVITE
User-Agent: My PBX
Max-Forwards: 70
Date: Tue, 27 Jan 2008 00:25:28 GMT
Allow: INVITE, ACK, CANCEL, OPTIONS, BYE, REFER, SUBSCRIBE, NOTIFY
Supported: replaces
Content-Type: application/sdp
Content-Length: 281
v=0
o=root 10871 10871 IN IP4 10.0.0.10
s=session
c=IN IP4 10.0.0.10
t=0 0
m=audio 11690 RTP/AVP 0 3 8 101
a=rtpmap:0 PCMU/8000
a=rtpmap:3 GSM/8000
a=rtpmap:8 PCMA/8000
a=rtpmap:101 telephone-event/8000
a=fmtp:101 0-16
a=silenceSupp:off - - - -
a=ptime:20
a=sendrecv
```

Table 2.5 shows the INVITE as the proxy has forwarded it. This message, from the proxy to the called party, 7020, looks similar to the original INVITE. There is one major difference, however, besides the Via header. The PBX in use is configured to direct the bearer channel through itself, where it will bridge the two legs of the call, one leg from each phone to the PBX. Therefore, it has substituted its own session description, in SDP, requesting the voice connection to come back to its own IP address. Also note that the CSeq for this invite is 102, not 20. The proxy maintains its own sequence. Finally, note that the URI for the To header has now changed to have the part after the @ sign refer to the phone's IP.

The called party's phone responds to the proxy with the next message, in Table 2.6.

The called party's phone will also start to ring. This may not happen if the phone is busy or not available. But, in this case, the phone is available, and is ringing.

Table 2.6: SIP Ringing Response from Called Party to Proxy

```
SIP/2.0 180 Ringing
Via: SIP/2.0/UDP 10.0.0.10:5060;branch=z9hG4bK51108ed9;rport=5060
From: "7010"(sip:7010@corp.com);tag=as13a69dc0
To: ⟨sip:7020@172.27.0.20⟩;tag=807333543
Call-ID: 524f6a41059ad64d43dd4cce4b001b69@10.0.0.10
CSeq: 102 INVITE
Contact: ⟨sip:7020@192.168.0.20⟩
Content-Length: 0
```

Table 2.7: SIP Ringing Response from Proxy to Caller

```
SIP/2.0 180 Ringing SIP/2.0
Via: SIP/2.0/UDP ⟹
 192.168.0.10:5060;branch=z9hG4bK922648023;received=192.168.0.10
From: "7010"(sip:7010@corp.com);tag=1687298419
To: ⟨sip:7020@corp.com⟩;tag=as500634a6
Call-ID: 2114455679@192.168.0.10
CSeq: 20 INVITE
User-Agent: My PBX
Allow: INVITE, ACK, CANCEL, OPTIONS, BYE, REFER, SUBSCRIBE, NOTIFY
Supported: replaces
Contact: ⟨sip:7020@10.0.0.10⟩
Content-Length: 0
```

Table 2.7 shows the ringing message sent from the proxy, back to the caller. The caller receives this message, and starts playing a ringback tone for the caller to listen to, as the other party decides whether to answer. Some amount of time passes, and then the called party answers.

Table 2.8 shows the "200 OK" message, reflecting that the call has been answered. In the OK message, we can see that the called party states where it wants its leg of the call to arrive to, and how it would like to set up the bearer channel.

The proxy will receive this message and patch together the second leg of the call from it to the called party. To do so, the proxy, which has inserted itself between the two endpoints of the call, needs to send an acknowledgment (ACK), as shown in Table 2.9. This is needed because UDP is a lossy protocol, and there is a possibility that one of the messages didn't get through. If the original INVITE did not get through, the sender will continue to send duplicate INVITEs, every second or so, until the other side gets it. Now, with the OK having been sent, the called party needs to know that the caller knows and is in the call.

Table 2.8: SIP OK Response from Called Party to Proxy

```
SIP/2.0 200 OK
Via: SIP/2.0/UDP 10.0.0.10:5060;branch=z9hG4bK51108ed9;rport=5060
From: "7010"(sip:7010@corp.com);tag=as13a69dc0
To: (sip:7020@192.168.0.20);tag=807333543
Supported: replaces
Call-ID: 524f6a41059ad64d43dd4cce4b001b69@10.0.0.10
CSeq: 102 INVITE
Contact: (sip:7020@192.168.0.20)
Content-Type: application/sdp
Content-Length: 160
v=0
o=9002 1367913267 845466921 IN IP4 192.168.0.20
s=A_conversation
c=IN IP4 192.168.0.20
t=0 0
m=audio 9000 RTP/AVP 0
a=rtpmap:0 PCMU/8000
a=ptime:20
```

Table 2.9: SIP ACK Request from Proxy to Called Party

```
ACK sip:7020@192.168.0.20 SIP/2.0
Via: SIP/2.0/UDP 10.0.0.10:5060;branch=z9hG4bK632b9079;rport
From: "7010"(sip:7010@corp.com);tag=as13a69dc0
To: (sip:7020@192.168.0.20);tag=807333543
Contact: (sip:7010@10.0.0.10)
Call-ID: 524f6a41059ad64d43dd4cce4b001b69@10.0.0.10
CSeq: 102 ACK
User-Agent: My PBX
Max-Forwards: 70
Content-Length: 0
```

The proxy also needs to send the OK through to the caller.

Table 2.10 shows the message the caller gets. This lets the caller stop playing the ringback tone, and connecting the phone call from itself to the PBX. Now, both callers can hear each other.

There is nothing to be done after the ACK, as it marks the end of the call setup protocol.

We can now ask how the proxy participated. The messages you have seen look as if they could have gone directly from party to party. And this is true. However, the SIP proxy and registrar work together to provide a way for clients to find out about other clients, and have

Table 2.10: SIP OK Response from Proxy to Caller

```
SIP/2.0 200 OK
Via: SIP/2.0/UDP ⇒
 192.168.0.10:5060;branch=z9hG4bK922648023;received=192.168.0.10
From: "7010"(sip:7010@corp.com);tag=1687298419
To: (sip:7020@corp.com);tag=as500634a6
Call-ID: 2114455679@192.168.0.10
CSeq: 20 INVITE
User-Agent: My PBX
Allow: INVITE, ACK, CANCEL, OPTIONS, BYE, REFER, SUBSCRIBE, NOTIFY
Supported: replaces
Contact: (sip:7020@10.0.0.10)
Content-Type: application/sdp
Content-Length: 202

v=0
o=root 10871 10871 IN IP4 10.0.0.10
s=session
c=IN IP4 10.0.0.10
t=0 0
m=audio 12482 RTP/AVP 0 8
a=rtpmap:0 PCMU/8000
a=rtpmap:8 PCMA/8000
a=silenceSupp:off - - - -
a=ptime:20
a=sendrecv
```

Table 2.11: SIP ACK Request from Caller to Proxy

```
ACK sip:7020@corp.com SIP/2.0
Via: SIP/2.0/UDP 192.168.0.10:5060;branch=z9hG4bK1189369993
From: "7010"(sip:7010@corp.com);tag=1687298419
To: (sip:7020@corp.com);tag=as500634a6
Call-ID: 2114455679@192.168.0.10
CSeq: 20 ACK
Max-Forwards: 70
Content-Length: 0
```

their calls routed to voicemail, the outside world, or to have any other advanced call feature performed. For this to work, the SIP proxy and registrar must work together to form a PBX. Clients are prevented from communicating directly. You may have noticed that at no point do the messages from the caller to the proxy have the IP addresses of the called party, and vice versa. The Contact and Via headers are always kept consistent, and all of the messages are forced to flow through the proxy. Not all proxies act this way, but most proxies with full PBX functionality do.

2.2.1.4 A Rejected SIP Call

Sometimes the called party does not want to answer the call. This happens when the phone call comes in and the user is busy. The process looks very similar to a successful SIP call, except that, instead of the called party sending a 200 OK message, it sends a different one.

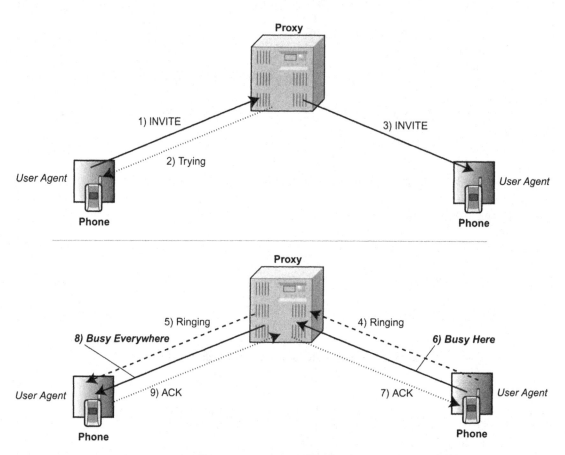

Figure 2.4:   SIP Call Rejection

Figure 2.4 shows the call setup flow. The change is that the OK is now replaced with a "486 Busy Here" message, which means that the called party is busy at the endpoint they were called at. The proxy could have used this status to forward the call onto a voicemail system, by either sending a new INVITE to the voicemail system (if it has a separate user agent) and bridging those two legs together, or by sending a "302 Moved Temporarily" response to the client and expecting the client to try the voicemail.

Table 2.12 shows the rejection that the called party will send if the user declines to take the call. The proxy will continue with its ACK, as before.

Table 2.12: SIP Busy Here Rejection from Called Party to Proxy

```
SIP/2.0 486 Busy Here
Via: SIP/2.0/UDP 10.0.0.10:5060;branch=z9hG4bK51108ed9;rport=5060
From: "7010"(sip:7010@corp.com);tag=as13a69dc0
To: ⟨sip:7020@192.168.0.20⟩;tag=807333543
Call-ID: 524f6a41059ad64d43dd4cce4b001b69@10.0.0.10
CSeq: 102 INVITE
Content-Length: 0
```

Table 2.13: SIP Busy Everywhere Rejection from Proxy to Caller

```
SIP/2.0 600 Busy Everywhere
Via: SIP/2.0/UDP ⇒
 192.168.0.10:5060;branch=z9hG4bK922648023;received=192.168.0.10
From: "7010"(sip:7010@corp.com);tag=1687298419
To: ⟨sip:7020@corp.com⟩;tag=as500634a6
Call-ID: 2114455679@192.168.0.10
CSeq: 20 INVITE
User-Agent: My PBX
Allow: INVITE, ACK, CANCEL, OPTIONS, BYE, REFER, SUBSCRIBE, NOTIFY
Supported: replaces
Content-Length: 0
```

Next, the proxy will send to the caller a busy message (Table 2.13), stating that the called party is busy at every user agent that it is aware of, which, for this example, is the one at 192.168.0.20. Because the proxy knows that the called party has no other alternatives, it changed the message code from "486 Busy Here" to "600 Busy Everywhere".

2.2.1.5 Hanging Up

Once the SIP call is going, there are no response codes that can hang the call up. Rather, the party who hangs up first sends a new request to the proxy, to disconnect the call. This type of message, called a BYE request, tells the receiver that the call is now over.

Table 2.14 shows the BYE message, sent when the caller hangs up the phone call. The proxy server will respond to the BYE with a "200 OK" response, which is similar to the ones we have seen before, except without any SDP description of the bearer channel. Then, the proxy server will send a BYE to the called party, and wait for the OK from there. Unlike with the call setup, the OK message is enough to mark the end of the teardown.

2.2.1.6 SIP Response Codes

The protocol has a number of response codes, some of which are more esoteric than others. We'll go through the main ones, one by one, to get a better sense of what can happen.

Table 2.14: SIP BYE from Caller to Proxy

```
BYE sip:7020@corp.com SIP/2.0
Via: SIP/2.0/UDP 192.168.0.10:5060;branch=z9hG4bK1818649196
From: "7010"<sip:7010@corp.com>;tag=1687298419
To: <sip:7020@corp.com>;tag=as500634a6
Call-ID: 2114455679@192.168.0.10
CSeq: 21 BYE
Max-Forwards: 70
Content-Length: 0
```

In-Progress Codes

• *100 Trying*: sent by any endpoint, when it needs to inform the requester that it is handling the request, but might not return quickly enough to prevent the sender from thinking the message was lost.

• *180 Ringing*: sent by the called party while it is ringing the phone.

• *181 Call is Being Forwarded*: might be sent by a proxy to the caller when the call is being forwarded to another destination, such as a second handset or voicemail. You may never see this one.

• *182 Queued*: sent by a server running a phone bank, informing the caller that the call is on hold to be answered. It can also be sent at a chokepoint when lines are busy but one of the proxies does not want to give up on establishing the call just yet.

Success Code

• *200 OK*: the request succeeded. If the request was an INVITE, the message has the session description for the audio channel.

Redirection Codes

• *301 Moved Permanently*: sent by a proxy (or possibly an endpoint) when a user has permanently changed his or her phone number to something else. The Contact header should have the forwarding address.

• *302 Moved Temporarily*: sent when the call is being forwarded. The Contact header will have where the call needs to be forwarded to, and the caller must start a new call to that forwarding address.

Request Failure Codes

• *400 Bad Request*: sent when the SIP message is not formatted properly. This is a good sign of interoperability issues between advanced features of different products.

- *401 Unauthorized*: sent by a device when it requires SIP authorization.

- *403 Forbidden*: sent for a number of reasons. Usually, this happens when a caller is not allowed to use a certain feature, perhaps because of access rights in the proxy. This also gets sent when unknown endpoints try to register and are not provisioned. Finally, this status can also be thrown in when devices, for whatever reason, do not wish to handle the request. Some phones or proxies send this when the other side is busy.

- *404 Not Found*: sent by a proxy for a number that does not exist.

- *405 Method Not Allowed*: sent by a proxy or an endpoint when they do not want to perform a SIP method for the called party. Devices which do not allow calls, such as registrars when the registrar is not a proxy, can send this, as can proxies that are not registrars and are being registered to. This is a good sign that the phone is misconfigured to use the wrong IP address for some PBX function.

- *407 Proxy Authentication Required*: sent by the proxy or PBX when it requires authentication. The calling phone might respond automatically by authenticating.

- *408 Request Timeout*: sent by an endpoint or proxy when the request just cannot be handled in time. This is often a sign of long delays or problems between specific SIP infrastructure resources, such as a proxy being unable to reach a separate proxy or registrar.

- *410 Gone*: sent when the called party did once exist, but no longer does, there is no forwarding information, and the proxy does not know why the extension is gone. Similar to "604 Does Not Exist Anymore," where the difference between the two will be described shortly.

- *480 Temporarily Unavailable*: sent when the called party's registration has expired, has not logged in, or is not yet electronically ready to receive the call.

- *482 Loop Detected*: sent by a proxy when it sees the same message twice. This is a sign that inter-proxy forwarding is not correctly configured. This concept is similar to an email forwarding loop.

- *483 Too Many Hops*: sent by a proxy when the Max-Forwards header field hits zero. This too is usually a sign that inter-proxy forwarding is not configured correctly, and that there is a forwarding loop.

- *486 Busy Here*: sent by a phone when it is already in a call or does not want to take the call because the user is busy. The proxy receiving the message can try the next registration for the extension, or can try forwarding the call off.

- *491 Request Pending*: a rare message, which happens when INVITEs cross pass for the same two clients from opposite directions. If you see this message, it is usually because of a coincidence, rather than misconfiguration.

- *493 Undecipherable*: the SIP security model cannot decrypt the message.

Server Failure Codes

- *500 Server Internal Error*: sent when the SIP protocol is not being followed correctly by the requester, or when there is something wrong with the endpoint receiving the request, such as misconfiguration or an error.

- *501 Not Implemented*: the recipient of the request does not implement the method.

- *502 Bad Gateway*: one proxy did not like the response it got from another proxy, and wishes to report it to the caller. Usually a sign of misconfiguration or incompatibility between the gateways.

- *503 Service Unavailable*: sent when the proxy or endpoint is not fully configured or is undergoing maintenance.

- *504 Server Time-out*: sent by a proxy when it cannot access a non-SIP service, such as DNS, in a timely manner.

- *505 Version Not Supported*: the requester is using a version of SIP too old or too new for the other end.

Global Failure Codes

- *600 Busy Everywhere*: sent by a proxy when it has tried every endpoint for the extension, they are all busy, and there is no place to route the call. This should generate a busy tone or busy indication, for the caller.

- *603 Decline*: the called party does not want to participate, and does not want to explain why.

- *604 Does Not Exist Anymore*: similar to "410 Gone," but is definitive. The called party no longer exists, because the extension was deleted.

- *606 Not Acceptable*: sent by the endpoint when it cannot handle the media stream being requested. This can happen when an IPv6 phone (with no IPv4 address) calls an IPv4 phone, or when a videophone calls a standard audio phone, for example.

This list is not exhaustive. Moreover, SIP is a very flexible protocol, subject to a number of different interpretations. Because of this, different phones and different proxies might send different messages, depending on how they interpret the standard and what devices they interoperate with will do when they receive the message.

2.2.1.7 Authentication

SIP authentication is based on the challenge-response protocol used in HTTP, and allows username and password authentication into the registrar, proxies, and services. With SIP, the concept for authentication is that the registrar or proxy will send one of the status messages that requires authentication (401 or 407), and the client or upstream server will respond.

The authentication method is based on the WWW-Authenticate header. The more common authentication method is using the digest scheme. The response comes in an Authorization header in the retransmitted request. The digest method uses an MD5 hash over the username, password, realm, and resource that is being accessed, along with a nonce from the responder and a nonce from the requester.

Table 2.15: Example SIP Authentication Challenge

```
WWW-Authenticate: Digest
realm="corp.com",
qop="auth,auth-int",
nonce="235987cb9879241b1424d4bea417b22f71",
opaque="aca9787a9e87cae532a685e094ce8394"
```

Table 2.15 shows an example authentication challenge header. The realm is "corp.com". "qop" asks for the specific algorithm to be used to calculate the hashes. The nonce is, as expected, a random number from the server that proves that the client is alive and is not just a replay of an older, valid session. The opaque value is some server-specific information that the client is required to repeat, but has no cryptographic significance.

The client will perform a series of MD5 hashes on this information, as well as the resource being requested, and will add the results into the Authorization header shown in Table 2.16.

Table 2.16: Example SIP Authentication Response

```
Authorization: Digest username="7010",
 realm="corp.com",
 nonce="235987cb9879241b1424d4bea417b22f71",
 uri="sip:7020@corp.com ",
 qop=auth,
 nc=00000001,
 cnonce="3a565253",
 response="d0ea98700991bce0bd0ba0eab1becc01",
 opaque="aca9787a9e87cae532a685e094ce8394"
```

Authentication, by itself, does not provide any privacy or encryption for the traffic or prevent modifications of the parameters to any message. But it does prevent accidental misuse of resources.

2.2.1.8 Secure SIP

Full hop-by-hop security is possible with SIP, using what is called *SIPS*. This extension uses TLS in essentially the same way as HTTPS does. URIs are modified from "sip:" to "sips:", just as "http:" is modified to "https:" with SSL (noting that SSL is just an older version of TLS).

Using SIPS requires that both endpoints communicate using TCP. When the TCP connection is connected, the server will use TLS to request that the client authenticate. The client will then use its TLS credentials to authenticate, including using no credentials as an option, akin to the typical use of HTTPS in which only the server needs to authenticate using TLS. Once TLS finishes, both the requester and responder have a master key, which they use to encrypt the session. Within the session, password-based authentication can be used using the SIP digest authentication method. With the connection being encrypted with TLS, both parties are provided confidentiality and protection from forgeries or modification.

2.2.2 H.323

H.323 is an *International Telecommunication Union* (ITU) specification that defines how to establish both the signaling and the bearer channels. The goal was to use the call signaling from ISDN, a call control protocol for renegotiating calls as they are ongoing, and a registration protocol, to provide for an all-encompassing solution. (Compare how SIP defines only registration, call signaling, and merely basic call control.) However, the process was different, and the H.323 technology suffers from the typical emphasis on layering and precise botanical definitions of technologies that haunts the world of telecommunications.

H.323's major advantage, compared to SIP, is that it contains the complete protocol definition for the application, covering features such as media reservation and conference negotiation that SIP leaves alone. H.323 is also able to pull together a number of other ITU definitions and technologies, into one larger umbrella. Because of the ITU protocols' amount of definition for media applications, H.323 is still the signaling protocol of choice for many videoconferencing applications.

That being said, H.323's relevance for voice is waning. For that reason, we will stick to H.323 at a higher level than we did with SIP.

2.2.2.1 H.323 Architecture

H.323 has a somewhat similar architecture to SIP. Figure 2.5 shows this architecture.

The endpoints are known as *terminals*. The terminals must register with the registrar, which is now in a function known as the *gatekeeper*. The gatekeeper is the PBX, and has complete responsibility for administration, user definitions, registration, and routing. Gateways are

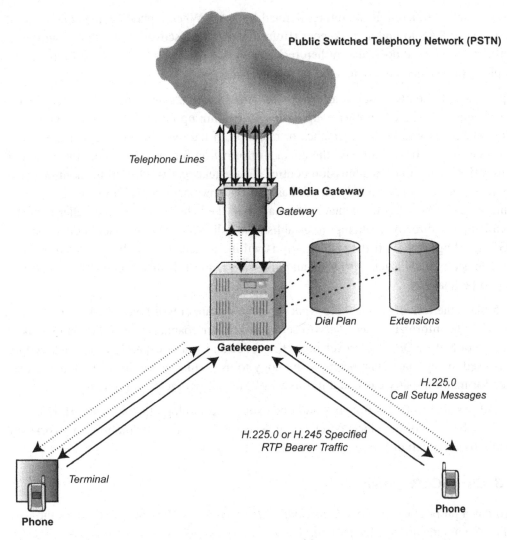

Figure 2.5: H.323 Architecture

now special devices that are specifically called out for bridging signaling and media between two different networks.

H.323 uses a protocol known as H.225.0 for call setup signaling. H.225.0 itself is a package that refers to Q.931 for call signaling definitions. (This sort of nesting is typical for telecommunications definitions). The good news is that it stops at Q.931, and we can identify it. ITU Q.931 is the call signaling protocol used in ISDN lines, and is referred to in Section 2.2.6. H.225.0 also includes the *Registration, Admission, and Status* (RAS) protocol, used by the client to register with the network.

The registration function is, therefore, defined by RAS. When a phone comes online, its first task is to use RAS to find a gatekeeper (from a known or discovered list) that is willing to let it register. Once it does that, it then requests to register. After the registration is complete, the phone is ready to send or receive calls.

To place a call, the phone sends an admission request to the gatekeeper. The gatekeeper's job is to find out where the other endpoint is, by looking up in its extension or routing tables. The result will be an acceptance or rejection. If the result is an acceptance, the gatekeeper will also respond with the contact information for the other endpoint. Notice that the model here is based on admission control. The gatekeeper is allowed to monitor voice resources, and reject calls purely on the basis of there not being enough resources. In any event, the caller now has the contact information of the called party, so the caller contacts the called party directly to attempt to establish the call. This direct contact is done using Q.931 signaling over IP. If the called party is willing to accept the call, the called party must contact its local gatekeeper with an admission request. If that is granted, the call is ready to be finalized.

H.245 plays the role of establishing what the bearer channel will hold. H.245 was designed to provide the information necessary to set up the bearer channels over RTP, and so takes the place of SIP's SDP. H.245 exchanges the codec and bearer capabilities of each endpoint, and is used to negotiate what bearer technology to use. This can be done in a manner that works for multiple-party calls, and in this way is useful with teleconferencing.

It is still possible to find softphones and open source technology that supports H.323, especially because of the videoconferencing aspect. However, voice mobility networks are unlikely to see much of H.323.

2.2.3 Cisco SCCP: "Skinny"

Cisco has a proprietary *Skinny Client Control Protocol* (SCCP), which is used by the Cisco Unified Communications Manager and Cisco phones as their own signaling protocol. SCCP requires the Cisco Unified Communications Manager or open-source PBXs to operate. Given the downside of proprietary protocols, the main reason for discussing SCCP within the context of voice mobility is only that Cisco's Wi-Fi-only handsets support SCCP, and so SCCP may be seen in some voice mobility networks. Unfortunately, SCCP internal documentation is not widely available or as well understood as an open protocol is, and so enterprise-grade implementations tend to lock the user into one vendor only.

SCCP runs on TCP, using port 2000. The design goal of SCCP was to keep it "skinny," to allow the phone to have as little intelligence as needed. In this sense, the Cisco Unified Communications Manager (or older Cisco Call Manager) is designed to interface with other telephone technologies as a proxy, leaving the phone to deal with supporting the one proprietary protocol.

SCCP has a markedly different architecture from what we have seen already. SCCP is still an IP-based protocol, and there is the one point of contact that the phone uses for all of its signaling. However, the signaling design of SCCP has the remarkable property, unlike with SIP or H.323, that the phone is not self-contained as an extension. Rather, SCCP is entirely user event–based. The phone's job is to report back to the call manager, in real time, whenever a button is pressed. The call manager then pushes down to the phone any change in state that should accompany the button press. In this way, the entire logic as to what buttons mean is contained in the call manager, which locally runs the various telephone endpoint logic. In this way, SCCP has more in common with Remote Desktop than it has with telephone signaling protocols: the phone's logic really runs in some centralized terminal server, which is called the call manager. To emphasize this point, Table 2.17 lists a typical sequence of events between a phone and a call manager, from when the phone is taken off the hook.

As you can see, the phone's entire personality—the meaning of the buttons, what the display states, which lights are lit, the tones generated—are entirely controlled by the call manager.

Overall, this is a marked difference from true telephone signaling protocols. In this sense, then, one can consider SCCP to be mostly a remote control protocol for phones, and the call manager is thus left with the burden of implementing the true telephone protocol. Unfortunately, however, when SCCP is used with a packet-based voice mobility network, the protocol going over the wireless or edge network is going to be SCCP, and not whatever protocol the call manager is enabled with.

Bearer traffic, on the other hand, still uses RTP, as do the other protocols we have looked at so far. Therefore, most of the discussion on bearer traffic, and on voice traffic in general, holds for SCCP networks.

2.2.4 Skype

Skype is mentioned here because it is such an intriguing application. Famous for its resiliency when running over the Internet, or any other non-quality-of-service network, as well as for its chat feature and low-cost calls, questions will always come up about Skype. Undoubtedly, Skype has helped many organizations reduce long distance or international phone bills, and many business travelers have favored it when on the road and in a hotel, to avoid room and cell charges for telephone use.

Skype is a completely proprietary peer-to-peer protocol, encrypted hop-by-hop to prevent unauthorized snooping. There are plenty of resources available on how to use Skype, so it will be appropriate for us to stick with just the basics on how it applies for voice mobility.

The most important issue with Skype is that it is not manageable in an enterprise sense. Not only is it a service hosted outside the using enterprise, but the technology itself is encrypted

Table 2.17: Example SCCP Call Setup Event Flow

#	Direction	Event Name	State	Meaning
1	Phone → Call Manager	*Offhook*	Dialing	User has taken the phone off the hook.
2	Call Manager → Phone	*StationOutputDisplayText*		Displays a prompt that the phone is off hook and waiting for digits.
3	Call Manager → Phone	*SetRinger*		Turns off the ringer.
4	Call Manager → Phone	*SetLamp*		Turns on the light for the line that is being used.
5	Call Manager → Phone	*CallState*		Sets the phone up so that the user can hear audio and press buttons.
6	Call Manager → Phone	*DisplayPromptStatus*		The phone is not connected to any other extension yet.
7	Call Manager → Phone	*SelectSoftKeys*		
8	Call Manager → Phone	*ActivateCallPlane*		
9	Call Manager → Phone	*StartTone*		Starts a dial tone.
10	Phone → Call Manager	*KeypadButton* (dialed 7)		The user dialed the number 7.
11	Call Manager → Phone	*StopTone*		Stops the dial tone, acknowledging that a digit has been dialed.
12	Call Manager → Phone	*SelectSoftKeys*		Changes the keys of interest to just the number pad (no redial buttons, etc.).
13	Phone → Call Manager	*KeypadButton* (dialed 0)		The user dialed the number 0.
14	Phone → Call Manager	*KeypadButton* (dialed 2)		The user dialed the number 2.
15	Phone → Call Manager	*KeypadButton* (dialed 0)		The user dialed the number 0.
16	Call Manager → Phone	*SelectSoftKeys*	Ringing	Changes the keys of interest.
17	Call Manager → Phone	*CallState*		Changes the state of the phone.
18	Call Manager → Phone	*CallInfo*		
19	Call Manager → Phone	*DialedNumber*		Reports that 7020 has been dialed.
20	Call Manager → Phone	*StartTone*		Starts playing a ringback tone.
21	Call Manager → Phone	*DisplayPromptStatus*		Changes the prompt to show that the other side of the phone is ringing.
22	Call Manager → Phone	*CallInfo*		The call is still ringing.

Table 2.17: *Continued*

#	Direction	Event Name	State	Meaning
23	Call Manager → Phone	*StopTone*	Connected	Stops playing the ringback tone.
24	Call Manager → Phone	*DisplayPromptStatus*		Displays that the phone call was answered.
25	Call Manager → Phone	*OpenReceiveChannel*		Prepares for the downward leg of the call.
26	Phone → Call Manager	*OpenReceiveChannelAck*		Acknowledges the downward leg.
27	Call Manager → Phone	*StartMediaTransmission*		The call's bearer channel starts flowing.
28	Phone → Call Manager	*OnHook*	Hanging Up	The caller hung up.
29	Call Manager → Phone	*CloseReceiveChannel*		Tears down the receive leg.
30	Call Manager → Phone	*StopMediaTransmission*		Stops the bearer channel entirely.
31	Call Manager → Phone	*SetSpeakerMode*		Restores the phone to the original state.
32	Call Manager → Phone	*ClearPromptStatus*		
33	Call Manager → Phone	*CallState*		
34	Call Manager → Phone	*DisplayPromptStatus*		
35	Call Manager → Phone	*ActivateCallPlane*		
36	Call Manager → Phone	*SetLamp*		Turns off the light for the line that was in use.

to prevent even basic understanding or diagnosis. Furthermore, it cannot be run independent of Internet connectivity, and it is designed to find ways around firewalls. As a primarily consumer-oriented technology, Skype does not yet have the features necessary for enterprise deployments, and thus is severely limited in a sense useful for large-scale voice mobility.

Another main issue with Skype is that it does not take advantage of quality-of-service protocols to provide reliable or predictable, or even prioritized, voice quality. Traffic engineering with Skype is incredibly difficult, especially if one tries to predict how Skype will consume resources if large portions of the networked population choose to use it, inside or outside the office.

On the other hand, Skype comes with better, high-bitrate codecs that make voice sound much less tinny than the typical low-bitrate codecs used by telephones that may have to access the public switched telephone network (PSTN). Skype's ability to free itself from

PSTN integration as the standard case (Skype's landline telephone services can be thought of more as special cases) has allowed it to be optimized for better voice quality in a lossy environment.

Skype is unlikely to be useful in current voice mobility deployments, so it will not be mentioned much further in this book. However, Skype will always be found performing somewhere within the enterprise, and so its usage should be understood. As time progresses, it may be possible that people will have worked out a more full understanding of how to deploy Skype in the enterprise.

2.2.5 Polycom SpectraLink Voice Priority (SVP)

Early in the days of voice over Wi-Fi, a company called SpectraLink—now owned by Polycom—created a Wi-Fi handset, gateway, and a protocol between them to allow the phones to have good voice quality, when Wi-Fi itself did not yet have Wi-Fi Multimedia (WMM) quality of service. SVP runs as a self-contained protocol, for both signaling and bearer traffic, over IP, using a proprietary IP type (neither UDP nor TCP) for all of the traffic.

SVP is not intended to be an end-to-end signaling protocol. Rather, like Cisco's SCCP, it is intended to bridge between a network server that speaks the real telephone protocol and the proprietary telephone. Therefore, SCCP and SVP have a roughly similar architecture. The major difference is that SVP was designed with wireless in mind to tackle the early quality-of-service issues over Wi-Fi, whereas SCCP was designed mostly as a way of simplifying the operation of phone terminals over wireline IP networks.

Figure 2.6 shows the SVP architecture. The SVP system integrates into a standard IP PBX deployment. The SVP gateway acts as the location for the extensions, as far as the PBX is concerned. The gateway also acts as the coordinator for all of the wireless phones. SVP phones connect with the gateway, where they are provisioned. The job of the SVP gateway is to perform all of the wireless voice resource management of the network. The SVP performs the admission control for the phones, being configured with the maximum number of phones per access point and denying phones the ability to connect to it through access points that are oversubscribed. The SVP server also engages in performing timeslice coordination for each phone on a given access point.

This timeslicing function makes sense in the context of how SVP phones operate. SVP phones have proprietary Wi-Fi radios, and the protocol between the SVP gateway and the phone knows about Wi-Fi. Every phone reports back what access point it is associated to. When the phone is placed into a call, the SVP gateway and the phone connect their bearer channels. The timing of the packets sent by the phone is such that it is directly related to the timing of the phone sent by the gateway. Both the phone and the gateway have specific requirements on how the packets end up over the air. This, then, requires that the access points also be modified to be compatible with SVP. The role of the access point is to

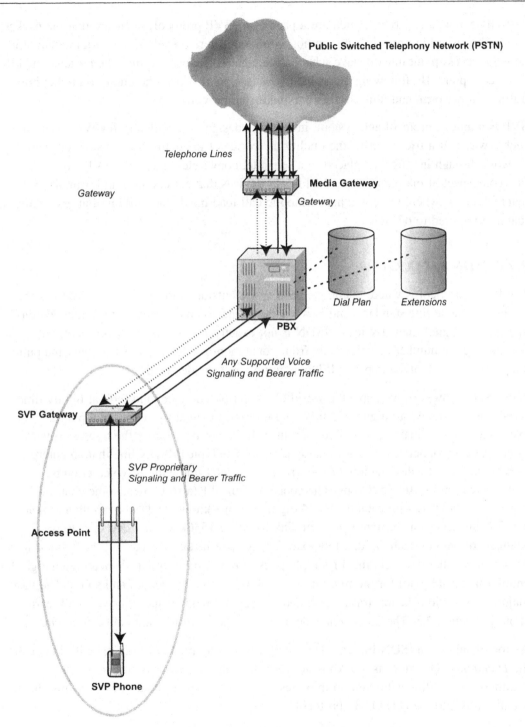

Figure 2.6: SVP Architecture

dutifully follow a few rules which are a part of the SVP protocol, to ensure that the packets access the air at high priority and are not reordered. There are additional requirements for how the access point must behave when a voice packet is lost and must be retransmitted by the access point. By following the rules, the access point allows the client to predict how traffic will perform, and thus ensures the quality of the voice.

SVP is a unique protocol and system, in that it is designed specifically for Wi-Fi, and in such a way that it tries to drive the quality of service of the entire SVP system on that network through intelligence placed in a separate, nonwireless gateway. SVP, and Polycom SpectraLink phones, are Wi-Fi-only devices that are common in hospitals and manufacturing, where there is a heavy mobile call load inside the building but essentially no roaming required to outside.

2.2.6 ISDN and Q.931

The ISDN protocol is where telephone calls to the outside world get started. ISDN is the digital telephone line standard, and is what the phone company provides to organizations that ask for digital lines. By itself, ISDN is not exactly a voice mobility protocol, but because a great number of voice calls from voice mobility devices must go over the public telephone network at some point, ISDN is important to understand.

With ISDN, however, we leave the world of packet-based voice, and look at tightly timed serial lines, divided into digital circuits. These circuits extend from the local public exchange—where analog phone lines sprout from before they run to the houses—over the same types of copper wires as for analog phones. The typical ISDN line that an enterprise uses starts from the designation *T1*, referring to a digital line with 24 voice circuits multiplexed onto it, for 1536kbps. The concept of the T1 (also known, somewhat more correctly, as a DS1, with each of the 24 digital circuits known as DS0s) is rather simple. The T1 line acts as a constant source or sink for these 1536kbps, divided up into the 24 channels of 64kbps each. With a few extra bits for overhead, to make sure both sides agree on which channel is which, the T1 simply goes in round-robin order, dedicating an eight-bit chunk (the actual byte) for the first circuit (channel), then the second, and so on. The vast majority of traffic is bearer traffic, encoded as standard 64kbps audio, as you will learn about in Section 2.3. The 23 channels dedicated for bearer traffic are called *B channels*.

As for signaling, an ISDN line that is running a signaling protocol uses the 24th line, called the *D channel*. This runs as a 64kbps network link, and standards define how this continuous serial line is broken up into messages. The signaling that goes over this channel usually falls into the ITU Q.931 protocol.

Q.931's job is to coordinate the setting up and tearing down of the independent bearer channels. To do this, Q.931 uses a particular structure for their messages. Because Q.931

can run over any number of different protocols besides ISDN, with H.323 being the other major one, the descriptions provided here will steer clear of describing how the Q.931 messages are packaged.

Table 2.18: Q.931 Basic Format

Protocol Discriminator	Length of Call Reference	Call Reference	Message Type	Information Elements
1 byte	1 byte	1–15 bytes	1 byte	*variable*

Table 2.18 shows the basic format of the Q.931 message. The *protocol discriminator* is always the number 8. The call reference refers to the call that is being referred to, and is determined by the endpoints. The information elements contain the message body, stored in an extensible yet compact format.

The message type is encompasses the activities of the protocol itself. To get a better sense for Q.931, the message types and meanings are:

- *SETUP*: this message starts the call. Included in the setup message is the dialed number, the number of the caller, and the type of bearer to use.

- *CALL PROCEEDING*: this message is returned by the other side, to inform the caller that the call is underway, and specifies which specific bearer channel can be used.

- *ALERTING*: informs the caller that the other party is ringing.

- *CONNECT*: the call has been answered, and the bearer channel is in use.

- *DISCONNECT*: the phone call is hanging up.

- *RELEASE*: releases the phone call and frees up the bearer.

- *RELEASE COMPLETE*: acknowledges the release.

There are a few more messages, but it is pretty clear to see that Q.931 might be the simplest protocol we have seen yet! There is a good reason for this: the public telephone system is remarkably uniform and homogenous. There is no reason for there to be flexible or complicated protocols, when the only action underway is to inform one side or the other of a call coming in, or choosing which companion bearer lines need to be used. Because Q.931 is designed from the point of view of the *subscriber*, network management issues do not need to be addressed by the protocol. In any event, a T1 line is limited to only 64kbps for the entire call signaling protocol, and that needs to be shared across the other 23 lines.

Digital PBXs use IDSN lines with Q.931 to communicate with each other and with the public telephone networks. IP PBXs, with IP links, will use one of the packet-based signaling protocols mentioned earlier.

2.2.7 SS7

Signaling System #7 (SS7) is the protocol that makes the public telephone networks operate, within themselves and across boundaries. Unlike Q.931, which is designed for simplicity, SS7 is a complete, Internet-like architecture and set of protocols, designed to allow call signaling and control to flow across a small, shared set of circuits dedicated for signaling, freeing up the rest of the circuits for real phone calls.

SS7 is an old protocol, from around 1980, and is, in fact, the seventh version of the protocol. The entire goal of the architecture was to free up lines for phone calls by removing the signaling from the bearer channel. This is the origin of the split signaling and bearer distinction. Before digital signaling, phone lines between networks were similar to phone lines into the home. One side would pick up the line, present a series of digits as tones, and then wait for the other side to route the call and present tones for success, or a busy network. The problem with this method of in-band signaling was that it required having the line held just for signaling, even for calls that could never go through. To free up the waste from the in-band signaling, the networks divided up the circuits into a large pool of voice-only bearer lines, and a smaller number of signaling-only lines. SS7 runs over the signaling lines.

It would be inappropriate here to go into any significant detail into SS7, as it is not seen as a part of voice mobility networks. However, it is useful to understand a bit of the architecture behind it.

SS7 is a packet-based network, structured rather like the Internet (or vice versa). The phone call first enters the network at the telephone exchange, starting at the *Service Switching Point* (SSP). This switching point takes the dialed digits and looks for where, in the network, the path to the other phone ought to be. It does this by sending requests, over the signaling network, to the *Service Control Point* (SCP). The SCP has the mapping of user-understandable telephone numbers to addresses on the SS7 network, known as *point codes*. The SCP responds to the SSP with the path the call ought to take. At this point, the switch (SSP) seeks out the destination switch (SSP), and establishes the call. All the while, routers called *Signal Transfer Points* (STPs) connect physical links of the network and route the SS7 messages between SSPs and SCPs.

The interesting part of this is that the SCP has this mapping of phone numbers to real, physical addresses. This means that phone numbers are abstract entities, like email addresses or domain names, and not like IP addresses or other numbers that are pinned down to some location. Of course, we already know the benefit of this, as anyone who has ever changed cellular carriers and kept their phone number has used this ability for that mapping to be changed. The mapping can also be regional, as toll-free 800 numbers take advantage of that mapping as well.

2.3 Bearer Protocols in Detail

The bearer protocols are where the real work in voice gets done. The bearer channel carries the voice, sampled by microphones as digital data, compressed in some manner, and then placed into packets which need to be coordinated as they fly over the networks.

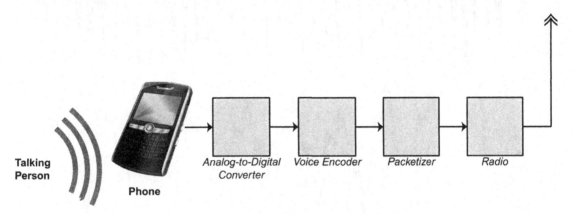

Figure 2.7: Typical Voice Recording Mechanisms

Voice, as you know, starts off as sound waves (Figure 2.7). These sound waves are picked up by the microphone in the handset, and are then converted into electrical signals, with the voltage of the signal varying with the pressure the sound waves apply to the microphone.

The signal (see Figure 2.8) is then sampled down into digital, using an *analog-to-digital converter*. Voice tends to have a frequency around 3000 Hz. Some sounds are higher—music especially needs the higher frequencies—but voice can be represented without significant distortion at the 3000Hz range. Digital sampling works by measuring the voltage of the signal at precise, instantaneous time intervals. Because sound waves are, well, wavy, as are the electrical signals produced by them, the digital sampling must occur at a high enough rate to capture the highest frequency of the voice. As you can see in the figure, the signal has a major oscillation, at what would roughly be said is the pitch of the voice. Finer variations, however, exist, as can be seen on closer inspection, and these variations make up the depth or richness of the voice. Voice for telephone communications is usually limited to 4000 Hz, which is high enough to capture the major pitch and enough of the texture to make the voice sound human, if a bit tinny. Capturing at even higher rates, as is done on compact discs and music recordings, provides an even stronger sense of the original voice.

Sampling audio so that frequencies up to 4000 Hz can be preserved requires sampling the signal at twice that speed, or 8000 times a second. This is according to the *Nyquist Sampling Theorem*. The intuition behind this is fairly obvious. Sampling at regular intervals is choosing which value at those given instants. The worst case for sampling would be if

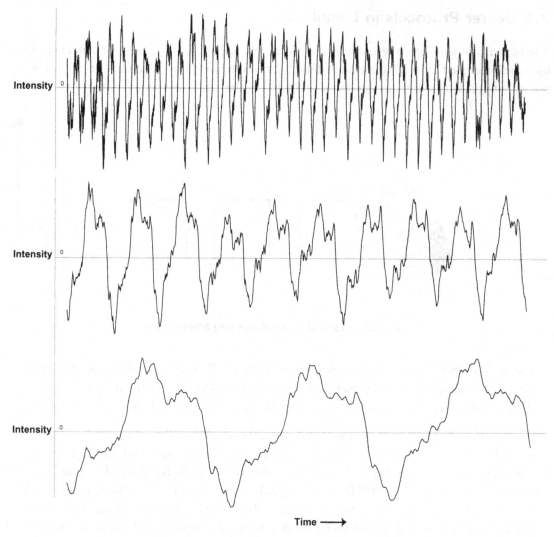

Figure 2.8:   Example Voice Signal, Zoomed in Three Times

one sampled a 4000 Hz, say, sine wave at 4000 times a second. That would guarantee to provide a flat sample, as the top pair of graphs in Figure 2.9 shows. This is a severe case of undersampling, leading to *aliasing* effects. On the other hand, a more likely signal, with a more likely sampling rate, is shown in the bottom pair of graphs in the same figure. Here, the overall form of the signal, including its fundamental frequency, is preserved, but most of the higher-frequency texture is lost. The sampled signal would have the right pitch, but would sound off.

The other aspect to the digital sampling, besides the 8000 samples-per-second rate, is the amount of detail captured vertically, into the intensity. The question becomes how many bits

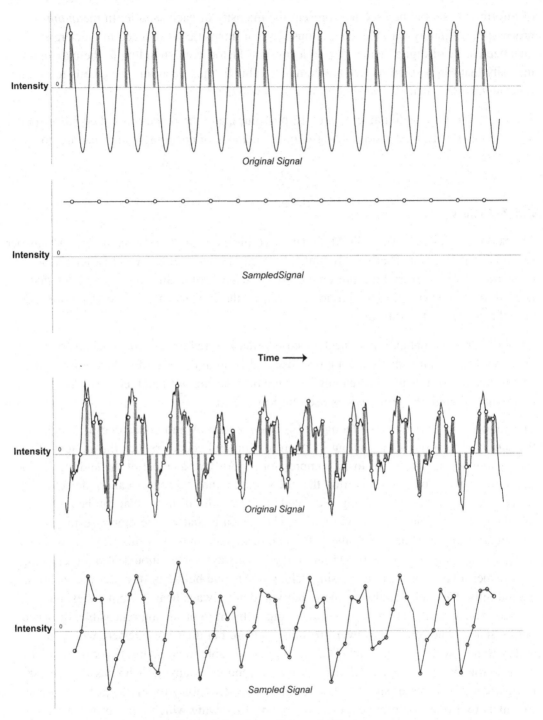

Figure 2.9: Sampling and Aliasing

of information should be used to represent the intensity of each sample. In the *quantization* process, the infinitely variable, continuous scale of intensities is reduced to a discrete, quantized scale of digital values. Up to a constant factor, corresponding to the maximum intensity that can be represented, the common value for quantization for voice is to 16 bits, for a number between $-2^{15} = -32,768$ to $2^{15} - 1 = 32,767$.

The overall result is a digital stream of 16-bit values, and the process is called *pulse code modulation* (PCM), a term originating in other methods of encoding audio that are no longer used.

2.3.1 Codecs

The 8000 samples-per-second PCM signal, at 16 bits per sample, results in 128,000 bits per second of information. That's fairly high, especially in the world of wireline telephone networks, in which every bit represented some collection of additional copper lines that needed to have been laid in the ground. Therefore, the concept of audio compression was brought to bear on the subject.

An audio or video compression mechanism is often referred to as a *codec*, short for *coder-decoder*. The reason is that the compressed signal is often thought of as being in a code, some sequence of bits that is meaningful to the decoder but not much else. (Unfortunately, in anything digital, the term *code* is used far too often.)

The simplest coder that can be thought of is a *null* codec. A null codec doesn't touch the audio: you get out what you put in. More meaningful codecs reduce the amount of information in the signal. All lossy compression algorithms, as most of the audio and video codecs are, stem from the realization that the human mind and senses cannot detect every slight variation in the media being presented. There is a lot of noise that can be added, in just the right ways, and no one will notice. The reason is that we are more sensitive to certain types of variations than others. For audio, we can think of it this way. As you drive along the highway, listening to AM radio, there is always some amount of noise creeping in, whether it be from your car passing behind a concrete building, or under power lines, or behind hills. This noise is always there, but you don't always hear it. Sometimes, the noise is excessive, and the station becomes annoying to listen to or incomprehensible, drowned out by static. Other times, however, the noise is there but does not interfere with your ability to hear what is being said. The human mind is able to compensate for quite a lot of background noise, silently deleting it from perception, as anyone who has noticed the refrigerator's compressor stop or realized that a crowded, noisy room has just gone quiet can attest to. Lossy compression, then, is the art of knowing which types of noise the listener can tolerate, which they cannot stand, and which they might not even be able to hear.

(Why noise? Lossy compression is a method of deleting information, which may or may not be needed. Clearly, every bit is needed to restore the signal to its original sampled state. Deleting a few bits requires that the decompressor or the decoder restore those deleted bits' worth of information on the other end, filling them in with whatever the algorithm states is appropriate. That results in a difference of the signal, compared to the original, and that difference is distortion. Subtract the two signals, and the resulting difference signal is the noise that was added to the original signal by the compression algorithm. One only need amplify this noise signal to appreciate how it sounds.)

2.3.1.1 G.711 and Logarithmic Compression

The first, and simplest, lossy compression codec for audio that we need to look at is called *logarithmic compression*. Sixteen bits is a lot to encode the intensity of an audio sample. The reason why 16 bits was chosen was that it has fine enough detail to adequately represent the variations of the softer sounds that might be recorded. But louder sounds do not need such fine detail while they are loud. The higher the intensity of the sample, the more detailed the 16-bit sampling is *relative* to the intensity. In other words, the 16-bit resolution was chosen conservatively, and is excessively precise for higher intensities. As it turns out, higher intensities can tolerate even more error than lower ones—in a relative sense, as well. A higher-intensity sample may tolerate four times as much error as a signal half as intense, rather than the two times you would expect for a linear process. The reason for this has to do with how the ear perceives sound, and is why sound levels are measured in decibels. This is precisely what logarithmic compression does. Convert the intensities to decibels, where a 1 dB change sounds roughly the same at all intensities, and a good half of the 16 bits can be thrown away. Thus, we get a 2:1 compression ratio.

The ITU G.711 standard is the first common codec we will see, and uses this logarithmic compression. There are two flavors of G.711: *μ-law* and *A-law*. μ-law is used in the United States, and bases its compression on a discrete form of taking the logarithm of the incoming signal. First, the signal is reduced to a 14-bit signal, discarding the two least-significant bits. Then, the signal is divided up into ranges, each range having 16 intervals, for four bits, with twice the spacing as that of the next smaller range. Table 2.19 shows the conversion table.

The number of the interval is where the input falls within the range. 90, for example, would map to 0xee, as 90 − 31 = 59, which is 14.75, or 0xe (rounded down) away from zero, in steps of four. (Of course, the original 16-bit signal was four times, or two bits, larger, so 360 would have been one such 16-bit input, as would have any number between 348 and 363. This range represents the loss of information, as 363 and 348 come out the same.)

A-law is similar, but uses a slightly different set of spacings, based on an algorithm that is easier to see when the numbers are written out in binary form. The process is simply to take the binary number and encode it by saving only four bits of significant digits (except the

Table 2.19: μ-Law Encoding Table

Input Range	Number of Intervals in Range	Spacing of Intervals	Left Four Bits of Compressed Code	Right Four Bits of Compressed Code
8158 to 4063	16	256	0x8	*number of interval*
4062 to 2015	16	128	0x9	*number of interval*
2014 to 991	16	64	0xa	*number of interval*
990 to 479	16	32	0xb	*number of interval*
478 t o 223	16	16	0xc	*number of interval*
222 to 95	16	8	0xd	*number of interval*
94 to 31	16	4	0xe	*number of interval*
30 to 1	15	2	0xf	*number of interval*
0	1	1	0xf	0xf
−1	1	1	0x7	0xf
−31 to −2	2	15	0x7	*number of interval*
−32 to −95	4	16	0x6	*number of interval*
−223 to −96	8	16	0x5	*number of interval*
−479 to −224	16	16	0x4	*number of interval*
−991 to −480	32	16	0x3	*number of interval*
−2015 to −992	64	16	0x2	*number of interval*
−4063 to −2016	128	16	0x1	*number of interval*
−8159 to −4064	256	16	0x0	*number of interval*

leading one), and to record the base-2 (binary) exponent. This is how floating-point numbers are encoded. Let's look at the previous example. The number 360 is encoded in 16-bit binary as

```
0000 0001 0110 1000
```

with spaces placed every four digits for readability. A-law only uses the top 13 bits. Thus, as this number is unsigned, it can be represented in floating point as

```
1.01101 (binary) × 2⁵.
```

The first four significant digits (ignoring the first 1, which must be there for us to write the number in binary scientific notation, or floating point), are "0110", and the exponent is 5. A-law then records the number as

```
0001 0110
```

where the first bit is the sign (0), the next three are the exponent, minus four, and the last four are the significant digits.

A-law is used in Europe, on their telephone systems. For voice over IP, either will usually work, and most devices speak in both, no matter where they are sold. The distinctions are now mostly historical.

G.711 compression preserves the number of samples, and keeps each sample independently of the others. Therefore, it is easy to figure out how the samples can be packaged into packets or blocks. They can be cut arbitrarily, and a byte is a sample. This allows the codec to be quite flexible for voice mobility, and should be a preferred option.

Error concealment, or *packet loss concealment* (PLC), is the means by which a codec can recover from packet loss, by faking the sound at the receiver until the stream catches up. G.711 has an extension, known as G.711I, or G.711, Appendix I. The most trivial error concealment technique is to just play silence. This does not really conceal the error. An additional technique is to repeat the last valid sample set—usually, a 10ms or 20ms packet's worth—until the stream catches up. The problem is that, should the last sample have had a plosive—any of the consonants that have a stop to them, like a *p, d, t, k*, and so on—the plosive will be repeated, providing an effect reminiscent of a quickly skipping record player or a 1980s science-fiction television character.* Appendix I states that, to avoid this effect, the previous samples should be tested for the fundamental wavelength, and then blocks of those wavelengths should be cross-faded together to produce a more seamless recovery. This is a purely heuristic scheme for error recovery, and competes, to some extent, with just repeating the last segment then going silent.

In many cases, G.711 is not even mentioned when it is being used. Instead, the codec may be referred to as PCM with μ-law or A-law encoding.

2.3.1.2 G.729 and Perceptual Compression

ITU G.729 and the related G.729a specify using a more advanced encoding scheme, which does not work sample by sample. Rather, it uses mathematical rules to try to relate neighboring samples together. The incoming sample stream is divided into 10ms blocks (with 5ms from the next block also required), and each block is then analyzed as a unit. G.729 provides a $16:1$ compression ratio, as the incoming 80 samples of 16 bits each are brought down to a ten-byte encoded block.

The concept around G.729 compression is to use *perceptual compression* to classify the type of signal within the 10ms block. The concept here is to try to figure out how neighboring samples relate. Surely, they do relate, because they come from the same voice and the same pitch, and pitch is a concept that requires time (thus, more than one sample). G.729 uses a couple of techniques to try to figure out what the sample must "sound like," so it can then throw away much of the sample and transmit only the description of the sound.

To figure out what the sample block sounds like, G.729 uses *Code-Excited Linear Prediction* (CELP). The idea is that the encoder and decoder have a *codebook* of the basics

---

* *Max Headroom* played for one season on ABC during 1987 and 1988. The title character, an artificial intelligence, was famous for stuttering electronically.

of sounds. Each entry in the codebook can be used to generate some type of sound. G.729 maintains two codebooks: one fixed, and one that adapts with the signal. The model behind CELP is that the human voice is basically created by a simple set of flat vocal chords, which *excite* the airways. The airways—the mouth, tongue, and so on—are then thought of as signal filters, which have a rather specific, predictable effect on the sound coming up from the throat.

The signal is first brought in and *linear prediction* is used. Linear prediction tries to relate the samples into the block to the previous samples, and finds the optimal mapping. ("Optimal" does not always mean "good," as there is almost always an optimal way to approximate a function using a fixed number of parameters, even if the approximation is dead wrong. Recall Figure 2.9.) The excitation provides a representation that represents the overall type of sound, a hum or a hiss, depending on the word being said. This is usually a simple sound, an "uhhh" or "ahhh." The linear predictor figures out how the humming gets shaped, as a simple filter. What's left over, then, is how the sound started in the first place, the excitation that makes up the more complicated, nuanced part of speech. The linear prediction's effects are removed, and the remaining signal is the *residue*, which must relate to the excitations. The nuances are looked up in the codebook, which contains some common residues and some others that are adaptive. Together, the information needed for the linear prediction and the codebook matches are packaged into the ten-byte output block, and the encoding is complete. The encoded block contains information on the pitch of the sound, the adaptive and fixed codebook entries that best match the excitation for the block, and the linear prediction match.

On the other side, the decoding process looks up the codebooks for the excitations. These excitations get filtered through the linear predictor. The hope is that the results sound like human speech. And, of then, it does. However, anyone who has used cellphones before is aware that, at times, they can render human speech into a facsimile that sounds quite like the person talking, but is made up of no recognizable syllables. That results from a CELP decoder struggling with a lossy channel, where some of the information is missing, and it is forced to fill in the blanks.

G.729a is an annex to G.791, or a modification, that uses a simpler structure to encode the signal. It is compatible with G.729, and so can be thought of as interchangeable for the purposes of this discussion.

2.3.1.3 Other Codecs

There are other voice codecs that are beginning to appear in the context of voice mobility. These codecs are not as prevalent as G.711 and G.729—some are not available in more than softphones and open-source implementations—but it is worth a paragraph on the subject. These newer coders are focused on improving the error concealment, or having better delay or jitter tolerances, or having a richer sound. One such example is the *Internet Low Bitrate Codec* (iLBC), which is used in a number of consumer-based peer-to-peer voice applications such as Skype.

Because the overhead of packets on most voice mobility networks is rather high, finding the highest amount of compression should not be the aim when establishing the network. Instead, it is better to find the codec which are supported by the equipment and provide the highest quality of voice over the expected conditions of the network. For example, G.711 is fine in many conditions, and G.729 might not be necessary. Chapter 3 goes into some of the factors that can influence this.

2.3.2 RTP

The codec defines only how the voice is compressed and packaged. The voice still needs to be placed into well-defined packets and sent over the network.

The *Real-time Transport Protocol* (RTP), defined in RFC 3550, defines how voice is packetized on most IP-based networks. RTP is a general-purpose framework for sending real-time streaming traffic across networks, and is used for nearly all media streaming, including voice and video, where real-time delivery is essential.

RTP is usually sent over UDP, on any port that the applications negotiate. The typical RTP packet has the structure given in Table 2.20.

Table 2.20: RTP Format

Flags	Sequence Number	Timestamp	SSRC	CSRCs	Extensions	Payload
2 bytes	2 bytes	4 bytes	4 bytes	4 bytes × number of contributors	variable	*variable*

The idea behind RTP is that the sender sends the timestamp that the first byte of data in the payload belongs to. This timestamp gives a precise time that the receiver can use to reassemble incoming data. The sequence number also increases monotonically, and can also establish the order of incoming data. The SSRC, for *Synchronization Source*, is the stream identifier of the sender, and lets devices with multiple streams coming in figure out who is sending. The CSRCs, for *Contributing Sources*, are other devices that may have contributed to the packet, such as when a conference call has multiple talkers at once.

The most important fields are the timestamp (see Table 2.20) and the payload type (see Table 2.21). The payload type field usually specifies the type of codec being used in the stream.

Table 2.21: The RTP Flags Field

	Version	Padding	Extension (X)	Contributor Count (CC)	Marked	Payload Type (PT)
Bit:	0–1	2	3	4–7	8	9–15

Table 2.22: Common RTP Packet Types

Payload Type	Encoded Name	Meaning
0	PCMU	G.711 with μ-law
3	GSM	GSM
8	PCMA	G.711 with A-law
18	G729	G.729 or G.729a

Table 2.22 shows the most common voice RTP types. Numbers greater than 96 are allowed, and are usually set up by the endpoints to carry some dynamic stream.

When the codec's output is packaged into RDP, it is done so to both avoid splitting necessary information and causing too many packets per second to be sent. For G.711, an RTP packet can be created with as many samples as desired for the given packet rate. Common values are 20ms and 30ms. Decoders know to append the samples across packets as if they were in one stream. For G.729, the RTP packet must come in 10ms multiples, because G.729 only encodes 10ms blocks. An RTP packet with G.729 can have multiple blocks, and the decoder knows to treat each block separately and sequentially. G.729 phones commonly stream with RTP packets holding 20ms or larger, to avoid having too many packets in the network.

2.3.2.1 Secure RTP

RTP itself has a security option, designed to allow the contents of the RTP stream to be protected while still allowing the quick reassembly of a stream and the robustness of allowing parts of the stream to be lost on the network. *Secure RTP* (SRTP) uses the *Advanced Encryption Standard* (AES) to encrypt the packets. (AES will later have a starring role in Wi-Fi encryption, as well as for use with IPsec.) The RTP stream requires a key to be established. Each packet is then encrypted with AES running in *counter mode*, a mode where intervening packets can be lost without disrupting the decryptability of subsequent packets in the sequence. Integrity of the packets is ensured by the use of the *HMAC-SHA1* keyed signature, for each packet.

How the SRTP stream gets its keys is not specified by SRTP. However, SIPS provides a way for this to be set up that is quite logical. The next section will discuss how this key exchange works.

2.3.3 SDP and Codec Negotiations

RTP only carries the voice, and there must be some associated way to signal the codecs which are supported by each end. This is fundamentally a property of signaling, but, unlike call progress messages and advanced PBX features, is tied specifically to the bearer channel.

SIP (see Section 2.2.1) uses SDP to negotiate codecs and RTP endpoints, including transports, port numbers, and every other aspect necessary to start RTP streams flowing. SDP, defined in RFC 4566, is a text-based protocol, as SIP itself is, for setting up the various legs of media streams. Each line represents a different piece of information, in the format of *type = value*.

Table 2.23: Example of an SDP Description

```
v=0
o=7010 1352822030 1434897705 IN IP4 192.168.0.10
s=A_conversation
c=IN IP4 192.168.0.10
t=0 0
m=audio 9000 RTP/AVP 0 8 18
a=rtpmap:0 PCMU/8000/1
a=rtpmap:8 PCMA/8000/1
a=rtpmap:18 G729/8000/1
a=ptime:20
```

Table 2.23 shows an example of an SDP description. This description is for a phone at IP address 192.168.0.10, who wishes to receive RTP on UDP port 9000. Let's go through each of the fields.

- Type "v" represents the protocol version, which is 0.

- Type "o" holds information about the originator of this request, and the session IDs. Specifically, it is divided up into the *username, session ID, session version, network type, address type,* and *address*. "7010" happens to be the dialing phone number. The two large numbers afterward are identifiers, to keep the SDP exchanges straight. The "IN" refers to the address being an Internet protocol address; specifically, "IP4" for IPv4, of "192.168.0.10". This is where the originator is.

- Type "s" is the session name. The value given here, "A_conversation", is not particularly meaningful.

- Type "c" specifies how the originator must be reached at—its connection data. This is a repetition of the IP address and type specifications for the phone.

- Type "t" is the timing for the leg of the call. The first "0" represents the start time, and the second represents the end time. Therefore, there is no particular timing bounds for this call.

- The "m" line specifies the media needed. In this case, as with most voice calls, there is only one voice stream from the device, so there is only one media line. The next parameters are the *media type, port, application,* and then the list of RTP types, for RTP. This call is an "audio" call, and the phone will be listening on port 9000. This is a UDP port, because the application is "RTP/AVP", meaning that it is plain RTP. ("AVP" means that this is standard UDP with no encryption. There is an "RTP/SAVP" option, mentioned shortly.) Finally, the RTP formats the phone can take are 0, 8, and 18, as specified in Table 2.22.

- The next three lines are the codecs that are supported in detail. The "a" field specifies an attribute. The "a=rtpmap" attribute means that the sender wants to map RTP packet types to specific codec setups. The line is formatted as *packet type, encoded name/ bitrate/parameters*. In the first line, RTP packet type "0" is mapped to "PCMU" at 8000 samples per second. The default mapping of "0" is already PCM (G.711) with μ-law, so the new information is the sample rate. The second line asks for A-law, mapping it to 8. The third line asks for G.729, asking for 18 as the mapping. Because the phone only listed those three types, those are the only types it supports.

- The last line is also an attribute. "a=ptime" is requesting that the other party send 20ms packets. The other party is not required to submit to this request, as it is only a suggestion. However, this is a pretty good sign that the sender of the SDP message will also send at 20ms.

The setup message in Table 2.23 was originally given in a SIP INVITE message. The responding SIP OK message from the other party gave its SDP settings.

Table 2.24 shows this example response. Here, the other party, at IP address 10.0.0.10, wants to receive on UDP port 11690 an RTP stream with the three codecs PCMU, GSM, and PCMA. It can also receive a format known as "telephone-event." This corresponds to the RTP payload format for sending digits while in the middle of a call (RFC 4733). Some codecs, like G.729, can't carry a dialed digit as the usual audio beep, because the beep gets distorted by the codec. Instead, the digits have to be sent over RTP, embedded in the stream. The sender of this SDP is stating that they support it, and would like to be sent in RTP type 101, a dynamic type that the sender was allowed to choose without restriction. Corresponding to this is the attribute "a=fmtp", which applies to this 101-digit type. "fmtp" lines don't mean anything specific to SDP; instead, the request of "0–16" gets forwarded to the telephone event protocol handler. It is not necessary to go into further details here on what "0–16" means. The "a=silenceSupp" line would activate *silence suppression*, in which

Table 2.24: Example of an SDP Responding Description

```
v=0
o=root 10871 10871 IN IP4 10.0.0.10
s=session
c=IN IP4 10.0.0.10
t=0 0
m=audio 11690 RTP/AVP 0 3 8 101
a=rtpmap:0 PCMU/8000
a=rtpmap:3 GSM/8000
a=rtpmap:8 PCMA/8000
a=rtpmap:101 telephone-event/8000
a=fmtp:101 0-16
a=silenceSupp:off - - - -
a=ptime:20
a=sendrecv
```

packets are not sent when the caller is not talking. Silence suppression has been disabled, however. Finally, the "a=sendrecv" line means that the originator can both send and receive streaming packets, meaning that the caller can both talk and listen. Some calls are intentionally one-way, such as lines into a voice conference where the listeners cannot speak. In that case, the listeners may have requested a flow with "a=recvonly".

After a device gets an SDP request, it knows enough information to send an RTP stream back to the requester. The receiver need only choose which media type it wishes to use. There is no requirement that both parties use the same codec; rather, if the receiver cannot handle the codec, the higher-layer signaling protocol needs to reject the setup. With SIP, the called party will not usually stream until it accepts the SIP INVITE, but there is no further handshaking necessary once the call is answered and there are packets to send.

For SRTP usage with SIPS, SDP allows for the SRTP key to be specified using a special header:

```
a=crypto:1 AES_CM_128_HMAC_SHA1_32 ⇒
inline:c3bFaGA+Seagd117041az3g113geaG54aKgd50Gz
```

This specifies that the SRTP AES counter with HMAC_SHA1 is to be used, and specifies the key, encoded in base-64, that is to be used. Both sides of the call send their own randomly generated keys, under the cover of the TLS-protected link. This forms the basis of RTP/SAVP.

CHAPTER 3

Elements of Voice Quality

3.0 Introduction

This chapter examines the factors that go into voice quality. First, we will look at how voice quality was originally, and introduce the necessary measurement metrics (MOS and R-Value). After that, we will move on to describing the basis for repeatable, objective metrics, by a variety of models that start out taking actual voice samples into account, but then turn into guidelines and formulas about loss and delay that can be used to predict network quality. Keep in mind that the point of the chapter is not to substitute for thousand-page telephony guidelines, but to introduce the reader to the basics for what it takes, and argue that perhaps—with mobility—the exactitude typically expected in static voice deployments is not that useful.

3.1 What Voice Quality Really Means

Chapter 2 laid the groundwork for how voice is carried. But what makes some phone calls sound better than others? Why do some voice mobility networks sound tinny and robotic, where others sound natural and clear? In voice, there are two ways to look at voice quality: gather a number of people and survey them about the quality of the call, or try to use some sort of electronic measurement and deduce, from there, what a real person might think.

3.1.1 Mean Opinion Score and How It Sounds

The *Mean Opinion Score*, or MOS (sometimes redundantly called the MOS score), is one way of ranking the quality of a phone call. This score is set on a five-point scale, according to the following ranking:

5. Excellent

4. Good

3. Fair

2. Poor

1. Bad

MOS never goes below 1, or above 5.

doi:10.1016/B978-1-85617-508-1.00001-3.

There is quite a science to establishing how to measure MOS based on real-world human studies, and the depth they go into is astounding. ITU P.800 lays out procedures for measuring MOS. Annex B of P.800 defines listening tests to determine quality in an absolute manner. The test requirements are spelt out in detail. The room to be used should be between 30 and 120 cubic meters, to ensure the echo remains within known values. The phone under test is used to record a series of phrases. The listeners are brought in, having been selected from a group that has never heard the recorded sentence lists, in order to avoid bias. The listeners are asked to mark the quality of the played-back speech, distorted as it may be by the phone system. The listeners' scores, on the one-to-five scale, are averaged, and this becomes the MOS for the system. The goal of all of this is to attempt to increase the repeatability of such experiments.

Clearly, performing MOS tests is not something that one would imagine can be done for most voice mobility networks. However, the MOS scale is so well known that the 1 to 5 scale is used as the standard yardstick for all voice quality metrics. The most important rule of thumb for the MOS scale is this: a MOS of 4.0 or better is toll-quality. This is the quality that voice mobility networks have to achieve, because this is the quality that nonmobility voice networks provide every day. Forgiveness will likely offered by users when the problem is well known and entirely relatable, such as for bad-quality calls when in a poor cellular coverage area. But, once inside the building, enterprise voice mobility users expect the same quality wirelessly as they do when using their desk phone.

Thus, when a device reports the MOS for a call, the number you are seeing has been generated electronically, based on formulas that are thought to be reasonable facsimiles of the human experience.

3.1.2 PESQ: How to Predict MOS Using Mathematics

Therefore, we turn to how the predictions of voice quality can actually be made electronically. ITU P.862 introduces *Perceptual Evaluation of Speech Quality*, the PESQ metric. PESQ is designed to take into account all aspects of voice quality, from the distortion of the codecs themselves to the effects of filtering, delay variation, and dropouts or strange distortions. PESQ was verified with a number of real MOS experiments to make sure that the numbers are reasonable within the range of normal telephone voices.

PESQ is measured on a 1 to 4.5 scale, aligning exactly with the 1 to 5 MOS scale, in the sense that a 1 is a 1, a 2 is a 2, and so on. (The area from 4.5 to 5 in PESQ is not addressed.) PESQ is designed to take into account many different factors that alter the perception of the quality of voice.

The basic concept of PESQ is to have a piece of software or test measurement equipment compare two versions of a recording: the original one and the one distorted by the telephone equipment being measured. PESQ then returns with the expected mean opinion score a group of real listeners are likely to have thought.

PESQ uses a perceptual model of voice, much the same way as perceptual voice codecs do. The two audio samples are mapped and remapped, until they take into account known perceptual qualities, such as the human change in sensitivity to loudness over frequency (sounds get quieter at the same pressure levels as they get higher in pitch). The samples are then matched up in time, eliminating any absolute delay, which affects the quality of a phone call but not a recording. The speech is then broken up into chunks, called *utterances*, which correspond to the same sound in both the original and distorted recording. The delays and distortions are then analyzed, counted, and correlated, and a number measuring how far removed the distorted signal is from the original signal is presented. This is the PESQ score.

PESQ is our first entry into the area of mathematical, or algorithmic, determination of call quality. It is good for measuring how well a new codec works, or how much noise is being injected into the sample. However, because it requires comparing what the talker said and what the listener heard, it is not practical for real-time call quality measurements.

3.1.3 Voice Over IP: The E-Model

How can we have access to a way of measuring the quality of voice over IP networks, measuring the contribution to the distortion caused uniquely by the voice mobility network? Once again, the ITU is here to the rescue. ITU G.107 introduces the *E-model*, a computational model that takes into account measurable network effects to determine the call quality that should have been expected for the call as seen on the network.

The output of the E-model is what is known as an *R-value*, a number on a scale from 0–100, similar to that used to produce letter grades in high school. The structure is as follows:

90% and up:	Very Satisfied
80%–90%:	Satisfied
70%–80%:	Some Users Dissatisfied
60%–70%:	Many Users Dissatisfied
60%–70%:	Nearly All Users Dissatisfied
0%–50%:	Not Recommended

The E-model includes noise levels injected, distortion, packet loss probabilities, mean delays, and echo problems. Table 3.1 shows the entire list of values that are used in computing the R-value for the E-model, including the allowed values and the defaults. With all of the defaults in place, the R-value will come out as 93.2, an excellent result. When G.107 is used in standard telephone networks, all of these values need to be measured. However, when measuring a voice mobility network, reasonable assumptions can be made about the quality of the end devices, the loudness of the room, how much echo is cancelled, and so on, and what is left is the contribution made by the packet network.

(Take note of the Advantage Factor, which is a fudge factor that lets testers add bonus points for mobility or convenience.)

Table 3.1: Components that Go into Calculating the R-Value

Name	Default Value	Permitted Range
Send Loudness	8 dB	0 to 18 dB
Receive Loudness	2 dB	−5 to 15 dB
Sidetone Masking	15 dB	10 to 20 dB
Listener Sidetone	18 dB	13 to 23 dB
D-Value, Send Side	3	−3 to 3
D-Value, Receive Side	3	−3 to 3
Talker Echo Loudness	65 dB	5 to 65 dB
Weighted Echo Path Loss	110 dB	5 to 110 dB
Mean One-Way Delay	0 ms	0 to 500 ms
Round-trip Delay	0 ms	0 to 1000 ms
Absolute Delay	0 ms	0 to 500 ms
Quantization Distortion	1 unit	1 to 14 units
Equipment Impairment	0	0 to 40
Path-loss Robustness	1	1 to 40
Random Packet-loss Probability	0%	0% to 20%
Burst Ratio	1	1 to 2
Circuit Noise	−70 dBm0p	−80 to −40 dBm0p
Noise Floor	−64 dBmp	
Room Noise at Send Side	35 dB(A)	35 to 85 dB(A)
Room Noise at Receive Side	35 dB(A)	35 to 85 dB(A)
Advantage Factor	0	0 to 20

The input to the voice mobility–focused E-model become the network effects, and the choice of codec. The choice of codec is key, because codecs introduce both distortion and delay, and the delay needs to be known, to be added to network delay.

The R-value result of the E-model can be mapped directly to the MOS value that we see from PESQ and subjective sampling. The formula for this (which follows) is graphed in Figure 3.1. Don't feel the need to try to calculate this, though, as most good tools that report R-value will also map them back to MOS.

$$MOS = 1 + 0.0035\,R + R(R-60)(100-R)\cdot 7/10^6$$

The overall R-value is made up of the sum of a few components. Specifically,

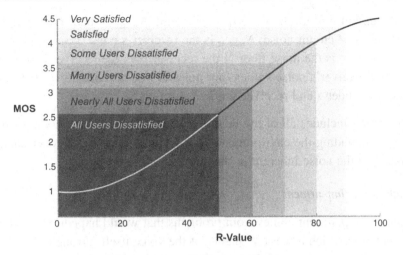

Figure 3.1: MOS from R-value

$$R = SNR - I_{\text{simultaneous}} - I_{\text{delay}} - I_{\text{loss-codec}} + A$$

where R is the R-value, not surprisingly; SNR is the signal-to-noise ratio for the voice, taking into account all of the background noise; $I_{\text{simultaneous}}$ is the impairment that happens simultaneously with the voice signal; I_{delay} is the impairment caused by delays in the voice stream; $I_{\text{loss-codec}}$ is the impairment caused by codec choice and packet loss; and A is the advantage factor that allows for hand-tuning the results to fit known MOS values, based on the perceived advantage the caller sees in the type of technology she is using. Each of these values is scaled so that the overall value can be in a range of 0 to 100.

Let's examine each value in turn.

3.1.3.1 Noise Impairment

The signal-to-noise ration is based on a loudness of the call as injected by the sender, and the noise levels which interfere with the call. The specific formula is

$$SNR = 15 - 1.5\,(SLR + N)$$

where SLR is the *send loudness*, and N is the sum of the noise values; both of these values are divided up into contributions from the circuit, room noise at the sender and receiver, and the receiver's noise floor. The send loudness is measured in decibels (dB) between the sender and a defined zero-point value. The noise sum N is composed, specifically, as

$$N = 10\log\left(10^{N_{\text{circuit}}/10} + 10^{N_{\text{sender}}/10} + 10^{N_{\text{receiver}}/10} + 10^{N_{\text{floor}}/10}\right)$$

where N_c is the circuit noise, relative to the zero-point; N_{sender} is the sender's noise, converted into units of circuit noise; $N_{receiver}$ is the receiver's noise, converted into units of circuit noise; and N_{floor} is the noise floor at the receiver plus the receiver loudness together. The sender's and receiver's noise values (not noise floors) are themselves basically the room noise at the sender's and receiver's side.

Together, this rating includes all of the factors that would affect the amount of background noise in the call, including the environmental noise both around the listener and picked up from the talker and the noise inherent in the circuits.

3.1.3.2 Simultaneous Impairment

The simultaneous impairment comes from problems that would happen no matter what the environment, and which affect the quality of the voice itself, through basic signal distortions. $I_{simultaneous}$ is made up of the sum of three factors. The first factor is the decrease in quality caused by there not being enough sender and receiver loudness together. Essentially, the call is too quiet. The second factor comes about from poor *sidetone*. Sidetone is, in this context, the sound of your own voice that comes back from the speaker in the handset. Sidetone is a natural extension of the normal act of speaking. When a person speaks, the vibration travels both through the person's head and through the environment, to the ears. When a person has a cold or is wearing an earplug, the natural feedback from the environment is deadened, and the person feels that she is speaking into a fog. This sidetone is how a caller can tell that he or she is speaking when the call is on mute: the caller will fail to hear any sound coming back, and the phone loses the effect of sounding "open." On landline phones, the lack of sidetone can be quite disturbing, and can give the speaker the impression that the phone is dead or that he or she is speaking too softly. On the other hand, the presence of too much sidetone can make the speaker stop talking, as the effect becomes one of shouting over one's own voice. Cellphones are notorious for having poor or nonexistent sidetone, and the result is that the speaker cannot effectively tell how loud she is speaking. The two sidetone values in Table 3.1 are weighted together, in a complex formula that looks for the optimal value. The third factor is caused by quantizing distortion, which is caused by the phone being digitally sampled into PCM, without regard to the codec.

3.1.3.3 Delay Impairment

The delay impairment factor I_{delay} stems from all of the sources of delays, and is itself the sum of three factors. The first factor is caused by the talker echo. Echo of reasonable loudness that comes back to the talker quickly is the sidetone mentioned previously, and is necessary. This is an example of *near-end echo*, because it originates in the talker's phone. However, if the echo is introduced too late from when the original sound was made, the echo ceases to be helpful and becomes a hindrance that usually gives the speaker some amount of pause, as he or she must compete with the delayed version of what is said. More often than not, this echo comes from the network itself, or the receiver's end, echoing the sounds back. This is called *far end echo*, because it comes from the far end of the call. Old-style acoustic handsets pick up near-end echo from the hollow tube between the microphone and the receiver, adding the comfort sidetone. All receivers pick up far-end echo from the crosstalk between the microphone and speaker at the other end. Every digital voice device has some amount of *echo cancellation*, which uses digital techniques to store the most recent sounds sent through the microphone and subtract them from the speaker when they come back. Sometimes, that is not enough, as anyone who has used a cellphone can attest to, as long echoes still come through now and again. Far-end echoes of this form result from long network round-trip delays that are not necessarily long enough to interrupt the conversation, but long enough to defeat the echo canceller. The problem is that echo cancellers can hold on to only so many milliseconds of recent voice and effectively cancel them out. If the echo is longer than that storage, the entirety of the echo will come through. The storage period is usually referred to as the *echo tail length*, for the reason that echoes do not usually come back as one reflection, but are spread out over time, and the amount of time the echo gets spread over is known as the echo tail.

One scenario where talker echo is prevalent is with conference calling. Many PBXs offer conference features, and many outside services exist to provide bridge number dialing. As conferences grow in size, the echo from each of the lines on the call increases the burden on the conference hosting service to filter out all of the echoes from those lines.

The second factor is caused by listener echo. Listener echo is a second-order echo: the sound goes from the talker to the listener, to the talker, and then back to the listener. It may also be caused by unusual problems, like buggy echo cancellers or line mixers, that introduce echo in the forward path. This is fairly rare.

The third factor is caused by absolute delay in the call, from the sender to the receiver. This is more noticeable in a two-way conversation than in a conference call.

3.1.3.4 Loss and Codec Impairment

The equipment impairment factor $I_{loss\text{-}codec}$ represents the joint impairment from the equipment—the choice of codec—and the loss rate of the network itself. The loss rate is measured in two methods: the random loss probability for each packet, and the average length of the burst loss. These rates are used to alter the impairment that the codec starts off with.

Codecs have different impairments because of how they compress. In order to represent the impairment by one number, the ITU did research into comparing the MOS value changes for each codec and used that to come up with a starting point. The codec impairment does not consider the base quantization error for converting to 8000 samples per second logarithmic

PCM, and so the impairment values are relative to G.711 PCM. Recommended impairments are given in ITU G.113 Appendix I, and for common codecs are 0 for G.711, 10 for G.729, 11 for G.729a, and anywhere from 5 to 20 for GSM, with no loss. Furthermore, the packet loss robustness for fully random packet loss can be set to 19 for G.729a, 25.1 for G.711 with Appendix I error concealment, and 4.3 for G.711 with no error concealment. With loss in place, and with error concealment turned on for the codecs, the values do go up. Using G.729 native error concealment on a 20ms packet, and G.711 error concealment on a 20ms packet where the first lost is covered up by repeating the previous 20ms sample, after which the call goes silent, the response for the codecs for loss is as follows. For six consecutive losses and a loss probability of 1.5%, G.729 provides an impairment of 9, and G.711 provides an impairment of 7. For eight consecutive packets, with a 2% packet loss, G.729 with error concealment provides an effective impairment of 11, and G.711 with the mentioned error concealment provides an impairment of 10. As mentioned in Chapter 2, G.711 generally performs better than G.729, and given that the overhead of most voice mobility networks exceeds the actual resource usage of the voice bearer payload, G.711 is often a better answer until the loss rates begin rise above a percent. After that point, the error concealment in the phones for each codec becomes the deciding factor.

Although the impairments vary with the loss rates, there are rules of thumb, and we will get to those in the next section.

3.2 What Makes Voice Over IP Quality Suffer

With the better understanding of what can be used to measure voice quality, and with the appropriate tools in our pocket, we can now look at the major factors that influence voice quality in a real voice mobility network. Thankfully, the properties that make the most difference are also the ones directly in the hands of those responsible for voice mobility networks.

3.2.1 Loss

Loss is the major contributor to poor voice quality in voice mobility networks. Loss comes in through all sorts of means. Wireless loss results when the phone is out of range, or when the network is congested with other traffic, or when the in-building coverage plan is spotty. Wherever it happens, loss removes words from people's sentences, making good communication impossible and stretching out the length of phone calls, as well as people's patience, to comic proportions.

Loss is one of the major factors in the E-model. Specifically, the E-model measures loss through the use of the *burst ratio* and *random packet-loss probability*. The reason for two metrics is simple. If the random packet-loss rate, or how often unrelated random packets are dropped, is low enough, the loss rate may be tolerable or not even noticeable, falling between pauses or breaths. However, if the losses all come in bursts, entire words can easily be lost or distorted, and the same loss rate can have a larger impact.

The burst ratio is defined as the average length of observed consecutive burst loss, divided by the average length of consecutive burst losses expected due to uniform random loss. In other words, randomness itself will drop some packets back-to-back, just as a coin flip can result in heads twice in a row. But, if the equipment is making this worse, by leading to back-to-back packet losses fairly often, the burst loss rate will show it. No introduced bursts lead to a burst rate of 1, which goes up as the equipment introduces burst loss.

The total ding to the equipment impairment is represented by the following formula:

$$I_{\text{loss-codec}} = I_{\text{codec}} + \frac{(95 - I_{\text{codec}})\,p}{B + p/r}$$

in which the overall impairment is $I_{\text{loss-codec}}$, the codec impairment itself is I_{codec}, the packet loss probability is p, the burst rate is B, and the packet loss robustness of the codec is r.

This formula leads us to a few rules of thumb for loss, which are quite handy. First, a ground rule. MOS can, in fact, be measured along the entire length of the call. However, because call quality varies over time, one common way to ascertain how well the network is doing with a call is to divide the call into n-second units (where n is usually 3), and to measure the MOS for each unit. Together, the average, minimum, and maximum MOS values can be looked at, to get a better understanding. The average number is more useful in that context, because the mobility of a phone tends to cause some fluctuations in most networks. (This is a way of introducing mobility advantage into the calculations in a more meaningful way than tacking on a number, although it is critical to keep an eye on the minimum MOS at all times.)

Let's assume that we are using a G.711 codec. The G.711 codec provides an impairment of 0 (reference) with no loss, and a packet loss robustness of 4.3, according to G.113. Assuming no burst loss greater than expected by the fixed probability distribution, we can calculate that the impairment due to packet loss will be around 10 at a 0.5% loss, 18 at 1% loss, 25 at 1.5% loss, and 30 at 2% loss. Remember that the impairment comes straight off the top. Assuming perfection in the rest of the system, no loss provides a 93.2 R-value, so a 0.5% loss results in around an 83.2 R-value, for a MOS that is still of toll quality, but a 1% loss drops to 75.2, for a MOS somewhere around 3.7.

For G.711 with Appendix I error concealment, the sensitivity to loss is mitigated substantially. A 3% loss can be taken until the R-value drops by 10, and a 4% loss rate can be taken until the R-value drops below toll quality.

G.729a takes a substantial hit up front. Without any loss, G.729a is approaching dropping below toll quality. However, its impairment curve matches the one for G.711 with error concealment nearly step for step. (The matching is not precise.)

Figure 3.2 shows the graph of impairments over packet error loss.

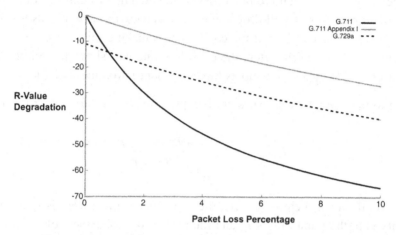

Figure 3.2:   R-Value Impairment over Packet Loss Rates

For G.711 with no error concealment, add about an extra 10 to the impairment for every half a percentage point of loss. With error concealment for G.711 or G.729, the add an extra 2 for every half a percentage point of loss. These values hold fairly tightly for any burst-loss ratio, so long as the packet loss rates are less than 2%. At 2% or higher, lower burst loss begins to help ease the fall.

On a grander scale, the simplest rule of thumb is the one currently used by the Wi-Fi Alliance for its voice certification efforts (see Chapter 5): a half a percentage point of true packet loss is about as far as you want to go with stock G.711, before the call quality begins to drop below toll grade.

3.2.2 Handoff Breaks

Handoffs cause consecutive packet losses. As mentioned in our previous discussion on packet loss, the impact of a handoff glitch can become large. The E-model does not make the best measurement of handoff break consternation, because it takes into account only the average burst length. Handoffs can cause burst loss far longer than the average, and these losses can delete entire words or parts of sentences.

Later chapters explore the details of where handoff breaks can occur. The two general categories are for *intratechnology handoffs*, such as Wi-Fi access-point to access-point, and *intertechnology handoffs*, such as from Wi-Fi to cellular. Both handoffs can cause losses ranging for up to a second, and the intertechnology handoff losses can be potentially far higher, if the line is busy or the network is congested when the handoff takes place.

The exact tolerance for handoff breaks depends on the mobility of the user, the density or cell sizes of the wireless technology currently in use, and the frequency of handoffs. Mobility tends to cut both ways: the more mobile the user is at the time of handoff, the more forgiving the user might be, so long as the handoff glitches stop when the user does. The density of the network base stations and the sizes of the cells determine how often a station hands off and how many choices a station has when doing so. These both add to the frequency of the glitches and the average delays the glitches see. Finally, the number of glitches a user sees during a call influences how they feel about the call and the technology.

There are no rules for how often the glitches should occur, except for the obvious one that the glitches should not be so many or for so long that they represent a packet loss rate beginning to approach a half of a percentage point. That represents one packet loss in a four second window, for 20ms packets. Therefore, a glitch of 100ms takes five packets, and so the glitch should certainly not occur more than once every 20 seconds. Glitches longer than that also run the risk of increasing the burst loss factor, and even more so run the risk of causing too many noticeable flaws in the voice call, even if they do not happen every few seconds. If, every two minutes, the caller is forced to repeat something because a choice word or two has been lost, then he would be right to consider that there is something wrong with the call or the technology, even though these cases do not fit well in the E-model.

Furthermore, handoff glitches may not always result in a pure loss, but rather in a loss followed by a delay, as the packets may have been held during the handoff. This delay causes the jitter buffer (jitter is explained in Section 3.2.4) to grow, and forces the loss to happen at another time, possibly with more delay accumulated.

A good rule of thumb is to look for technologies that keep handoff glitches less than 50ms. This keeps the delaying effect and the loss effect to reasonable limits. The only exception to this would be for handoffs between technologies, such as a fixed-mobile convergence handoff between Wi-Fi and cellular. As long as those events are kept not only rare but predictable, such as that they happen only on entering or exiting the building, the user is likely to forgive the glitch because it represents the convenience of keeping the phone call alive, knowing that it would otherwise have died. In this case, it is reasonable to not want the handoff break to exceed two seconds, and to have it average around a half of a second.

3.2.3 Delay

For voice mobility networks, we hope to already have an echo-free system. Digital handsets and PBXs have reasonable echo cancellation systems. The major source for problems with delay, then, is network delay alone. The E-model uses a very complicated formula to determine what that impairment would be:

$$I_{delay} = 25\left\{\left(1+X^6\right)^{1/6} - 3\left(1+[X/3]^6\right)^{1/6} + 2\right\}$$

where $X = lg(T/100)$, lg is the base-2 logarithm, and T is the end-to-end delay in milliseconds. This formula applies only when the delay is greater than 100ms; otherwise, the impairment is zero. The only way to get an appreciation of this is to view it plotted out, as in Figure 3.3.

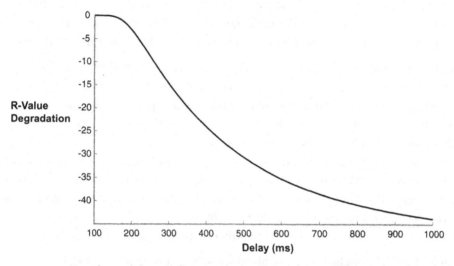

Figure 3.3:   Delay Impairment over Milliseconds

Delay impairment is measured independent of the codec, though the codec adds to the total delay. You may notice that the formula allows for up to 200ms of one-way, end-to-end delay, before any degradation is noticeable. Toll quality becomes challenged when, all else being perfect, the delay begins to cross 300ms.

Because loss and delay are present in networks together, it is best to avoid delays that get up to 200ms. Most of this delay budget should be considered to belong to the wireline network. End-to-end delays are added to by the codecs. The sending encoder for a 20ms G.711 stream will add 20ms, necessarily, to the delay: the frame comes out with the first sample delayed by the entire 20ms. G.729 adds an extra 5ms of delay for its encoder, on top of the 20ms for the packet rate typically used. The receiver will add a significant amount of delay for its reassembly *jitter buffer*, mentioned in the next section. This can easily be up to a couple of packets worth. Conference bridges or media gateways add an additional delay, starting at the packet size and going up from there. Therefore, the 200ms of end-to-end budget can get eaten into rather quickly. The well-known recommendation within the industry is to limit the delays added by the network itself, wireless and wired, to 50ms on top of whatever the phones and PBXs add.

3.2.4 Jitter

Jitter is the variation in delays that the receiver experiences. Jitter is a nuisance that the user does not hear directly, because the phones employ a jitter buffer to correct for any delays. Jitter can be defined in a number of ways. One way is to use the standard deviation or maximum deviation around the mean delay per packet. Another way is to use the known arrival intervals (such as 20ms), and subtract consecutive delays of packets that were not lost from the known arrival time, then take the standard deviation or the maximum deviation. Either way, the jitter, measured in times or percentages against the mean, tells how variable the network is.

Jitter is introduced by variable queuing delays within network equipment. Phones and PBXs are well known for having very regular transmission intervals. However, the intervening network may have variable traffic. As the queue depths change and the network loads fluctuate, and as contention-based media such as Wi-Fi links clog with density, packets are forced to wait. Wireless networks are the biggest culprit for introducing delay into an enterprise private network. This is because wireless packets can be lost and retransmitted, and the time it takes to retransmit a packet can usually be measured in units of a millisecond.

A jitter buffer's job is to sit on the receiver and prevent the jitter from causing an *underrun* of the voice decoder. An underrun is an awkward period of silence that happens when the phone has finished playing the previous packet and needs another packet to play, but one has not yet arrived. These underruns count as a form of error or loss, even if every packet does make it to the receiver, and loss concealment will work to disguise them. The problem with jitter becomes that an underrun must be followed by an increase in delay of the same amount, assuming no packets are lost. This can be seen by realizing that the delayed packet will hold up the line for packets behind it.

Here, the value of the jitter buffer can be seen. The jitter buffer lets the receiver build up a slight delay in the output. If this delay is greater than the amount of actual jitter on the network, the jitter buffer will be able to smooth things out without underruning.

In this sense, the jitter buffer converts jitter directly into delay. If the jitter becomes too large, the jitter buffer may have limited room, and start dropping earlier samples in the buffer to let the call catch up to be closer to real time. In this way, the jitter buffer can convert the jitter directly into loss.

Because jitter is always converted into delay first, then loss, it does not have a direct impact on the E-model by itself, but instead can be folded in to the other measures. However, the complication arises because the user or administrator does not usually know the exact parameters of the jitter buffer. How many samples, how much delay, will the jitter buffer take before it starts to drop audio? Does the jitter buffer start off with a fixed delay? Does it build up the delay as jitter forces it to? Or does it try to proactively build in some delay, which can grow or shrink as the underruns occur? These all have an impact on the E-model call quality.

As a result, a rule of thumb here is to match the jitter tolerance to the delay tolerance. The network, at least, should not introduce more than 50ms of jitter.

3.2.5 Non-IP Effects that Should Be Kept in Mind

The E-model makes plenty of room for non-IP effects on voice quality, and we would be wise to consider them here, even though the previous few sections chose to focus only on the network effects.

As mentioned earlier, echo is a problem to be tackled whenever calls are being tied together in conference bridges or are traversing through multiple media gateways. Analog lines introduce the problem of noise, as well as volume or gain control. Some analog lines may be tuned softer than others. Most of this requires reasonable end-to-end testing, however.

Then there are the intangibles. Is the network provisioned well enough that calls go through or are held predictably and reliably? Is the voice mobility network laid out well enough that users know that every point in the campus is a hot spot, or are some areas weak or dead? Cellular companies make entire marketing campaigns on the premise of the importance of coverage and dropped calls. (The number of bars on the phone or people standing behind the spokesman are both powerful examples of how important the predictability of the call quality is to callers.) This same concern needs to be applied to voice mobility networks produced within the enterprise. No amount of modeling will answer how much tolerance exists, but the general consensus is that voice mobility networks must work better than the cellular networks, when the callers are in the office. Mobility within the office does not generally count as a factor that can be used to increase the acceptance of the quality of the calls, and although mobility is a tremendous driving force to achieve higher productivity and less frustration, it is the sort of benefit that is hardly noticed until it is gone.

Keep in mind that the codec chosen can make an immediate ten-point difference in the R-value, in many cases.

3.3 How to Measure Voice Quality Yourself

The final section in this chapter is concerned with the ways in which administrators of voice mobility networks can directly ascertain the quality of the network.

3.3.1 The Expensive, Accurate Approach: End-to-End Voice Quality Testers

As mentioned in the discussion of PESQ (Section 3.1.2), existing tools can measure the quality of the voice network by directly pumping in prerecording voice samples and comparing the output. These tools are either expensive or home-grown, and are used to test large networks as a part of a planning or predeployment phase.

This sort of testing is more of a tuning exercise, and—much like how piano tuning is a rare and complicated enough exercise that it is not performed frequently—direct end-to-end testing is not diagnostic. Telephone equipment testing companies do make the sort of equipment to perform this end-to-end inspection, and these tools can be rented. Unfortunately, it is very difficult to know where to invest in this sort of heavily proactive effort.

More likely, the voice quality is measured by having administrators walk around the network with some number of phones in question, ensuring themselves that whatever problems they may face will likely be manageable. The problem with both forms of proactive testing is that they normally occur on only lightly loaded networks, and thus are not able to measure the effect of network load on voice quality. Network load is generally the largest impact on voice quality, in fact, partly because voice mobility network managers do a good job of testing their networks before they launch them for basic problems, which they quickly correct, and partly because voice mobility networks are more likely to be robust enough out of the box for basic voice connectivity.

3.3.2 Network Specific: Packet Capture Tests

Most of the major packet capture tools, for wireline and for wireless, make modules that are able to indirectly infer the MOS values using E-model calculations. Sometimes, these work by tracing the voice setup protocols, such as SIP, and determining what RTP flows map to phone calls and the properties of the phone calls. Other times, these tools will just look directly at the RTP streams, and not try to find out what phone numbers the streams map to In both cases, the tools then use the sequence number and timestamp fields in the RTP stream to determine values such as loss, delay, and jitter. Using assumed values for the jitter buffer, with the option of having the user overwrite them, the tools then model the expected effect and produce a score.

The major issue with these tools is that they show quality only up to the point where they are inserted. An easy example of the problem is to look at wireless networks. On a Wi-Fi network, a packet capture tool may be able to directly determine what packets it sees and come up with a score. By looking at the Wi-Fi protocol, the tool may do a good job of inferring whether the mobile phone received the packet from the access point, and at what time, and may produce a reasonably close call quality number. On the other hand, the upstream flow is likely to look quite good from the point of view of the test tool, because there is only one network in between the client and the tool. The entirety of the network upstream from the client goes missing, and the upstream MOS value can be entirely misleading.

Some network infrastructure devices are able to do these inferences within themselves, as they pass the data through. This may be a reasonable thing to do, again depending on the point of insertion and how well they are able to capture information as late into the network as possible. It is important, when using all of these tools, for you to consult with the vendor or maker of the tools to find out where the tools are measuring. For a wireless controller with voice metric capabilities, for example, make sure that the downstream metrics are measured on the access point, based on what happened over the air, and not just passing through the controller. For wireless overlay monitoring, make sure that there is an option to do a similar capture using a wired mirror port on one of the switches, for cases in which voice quality might begin to suffer and the network needs direct attention. Overall, do not rely on just one tool, and believe what the users say—no matter what the tool tells you.

3.3.3 The Device Itself

The most accurate and reasonable way to measure voice quality is from the endpoints themselves. Both some handsets and PBXs offer the ability for the device to produce the one-way MOS value or R-value for the receive side at the device itself. These numbers are based entirely on E-model calculations, assuming best-case or known-default scenarios for the rest of the system, but are likely to be the most accurate. Of course, it is difficult to ask a user to determine what the voice quality is of a call while on it, especially given that voice quality is not something a user wants to measure. However, for diagnosing locations that are having troubles, this tool is valuable for the administrator herself, who is able to avoid having to guess as to whether the call sounds reasonable, and may be able to detect variations in the MOS value or R-value.

In the end, keep in mind that the absolute values produced by any of the methods deserve being taken with a grain of salt. As time goes on, the administrator of a voice mobility network should be able to learn what the real quality means for any given value the tool suggests, even when the tool is placing results a half a MOS point too high or too low. However, the *variation* of the scores, especially when the network has changed, can be a valuable tool for point the way towards the solution.

CHAPTER 4

Voice Over Ethernet

4.0 Introduction

This chapter introduces the technologies necessary to carry voice over wireline packet networks. The first half of the chapter is a basic review of the concepts within packet networks, including IP and Ethernet. The second half takes a look directly at voice over these networking technologies.

4.1 The IP-Based Voice Network

The previous chapters explored the basics of how calls are set up and voice is carried over packet-based IP networks. However, the details about what makes the IP network itself work have not yet been addressed.

Voice started out on analog phone lines. Each pair of copper wires was dedicated to one specific phone, and to nothing else. This notion of a dedicated circuit has its advantages. It provides complete isolation of whatever might be going on with that line from the circumstances and problems of other phones in the network. No amount of calls being placed on a neighbor's line can make the original line itself become busy. This isolation and invariance is necessary for voice networks to function when unexpected circumstances occur, and ensures that the voice network is reliable in the face of massive fluctuations in the system. Provisioning is simple, as well, with one line per phone at the edge.

The problem with the concept of the dedicated line is that it is extremely wasteful. When the phone is not in use, the line stays empty. No other calls can be placed on that line. Even when a call is in place, the copper wire is fully occupied with carrying the voice traffic, a small bandwidth application, and a tremendous amount of excess signal capacity exists. Dedicated wires might make sense for short distances between the phone and some next-level aggregation equipment, but these dedicated lines were used as trunks between the aggregators, causing tremendous waste from both idleness and lost bandwidth. But probably the property that caused the most complications with wireline networking was that the dedicated line is not robust. If network problems occur—the bundle of cables is cut, or some intermediate equipment fails and can't do its job—all lines that are attached along that path are brought down with it.

doi:10.1016/B978-1-85617-508-1.00001-3.

Digital telephone networks started to eliminate some of the problems inherent to the one-line dedication of early circuit switching. By having digital processes encode and carry the voice, more voice calls could be multiplexed onto each line, better using the bandwidth available on the copper wire. Furthermore, by allowing for hop-by-hop switching with smarter switches between trunks, failures along one trunk could be accommodated. However, the network was still circuit-switched. A voice line could be used only for voice. Even where voice circuits were set aside for data links, the link is either fully in use or not at all. The granularity of the 64kbps audio line, the DS0, became a burden. Running applications that are not always on and have massive peak throughput but equally meek average throughput requirements meant that provisioning was always an expensive proposition: either dedicate enough lines to cover the peak requirement case, and pay for all of the unused capacity, or cap the capacity offered to the application. Furthermore, these circuits needed to be considered, managed, and monitored rather separately. The hard divisions between two circuits became a hard division between applications. Voice networks were famous for their reliability, strict clockwork operation—and complexity. They were not for easy-to-set-up, easy-to-move operations. The wires are drawn once and carefully, and the switches and intermediate equipment is set up by a team of dedicated and expensive experts who do nothing but voice all day. If you were serious about voice, you operated your own little phone company, complete with dedicated operators. If not, your only option was to have the phone company run your phone network for you.

Along came packet-switched networks. Sending small, self-contained messages between arbitrary endpoints on a network inherently made sense for computers. The idea of sending a message quickly, without tying up lines or going through cumbersome setup and teardown operations removed the restrictions on wasted lines. Although it was still true that lines could remain idle when not being used, the notion of allowing these *packets* of information into the line as the fundamental concept, rather than requiring continuous occupation and streaming, meant that lines that carried aggregated traffic from multiple users and multiple messages could be used more efficiently. If the messages were short enough, one line might do. No concerns about running out of lines and having the needed, or only, path to the receiver blocked. Instead, these messages could just be queued until space was available.

Along with this whole new way of thinking about occupying the resources came a different way of thinking about addressing and connecting the resources. In the early days, a phone number used to encode the exact topological location of the extension. Each exchange, or switch with switchboard operator, had a name and number, and calls were routed from exchange to exchange based on that number first. Changes to the structure or layout of the telephone system would require changes to the numbers. Packet-switching technologies changed that. Lines themselves lost their names and numbers. Instead, those names and numbers were moved to the equipment that glued the lines together. Every device itself now had the address. The binding of the addresses to the topology of the network remained, at some level. Devices could not be given any arbitrary address. Rather, they needed to have addresses that were similar to their neighbors. The notion of exchange-to-exchange *routing* was retained.

This notion, though, proved to be a burden. Changes to the network were quite possible, as either more devices needed addresses, or more new "exchanges" were added to the network. Either way, the problem of figuring out how to route messages through the network remained. The original design had each router know which lines needed to be used to send the messages along their way. The router might not know how the message should get to the final destination, but it always knew the next step, and could direct traffic along the right roads to the next intersection, where the next router took over. As the number of intersections increased, and the number of devices expanded, the complexity of maintaining these *routing tables* exploded. A way was needed for neighboring routers to find out about each other, and more importantly, to find out about what end devices they knew routes to. Thus, the *routing protocol* was born. These protocols spoke from router to router, exchanging information on a regular basis, ensuring that routers always had recent information on what destinations were valid and how to get there from here. But another thing happened. This idea of exchanging the routes had another benefit, in that it allowed the network itself to be restructured, or to fail in spots, and yet still be able to send traffic. Routers did not need to know the entire path to the destination, only the next hop. If a router knew two, different next hops for the same message, and one of the routes went down, the router could try the second one. If the router lost all of its paths to a particular set of destinations, the router before it could learn about that, and avoid using that path to get the messages through. If there was a way to get the message there, the network would find it, through the process of *convergence*, or agreement over time on the consistency of whether and how messages could be sent. The network became resilient, and point failures would not stop traffic from flowing.

This is the story of the Internet, and of all the protocols that make it work. Clearly, the story is simplified (and perhaps romanticized to highlight the point at hand), but the fundamentals are there. Circuit switching is difficult to manage, because it is incredibly wasteful and inflexible. Packet switching is much simpler to manage, and can recover from failures.

The Internet grew up on top of the lines offered by the circuit-switched technologies, but used a better way to dedicate the resources. It wasn't long before someone realized that voice itself could be put over these packet-switched lines. At first, that might sound wasteful, as using a digital line to carry a packet containing voice can never be more efficient than using that line to carry the same bits of voice directly because of the packet overhead. But packet networking technologies matured, and the throughputs offered on simple point-to-point links grew much faster than did the corresponding uses of the same copper line for digital voice—at least, in the enterprise. And the advantages of using a multipurpose technology allowed these voice over IP pioneers to use the network's flexibility and lack of dedication to one purpose to add to the voice over IP offerings quickly, without requiring retooling of physical wires. The ways in which provisioning was thought about changed, and the idea that voice and data networks can perhaps use the same resources became a compelling reason to try to save deployment and management costs.

There are a tremendous number of resources available for understanding the intricacies of how IP networks operate, including details on how to manage routing protocols and large trunk lines. Here, we will explore how voice fits into the packet-based IP network.

4.1.1 Wireline Networking Technologies and Packetization

The wireline networking technologies range from the most basic definition of how electrical signals are encoded over the copper line to the higher-level ways that computer software endpoints ensure that messages do not flood the network.

4.1.1.1 Ethernet

Nearly all wireline voice mobility networks in the enterprise start with *Ethernet*. Ethernet is a family of related networking technologies that establish how two machines that are physically connected can talk to each other. Ethernet was designed to be as simple to deploy as possible, so that it can be set up as an unmanaged network, where physically connecting two endpoints together, somehow, through the network is enough to allow them to find each other and communicate. (Note that this doesn't mean that higher-level protocols will work on this network without effort—just Ethernet itself.)

All of the Ethernet protocols belong to the IEEE 802.3 series and are based on the idea of encoding *frames*. A frame is a well-defined packet message, with a source, a destination, a length, and a type. The logical format of the Ethernet frame is shown in Table 4.1.

Table 4.1: Ethernet Frame Format

Destination	Source	Ethertype	Frame Body	FCS
6 bytes	6 bytes	2 bytes	*n* bytes	4 bytes

In Ethernet, links are anonymous. Endpoints, however—the line cards that the Ethernet cables plug into—are given addresses. These addresses are assigned at the time the device is built, and are permanently associated with the device. The Ethernet address is a 48-bit (6-byte) address, as shown in Table 4.2. The first three bytes, or 24 bits, is called the *Organizationally Unique Identifier* (OUI). Each manufacturer of Ethernet equipment is assigned one or more of these OUIs by the *Institute of Electrical and Electronics Engineers* (IEEE) *Registration Authority*. The manufacturer chooses the second 24 bits from a unique pool, often in order starting from 00:00:01. Together, the scheme guarantees that this address will never be accidentally taken by another device.

Ethernet also defines two special flags in the address. The L bit specifies a local address, which is dynamic and invented by a device for temporary usage. This has an application in Wi-Fi (see Chapter 5), but is otherwise not common. The G bit is for group-addressed frames—either broadcast or multicast. A group-addressed frame is meant to go out to multiple devices at once, for all of them to receive. Multicast transmissions use this mechanism. The special group address FF:FF:FF:FF:FF:FF (all 1s) is the broadcast address, and specifically requests to go to every device, whether they are in a multicast group or not.

Table 4.2: The Ethernet Address Format

	OUI	Manufacturer-Defined
Bit:	0–23	24–47

	L	G
Bit:	6	7

This is one way by which Ethernet guarantees that it does not require management to add or remove devices from the network. When a device wants to transmit over a wire to another device, it has no way of knowing if that second device is there. Ethernet was intentionally designed to be as simple as possible, so senders have to transmit and hope that the other device is there. When the sender creates a frame, it places the destination Ethernet address first in the frame, followed by its own address. Then comes the type of the frame, used to figure out what network protocol is running on top of Ethernet. An arbitrary frame body follows, subject to size restrictions: the body of the frame cannot be greater than 1500 bytes, usually, and cannot be less than 64 bytes. (Shorter frames must be padded.) Finally, Ethernet provides a way to determine whether noise on the Ethernet line causes any bit errors, by using a *frame check sequence* (FCS), a mathematical checksum of the bits in the frame that will generally not match the contents of a frame if there are any errors. Ethernet uses a CRC-32 checksum.

Ethernet itself is a serial protocol, much like serial lines used to connect modems together, but operating with much more sophistication and at a faster rate. Most Ethernet types today fall into two categories: copper and fiber. The commercially available copper Ethernet technologies all use a modified version of a telephone cable, made out of copper wires. Each cable carries eight small, insulated copper wires, twisted into pairs as is done for analog telephone lines. The plastic connectors at each end also look like telephone connectors, but have eight pins, rather than the usual six. These connectors, often referred to as *RJ45*, a specification in which the connectors figure prominently, snap into the corresponding sockets on all Ethernet devices. Differing numbers of the pairs within the four-pair cable may be used for different Ethernet technologies.

The first RJ45-based Ethernet is called *10BASE-T*, or simply original *Ethernet*. Devices that support 10BASE-T run at 10Mbps, across just two of the pairs within the cable, one for reception, and one for transmission. (The other pairs are not used for data.) These Ethernet lines run a serial protocol, where the voltage on the line is flipped to signal a one or a zero in the bits used to encode the frame. However, these serial lines do not constantly transmit. Instead, the line is usually idle. But when a device wants to transmit on the line, it simply starts transmitting. The transmission itself is the frame, just described. Before the frame itself is sent, a few bits are prepended to it. These bits, known as the *preamble*, are used to alert the device at the other end that the transmission is going to begin. The preamble is a 64-bit sequence of alternating ones and zeros, except for the last two bits, which are both ones. The receiving device detects that a transmission comes in, by looking for the sharp swings in voltage in the line from idle, representing the preamble's bits. By the time the preamble is done, the receiver will have figured out the timing of the bit patterns, in case the receiver's clock is slightly off from the sender's. The full bits of the frame proper come in, including the checksum. At the end of the transmission, the sender and receiver have to wait for a few microseconds, and then the line becomes idle and ready to be transmitted on again.

Given that 10BASE-T is a point-to-point physical system, as there can only be one transmitter on one twisted pair, and the other transmitter on the second, there needed to be some way to interconnect multiple lines and thus multiple devices together. The solution to that is the Ethernet *hub*. The hub works by connecting the twisted pair that is used by a device to transmit, to every link's twisted pair used to receive. This connection allows the transmission by one device to reach all of the others on the same *segment*, or other devices attached to the same hub. Hubs are purely electrical, and do not participate in the network itself. When a device transmits on an Ethernet hub, every device on that hub hears the signal. A receiver knows that the frame is for it by looking at the destination Ethernet address. If the address matches, then the frame is kept; otherwise, it is discarded unless the operating system on that device requests to receive all frames on the line. The use of hubs, and the definitions for 10BASE-T, require that the transmissions are all *half-duplex*, meaning that a reception and transmission cannot occur independently.

Adding multiple devices together on an Ethernet link introduces a problem. Two or more devices are capable of transmitting at the same time. If two devices do transmit at the same time, their signals will mix on the wire, and all of the receivers will receive the garbage created by the interference. Thankfully, there is a solution to avoid this. The overall concept is known by the unwieldy phrase *Carrier Sense Multiple Access with Collision Detection* (CSMA/CD). Let's break that phrase apart, starting from the end. The collision detection portion of Ethernet works rather simply. When the device starts transmitting, it watches the receive twisted pair for its own transmission to return to it. If it sees a signal that differs from its transmission, more than just by a delay, it knows that another device is also transmitting. To prevent wasting time by having both signals clobber each other, the transmitter stops transmitting the frame and sends out a jamming signal for a short burst to ensure that the line is dead. The senders then retry their frame, up to a certain number of times. If the signal does not get clobbered by the time the frame reaches the end, the sender knows that the frame must have made it down the wire safely. If that were all to it, however, two devices with data to send would never be able to avoid colliding. That is because they would both detect the collision at nearly the same time, stopping their transmissions and waiting the mandatory time for the line to come back to quiet, and then they would transmit their next frames immediately. The segment would be in constant collision. To avoid that, the CSMA portion of Ethernet is used. Carrier sense is the act of detecting that a transmission is already on the line. First, the devices check the receive line, to make sure that a signal is not coming in already, corresponding to a transmission already in progress. If there is a transmission in progress, the transmitter waits until it ends. Then it transmits. If a collision occurs, CSMA uses the notion of a random *backoff*. Instead of each device transmitting exactly after a fixed time from the previous frame after the collision, each device picks a random integer greater than or equal to zero and less than the maximum backoff for this transmission. They then wait that many of *slots*, each one measured in microseconds, before transmitting again. This step reduces the probability that the devices will collide a second time. If the collision occurs again, the maximum backoff doubles, starting from the first backoff of two slots. This process stops when the frame is successfully transmitted, or abandoned. The next frame will then go out with no backoff.

The problem with backoffs is that they lead to unstable behavior when the network is loaded. This congestion occurs because of excess collisions, and more and more of the time on the network becomes dedicated to retransmissions and less time to new data. The solution to the problem was with the introduction of the Ethernet *switch*. The Ethernet switch is similar in concept to the telephone switch. A telephone switch isolates the paths between two connections, allowing two devices to speak at a time as if they were directly connected, independent of the other traffic. An Ethernet switch doesn't directly connect circuits, being packet-based, but it does eliminate one device's dependance on the transmissions of the other devices.

The Ethernet switch works by terminating each Ethernet link. Whereas a hub ties the multiple links together into one interconnected *collision domain*, the switch acts as a separate receiver for each connected device. Two or more devices can transmit at once, on their individual ports, and the switch will independently receive and gather the frames. The frames are then analyzed, interpreted for their destination addresses, and the frame is then sent out on the link that has that address. Because the switch has to read and understand the

Ethernet frames, its job becomes one of a traffic director. The concept of simultaneous reception resolves collisions between two endpoint devices, but the switch goes further, by performing the above-mentioned detection of which device is on each link. This function is a crucial part of *bridging* traffic, and works by the switch maintaining a learning table, built up dynamically, of the Ethernet addresses that have been seen as *sources* on each link. This table is essentially *soft state*, meaning that the entries are not permanently recorded, and are built up or refreshed as needed. The last remaining problem is for when the switch gets a frame whose destination address has not yet been learned. In this case, the switch just forwards the frame on every port on the switch, except for the one the frame came in on.

With a switch, the transmit side of each Ethernet link can become a bottleneck. Multiple frames can come in destined to one link, especially if this link holds a common server of some sort. When this happens, the switch is forced to build a backlog of pending transmissions, known as a *queue*. This queue is a list of packets, usually ordered first-come, first-serve, or first-in, first-out (FIFO). One of the switch's major benefits is that it has translated the resource contention that occurred with hubs into an orderly, predictable wait for packets to get to a popular resource.

The other benefit of a switch is that the collision concept can now be removed entirely. Because, on a switch, there are only ever two possible transmitters on a link, and because there are separate transmit and receive pairs, there is no reason for the receiver to echo back the transmitter's signal. Instead, each device can operate the transmit and receive lines independently. This is known as *full duplex* operation.

Full duplex operation was introduced with *100BASE-TX*, or *Fast Ethernet*. Fast Ethernet runs at 100Mbps for each direction, greatly increasing the possible data rate on the line. Fast Ethernet uses the same cables and connectors as the original 10BASE-T Ethernet (though lesser-quality cables of the type cannot be used), and the standard defines Fast Ethernet to be backward-compatible with the original. The mechanics of the Fast Ethernet encoding are more advanced than that of the original Ethernet. I will not concern you with the details here, as Ethernet signals are rarely noisy and insight into the encoding is not necessary. The key is that the frame format remains the same, but the data rate is ten times that of Ethernet. Additionally, because Fast Ethernet can use a switch, the backoff procedure is no longer required when transmitting to a Fast Ethernet peer. However, Ethernet hubs are still allowed. Furthermore, 10BASE-T devices may still be connected to a Fast Ethernet port. To determine whether the device can use the 100BASE-TX standard, a protocol known as *autonegotiation* occurs. Autonegotiation starts as soon as two ports are plugged together. 100BASE-TX devices send out special signals on the line, establishing that it is using 100BASE-TX and communicating its duplex setting. If the other side is also 100BASE-TX, the link will use Fast Ethernet. On the other hand, if the other device does not respond with the other signal (and instead sends the usual 10BASE-T link detection pulses), the 100BASE-TX device will downgrade until the link is unplugged.

Gigabit Ethernet over copper is specified in the *1000BASE-T* standard. Gigabit Ethernet, again, is backward-compatible, and uses similar cables, though with tighter quality requirements than either of the previous standards. One major difference for Gigabit Ethernet is that it uses all four twisted pairs in the cable. Combined with using a more advanced bit coding, this produces the extra speed—100 times faster than the original Ethernet.

4.1.1.2 The Internet Protocol (IP)

Ethernet defines how devices can be physically connected. But users do not know the Ethernet addresses of the devices providing the services they wish to use. More intelligence is needed to separate out the physical addressing, replacing it with logical addressing that an administrator can decide on, and allow multiple physical networks to be connected. The *Internet Protocol* (IP) defines how this addressing and packet formatting is to occur.

IP was originally specified in RFC 791 and expanded upon later, and comes in two versions: version 4 (IPv4) and version 6 (IPv6). The two main concepts for IP are the *IP address* and *IP frame*.

4.1.1.2.1 IPv4

IPv4 is the version used most often on the Internet today, by a wide margin. IPv4 uses a four-byte address written out as dotted decimal numbers, such as 192.168.0.1. These addresses are given out by an international agency in blocks for large organizations to use. Generally, individual IP addresses are provided to organizations by their Internet service providers. Ranges of addresses tend to be specified using the *slash notation*. For example, 192.168.0.0/16 means that the upper 16 bits are what was written, and the rest are within the range defined by allowing the remaining lower bits to be set to any value.

Of the 32-bit address space, some of the addresses have special meanings. The 127.0.0.0/8 address range is for *loopback* networking, and, when used as a destination, are kept internally to the machine that is doing the sending. This allows an IP device to send packets to itself. The 169.254.0.0/16 range is for link-local addresses, meaning that their use cannot extend past the Ethernet switching network they are used on. In addition, 10.0.0.0/8, 172.16.0.0/12, and 192.168.0.0/16 are all private addresses. These are the addresses most commonly used in voice mobility networks within the enterprise. They are not valid on the public Internet itself, but are designed for private networks based on IP. The 224.0.0.0/4 network is used for multicast traffic: each address is a different multicast group. Finally, the 255.255.255.255 address is the link-local broadcast address, meant to go out to all devices on the Ethernet switching network (and using the FF:FF:FF:FF:FF:FF Ethernet broadcast address for the underlying packet).

IPv4 runs on Ethernet by setting the Ethernet type to 0x0800. The IP packet has a header and payload, as shown in Table 4.3.

Table 4.3: IPv4 Packet Format

Version/Header Length	TOS/DSCP	Total Length	Identification	Fragment	TTL	...
1 byte	1 byte	2 bytes	2 bytes	2 bytes	1 byte	

	Protocol	Header Checksum	Source	Destination	Options	Data
...	1 byte	2 bytes	4 bytes	4 bytes	variable	n bytes

Table 4.4: The Version/Header Length Field

	Version	Header Length
Bit:	0–3	4–7

The Version/Header Length field is specified in Table 4.4. For IPv4, the version is always 4. The header length measures how long the header is (up to the data field), in four-byte increments. The Type of Service/Diffserv Code Point (TOS/DSCP) field is used to specify the quality-of-service properties of the packet. The Total Length measures the entire length of the packet, and will come into play with fragmentation. The Identification field is used to track which fragments belong to the same overall packet; between separate packets, most devices tend to increment this by one, although this is not required. The Fragment field specifies what the offset is for this fragment in the entire packet. The *TTL* (*Time To Live*) field is used for forwarding, and specifies how many times this packet can be forwarded before it is dropped. The Protocol field specifies what higher-layer protocol is used on top of IP for this packet. The Header Checksum is a literal *one's complement* 16-bit sum of the header of the packet, and is used to detect if the underlying network flips a bit by mistake. (The Ethernet CRC is adequate for that purpose, so this field, although always set and always checked, is not terribly useful.) Finally, to the interesting information. The Source and Destination fields hold the IP addresses of the originator and final destination of this packet. The header ends here, and is followed by the next protocol's headers or data. This entire set of bytes is the payload of the underling Ethernet frame.

IP is designed to be relayed, or *forwarded*, between computers, across different network segments, and across the world if needed. This is the major distinction for IP, as it has allowed the Internet to be constructed from an assembly of smaller networks. The idea is that any IP-connected device that has multiple links can forward messages if configured to do so. Each link has its own IP address, as required. When an packet comes in for an IP address that is not that of the machine (how that happens will be mentioned in a moment), the device will look up a routing table to find out where the next machine is that this packet needs to go to. IP forwarding works on the concept of *longest-prefix matching*. Because there are too many IP addresses for a machine to know about, and because the IP address space tends to be organized in ranges, the forwarding device (a *router* in this context) looks up a series of routing rules that it has configured. Each routing rule is set up as a network prefix (as specifiable by slash notation), and the IP address of a machine that is on one of the links the router has. This address is the *next hop*. Because a destination address might match more than one rule, the one rule that matches the most leading bits—the longest prefix—will win, and that next hop will be used.

This concept of next hops explains why a router or machine may get a packet for a different destination IP address than it uses for itself. If another machine is set up to forward packets to it—and any machine can be set up to forward to any other, without restriction, so long as both are on the same switching network—then the first machine will get packets for other devices. The concept of prefix routing makes sense when you think of most enterprise routers. Enterprises, all but the largest, usually have a limited number of address ranges that are used locally. All of the rest, every other one, is out on the Internet. Connecting the enterprise to the Internet is one Internet router. The enterprise routers thus need only to have forwarding rules for the address ranges that they have in the enterprise, plus one route, called a *default route*, that tells the IP address of the Internet router. This default route uses a 0.0.0.0/0 prefix, meaning that every address matches, because the prefix is trivially short. Therefore, longest-prefix matching ensures that the default route matches last. Nonrouters will normally only have this default route, because they will not forward other devices' traffic. In this way, locally generated traffic is forwarded, even when other traffic will not be. The final bit of information to know is that not all traffic is forwarded on to next hops. Each link into the system has not only an IP address but a *subnet mask*, or a prefix that specifies what other IP addresses are directly on that link. For example, using the same slash notation, 192.168.10.20/24 states that the IP address of the link is 192.168.10.20, and all IP addresses starting with 192.168.10 are directly on the link, and do not need to be routed to the next hop. Those direct-link addresses belong to the same subnet. Every time the frame is forwarded, the TTL is reduced by one. Once it hits zero, the packet is dropped, rather than forwarded. Nothing else is modified while the packet is forwarded—the source and destination addresses are always those of the originator and the final destination of the packet.

Because IP runs on top of Ethernet, there must be a way to map IP addresses to Ethernet addresses. Every IP address has an Ethernet address—that of the Ethernet device the IP address was assigned to. When a sender needs to send out an IP packet, and it has used its forwarding logic to figure out which link the next hop or final destination is on, the sender needs to use a resolution protocol to ask the devices on the network for which one has the IP address it needs. The protocol is called the *Address Resolution Protocol* (ARP). ARP runs on a different Ethernet protocol type 0x0806. The idea is that each sender maintains an *ARP cache*. This cache stores the Ethernet address that is known for a given IP address. The cache is updated whenever another device sends a packet to the first one, as the Ethernet source address is assumed to be bound to the IP source address, so long as that IP address is on the same subnet. However, if the cache does not have an address mapping that is needed, the sender will send an ARP request to the network. These ARP requests are broadcasted using Ethernet, and any device that receives the ARP request and has that IP address is required to respond, unicast to the ARP sender, acknowledging the binding with another ARP message. The format of an ARP message is shown in Table 4.5.

Table 4.5: ARP Message Format

Hardware Type	Protocol	Hardware Size	Protocol Size	Opcode	...
2 bytes	2 bytes	1 byte	1 byte	1 byte	

	Sender Ethernet	Sender IP	Target Ethernet	Target IP
...	6 bytes	4 bytes	6 bytes	4 bytes

For Ethernet networks, the Hardware Type is always 1, and the Protocol Type is always 0x800. The Hardware size is the length of the Ethernet address, 6. The Protocol Size is the length of the IP address, 4. There are two opcodes: 1 is for a request, and 2 is for a reply. Finally, the addresses state the mapping that is requested or being answered for. When a machine requests to find out which other device has an IP address, it will send its Ethernet and IP addresses as Sender, and the IP address it is looking for as Target, with the Target Ethernet set to 0. The respondent will fill in its Ethernet and IP address as sender, the original requester's Ethernet and IP as target, and then send the response back.

With ARP, the binding of IP addresses to Ethernet addresses can be dynamic and changing.

Earlier, the concept of fragmentation was alluded to. IP provides a service that lets a packet be split across a number of smaller packets. The reason for this is that IP is meant to be carried over a wide variety of link-layer technologies, not just Ethernet, and those technologies may have a different maximum payload size. To make sure that a packet that is of a valid length that is sent in one network can arrive safely at the other, the concept of fragmentation was introduced. The router, or sender who has a packet which is too large, and which does not already have the "Do Not Fragment" bit in its Fragment field set, will divide the packet into two or more smaller ones, each with a copy of the original IP header.

The data fields will be the individual segments, with the offset of the first byte of the data field from the start of the original (or reassembled) packet being given in the Fragment field. The fragments are sent over the network, and the receiver is required to reassemble all of the fragments before sending it up to the higher layers. The receiver knows that it has reached the end of the fragment chain by looking at the "More Fragment" bit in the Fragment field. The last fragment will not have that bit set. All fragments of an original packet share that packet's original Identification field. The maximum size of an IP packet, including all headers, is 65,535 bytes.

IP, like most other packet networking technologies, makes no guarantees as to whether a packet will arrive at its destination. Packets may arrive with arbitrary delays, and may even come out of order (although this is to be discouraged). This best-effort delivery guarantee—the network will try, but will not commit resources up front—is key to IP's success. It, unfortunately, also runs counter to the goals of voice.

Clearly, IPv4 is the bread-and-butter protocol for voice mobility. What was presented here was a brief, high-level survey, and readers are encouraged to fill any major gaps in understanding before undertaking major roles in voice mobility networks.

4.1.1.2.2 IPv6

IPv6, specified in RFC 2460, was created to address a few design issues with the previous IPv4. The major issue to be addressed is the limited number of IPv4 addresses. As the Internet grew, many devices that were not counted on originally to have networking support were given it, and IP addresses were allocated in large chunks to organizations, whereas many of the later addresses in the chunks went unused, being reserved for future growth. The people behind IPv6 decided, not without controversy, that more addresses were needed. As a result, they created the most defining feature of IPv6.

Each address in IPv6 is 128 bits. The address fields are split up into very large ranges and subfields, with the understanding that these large fields are to be used to simplify network allocation. IPv6 addresses are written in hexadecimal notation, rather than decimal, and are separated every four digits by colons. For example, one address might be 1080:0:0:0:8:800:200C:407A, where it is understood that leading zeros can be omitted. There is a shortcut, as well, where long ranges of zeros can be written with the double colon, ::. Thus, 1080::8:800:200C:407A specifies the same address as the earlier one.

As with IPv4, there are a few ranges, specified in slash notation, which are set aside for other purposes. The address ::1 represents the loopback address. Addresses of the form FE80::/10 are link-local addresses. Addresses of the form FC00::/8 are private addresses. The multicast address space is of the form FF00::/120. Finally, for backward compatibility, IPv6 specifies how to embed IPv4 addresses into this space. If the left 96 bits of the address are left zero, the right 32 bits are the IPv4 address. This allows the machine using

192.168.0.10, say, to use the IPv6 address ::192.168.0.10 (they allow the dotted decimal notion just for this). This means that the machine ::192.168.0.10 understands and can receive IPv6, but was assigned only an IPv4 address by the administrator. On the other hand, machines that speak only IPv4 and yet have had packets converted to IPv6 by some router are also given an address. If 192.168.0.10 belonged to this group, it would receive the IPv6 address ::FFFF:192.168.0.10. The FFFF is used to signify that the machine cannot speak IPv6.

The IPv6 header is given in Table 4.6.

Table 4.6: IPv6 Packet Format

Version/ Flow	Payload Length	Next Header	Hop Limit	Source	Destination	Options	Data
4 bytes	2 bytes	1 byte	1 byte	16 bytes	16 bytes	*optional*	*variable*

Table 4.7: The Version/Flow Field

	Version	Traffic Class	Flow Label
Bit:	0–3	4–11	12–31

The Version/Flow field (Table 4.7) species important quality-of-service information about the flow. The version, of course, is 6. The Traffic Class specifies the priority of the packet. The Flow Label specifies which flow this packet belongs to. The Payload Length specifies how long the packet is from the end of the IPv6 header to the end. Thus, this is the length of the options and the data. (Note that, in IPv4, the options are counted in the header, not the payload.) The Next Header field specifies the type of the header following the IPv6 header, or if there is no IPv6 option following, then this specifies the protocol of the higher-layer unit this packet carriers. The Hop Limit is the TTL, but for IPv6. The Source and Destination addresses have the same meaning as in IPv4.

IPv6 is routed in the same way as IPv4 is, although there is a lot more definition in how devices learn of routes. In IPv6, devices are able to learn of routers by their own advertisements, using a special protocol for IPv6 administrative communications (ICMPv6, as opposed to ICMPv4 used with IPv4).

IPv6 is a major factor in government or public organization networks, and has an impact in voice mobility in those environments. Many private voice mobility networks, however, can still safely use IPv4.

4.1.1.3 UDP

The User Datagram Protocol, or UDP, is defined in RFC 768. The purpose of UDP is to provide a notion of *ports*, or mailboxes, on each IP device, so that multiple applications can exist on the same machine. A UDP port is a 16-bit value, assigned by the application opening the port. Packets arriving for a UDP port are placed into a queue used just for the application that has the port open: these queues, and the ports they are attached to, are generally called *sockets*. UDP-based applications often have well-known, assigned port numbers. Common UDP applications for voice mobility are SIP on port 5060, DNS on port 53, and RADIUS, on port 1812.

Every socket has a port, even those that do not need a well-known one. Ports can be assigned automatically, according to whatever might be free at the time, These are called *ephemeral ports*.

UDP embeds directly into an IPv4 or IPv6 packet. The format of the UDP header is shown in Table 4.8.

Table 4.8: UDP Packet Format

Source Port	Destination Port	Length	Checksum	Data
2 bytes	2 bytes	2 bytes	2 bytes	*variable*

The Source Port is a 16-bit value of the socket sending the UDP packet. It is allowed to be 0, although that is rarely seen; ephemeral ports are far more common. The Destination Port is that of the socket that needs to receive the packet. The length field specifies the entire length of the UDP datagram, from the Source Port to the end of the Data. This is redundant in IP, because IP records the length of its payload, and the UDP packet is the only thing that needs to fit. The checksum is an optional field for UDP, which covers the data of the packet, as well as the UDP header and a few fields of the IPv4 or IPv6 header (such as source, destination, protocol, and length).

UDP suffers from the same problems as the underlying IP technology does. Packets can get dropped or reordered. Applications that depend on UDP, such as SIP and RTP, need to make plans for when packets do get lost. This is a major portion of voice mobility.

4.1.1.4 TCP

The *Transmission Control Protocol* (TCP) is the heavy-duty older sibling of UDP. TCP, specified in a number of RFCs and other sources, is a protocol designed to correct for the vagaries of IP's underlying delivery, for use in data applications.

Unlike UDP and IP, TCP provides the view to the using application that it is a byte stream, not a packet datagram service. Of course, TCP is implemented with packets. This means that TCP must ensure that packet loss and reordering do not get revealed to the end application, and so some notion of reliable transport is necessary. Furthermore, because

TCP is the dominant protocol for data, it must deal with trying to avoid overwhelming the network that it is being used in. Therefore, TCP is also charged with *congestion control—* being able to avoid creating congestion that brings down a network—while finding the best throughput it can.

The header structure for TCP is given in Table 4.9.

Table 4.9: TCP Packet Format

Source Port	Dest. Port	Sequence	Ack.	Flags	Window	Checksum	Urgent	Options	Data
2 bytes	2 bytes	4 bytes	4 bytes	2 bytes	2 bytes	2 bytes	2 bytes	*Optional*	*variable*

Table 4.10: The TCP Flags Field

	Data Offset	Reserved	CWR	ECE	URG	ACK	PSH	RST	SYN	FIN
Bit:	0–3	4–7	8	9	10	11	12	13	14	15

The Source and Destination ports are similar to TCP, and well-known ports are allocated in the same range. No TCP port can be zero, however, and the UDP port and TCP port with the same number are actually independent; only convention suggests that an application use the same number for both. Examples of well-known TCP ports are SSH on 22 and HTTP on 80. The Sequence and the Acknowledgement fields are used for defining the flow state. The Window field specifies how many bytes of room the receiver has to hold onto out-of-order data. The checksum, mandatory, covers the data, TCP header, and certain fields of the IP header. The urgent field is almost always zero, but was conceived as a way that TCP could send important data in a side channel. Options are possible after that, and then the data comes. Unlike UDP, TCP does not provide the length explicitly, as IP already does.

The flags (see Table 4.10) are divided up into the Data Offset, which specifies how long the options will be by when the first bit of data will appear. The CWR and ECE flags are not often used, and are for network congestion notification. The URG flag is for whether the Urgent field is meaningful. The ACK flag is used for every packet that is a response to another. The PSH flag is set when this particular packet was the result of the application saying that it wants to flush its send buffer. Small writes to the sending TCP socket do not cause packets to come out right away, unless that feature is specifically requested. Rather, the sender's operating system holds on to the data for a bit, hoping to get a larger chunk, which is more efficient to send. The application can flush that holding on, however, and the resulting packet will have the PSH bit set. RST is set when the sender has know idea about the socket the packet is coming in for. SYN is used to set up a TCP flow, and FIN is used to tear it down.

TCP needs to keep track of flow state, in order to provide the appearance of a stream. The TCP stream is a two-way channel, symmetric in the sense that no one side is favored over the other. Each side keeps sender state and receiver state. A part of that state is that every byte in the TCP stream, since the stream began, is given an increasing sequence number. As packets are pushed out to the network by the sender, the sender keeps copies of those packets. When the receiver gets a packet, it acknowledges it by sending a packet with the ACK flag set, and the Acknowledgment field to the sequence number of the highest byte it has received before a break in the sequence occurs. This acknowledgment can come back as a part of the next return-direction data packet, but if none are queued, then ACKs are generated in their own, otherwise empty, packets every 200ms. This process is called *delayed acknowledgment*. If the acknowledgment is never received by the sender, the sender has to assume that either the original data packet, or the acknowledgment itself, got lost. The sender will then retry the packet some time later. Once a packet has an acknowledgment received for it, the sender will finally free up the packet. On the receive side, the receiver cannot send information back to the application unless there is a contiguous run of bytes at the head of the reassembly buffer. If not, the buffer holds onto the bytes, and advertises the hole in the next acknowledgment.

TCP uses sophisticated flow control techniques to prevent the sender from sending too much. The basic flow control technique is that the sender cannot have any more outstanding packets sent than the window size it hears on any given TCP packet in return. This prevents the receive buffer from being overrun. On top of that, however, TCP engages in congestion control. The sender specifically tries to measure the round-trip time of the network, and its loss rate. Because TCP is a handshaking protocol, and the sender cannot send when the window is full unless it receives an acknowledgment first, TCP will perform most optimally if it can stick enough packets in the wire to fill the round trip time. If, after a round trip time elapses, an acknowledgment does not come in for a packet, the sender can assume the packet did not arrive and retransmit it right away. However, what if the network is congested? In that case, switches and routers can start dropping packets. TCP reacts to that packet loss. Its first method is to avoid flooding the line to begin with. With TCP, past success begets future success. To make that work, TCP starts of slowly, sending one packet at a time. Every acknowledgement gives it more confidence, and a reason to send one more packet than before in the next round trip. This process, called *slow start*, continues until the network finally drops a packet. Once a packet is dropped, the sender will notice it, because subsequent packets that are sent to the receiver will cause *duplicate acknowledgments*, as the hole that the loss created prevents the receiver from acknowledging the later sequence numbers, and yet the receiver is required to send an acknowledgment. The back-to-back duplicates cause TCP to back off, by cutting its *congestion window*—the number of packets it thinks it can have outstanding every round trip—in half. The sender then tries to ease back in, by growing its congestion window once every round trip time. This process is finely tuned to ensure that the network does not become overly crowded by aggressive behavior. In the early days of the Internet, this did, in fact, happen and was the motivation behind introducing congestion control.

Because TCP refuses to allow any loss, it is required to block the sender and receiver until it can resolve outstanding packet matters. This makes TCP generally inappropriate for voice mobility. Interestingly, TCP can be used for the signaling protocols, such as with Secure SIP (Chapter 2), as long as the applications that use it are prepared to handle cases on lossy networks where the application gets stuck. Also, TCP is being used increasingly for video, mostly because of applications such as consumer-oriented video sharing services, which make the assumption that simplicity is best.

4.2 Quality of Service on Wired Networks

The benefit of packet networks is that they are incredibly flexible. The ironic thing about the transfer from circuit networks to packet networks, however, is that the best-effort nature of packet delivery requires that packet networks develop the quality-of-service sophistication that was not needed for circuits networks. On a circuit, there is always room for a high-quality call, if there is room at all. On a packet-based network, however, it is very difficult to tell whether that call can just be squeezed in, or whether problems will arise.

There are two general methods for solving the quality-of-service problem with packet networks. Both methods are based on notions designed for IP, to ensure the simplest network that can deliver on the promises.

4.2.1 Integrated Services

The concept of *integrated services* is that the quality-of-service mechanisms are integrated directly into the forwarding network. Integrated services are based on the notion of resource reservations, similar to with circuit networks. The difference between circuit-oriented resource reservations and integrated services reservations is that the latter can accommodate a nearly infinite variety of packet rates, sizes, types, and behaviors.

Integrated services is based on the protocol called *RSVP*, the *Resource Reservation Protocol* in RFC 2205 (though named to sound like the phrase *répondez, s'il vous plaît*, as appropriate for reservations). The idea is rather simple. A receiver that needs to send a flow of a certain type requests the ability to get that flow with a specific quality of service. The sender uses a multicast group to send out specific messages, called *PATH* messages, to announce the availability of a stream. When a listener wants to join, it sends a response called a *RESV* (for reservation) directly back to the sender. Along the way, the intervening routers on the path have the responsibility of listening in on that protocol, and trying to provide just that quality of service.

The crux of the mechanism is the voluntary announcing of the flow qualities required using a format known as a *traffic specification*, or *TSPEC*. The TSPEC for RSVP is shown in Table 4.11. The Token Bucket Rate is a floating-point number that represents the amount of bytes per second that the flow is expected to take. The Token Bucket Size is a floating-point number specifying the amount of tokens, in bytes, that a flow can accumulate. This is something like a measure of the backlog. Token buckets work to regulate a flow. The idea is that the flow is given tokens at a fixed rate, one token per allowed byte. Thus, a flow that was admitted for one kilobyte a second will get 1000 tokens a second. The bucket has a maximum capacity, the size, to prevent a flow from getting an infinite hall pass to transmit if it has been idle for a while and not using its resources. Every byte that passes by from that flow needs a token to continue, and takes it from the bucket. Once the buckets are done, the network can either buffer the packet until it gets a token, or police the flow by dropping that packet. RSVP works usually under the latter condition. However, RSVP also endeavors to ensure that the flows that are within their limits get to use the resources before best effort traffic.

The Minimum Policed Unit is set to the size of the smallest packet that is used in the flow: for voice or video, this is likely to be an RTP packet. The Maximum Policed Size field specifies the largest packet size that the sender may want to generate.

Table 4.11: RSVP TSPEC Format

Token Bucket Rate	Token Bucket Size	Peak Data Rate	Minimum Policed Unit	Maximum Policed Size
4 bytes	4 bytes	4 bytes	4 bytes	4 bytes

RSVP is a form of admission control. When the resource requests are made, if any one of the routers cannot support the flow because it has already exceeded its admissible capacity, it will inject a reject message, to inform the listener that its flow's quality of service will not be granted.

RSVP's greatest disadvantage is that it requires all of the routers to keep state—even soft state—on every flow, and to take action based on the behavior of the flow. Because of this, RSVP is not commonly used in voice mobility networks, and the concepts are not used for wireline. The same concepts behind RSVP, however, appear in the context of wireless networks, where the stakes are higher and the number of devices that must maintain state are dramatically reduced (to one base station and one client).

4.2.2 Differentiated Services

So how do wireline networks get quality of service? They do so through the use of prioritization. Instead of asking for, and accounting for, resources and reservations and policing, the network becomes very simple. Traffic is divided up into classes. Some classes are better than others, and will get special treatment. Most likely, this treatment is just to cut to the head of the line. Each packet, not flow, is independently marked with the priority or class it belongs to. Every router and switch along the way that understands the tags will provide that differentiation, and the ones that do not simply ignore the tags and treat the packet as best effort.

This is the concept of *differentiated services*. For IP networks, the TOS/DSCP field in IPv4 and Traffic Class field in IPv6 is expected to hold the specific class or priority that the packet belongs to. The sender self-marks the packet, and the network takes it from there.

Here, the two conflicting concepts of the IPv4 Type of Service (TOS) come in contact with the Differentiated Services Code Point (DSCP) definition, for the same byte in the header. Each is a mechanism that was created to try to classify packets on a per-packet basis. TOS is the older mechanism, and is now considered to have fallen out of use. However, for the purposes of voice mobility, a lot is similar about TOS and DSCP. TOS defined, among other things, eight priority levels.

The format of the now formally deprecated TOS field is shown in Table 4.12.

Table 4.12: The TOS Field in IPv4

	Precedence	Delay	Throughput	Reliability	Reserved
Bit:	0–2	3	4	5	6–7

The precedence value is a prioritization that is used within the network to determine its handling. The values run from 0 to 7, with 0 being the lower end of the range. The definitions originally conceived for this value is given in Table 4.13.

Table 4.13: The TOS Precedence

Value	Old Meaning	802.1p Meaning	WMM Meaning
7	Network Control	Network Management	Voice
6	Internetwork Control	Voice	Voice
5	CRITIC/ECP	Video	Video
4	Flash Override	Controlled Load	Video
3	Flash	Excellent Effort	Best Effort
2	Immediate	Undefined	Background
1	Priority	Background	Background
0	Routine	Best Effort	Best Effort

The table suggests a gradual rise in priority from 0 to 7. The problem with this definition is that different technologies use the 0–7 range for priorities. Most equipment endeavors to maintain a consistent mapping for the number to a priority level, no matter how the priority got to the packet. The three different meanings are shown in the columns. The second column is from IEEE 802.1p, which is a per-frame prioritization extension to Ethernet, and uses a special header to advertise the priority. The third column contains the meaning of the same eight values in WMM, the Wi-Fi prioritization standard. In general, it is best to assume the meaning of the final two columns. Note that the priority for values 1 and 2 are actually less than best effort in that case. When in doubt, do not use those priorities.

The remaining three flags in Table 4.12 represent extra information that may have been useful for the packet. Setting the delay bit meant to ask for low delays, whereas setting the throughput or reliability bit was meant to signal that throughput or reliability was a greater concern to the application.

TOS is considered to be replaced, and yet many modern devices in the world of IP telephones use the TOS meanings, and not the later DSCP meanings, in order to support older network configurations that may still be in use.

DSCP requires that the TOS meanings for the top three bits still be preserved, as long as the remaining bits are zero. However, DSCP looks at the one byte a different way. Table 4.14 shows the new meaning.

Table 4.14: The DSCP Field in IPv4 (Same Byte as TOS; Different Meaning)

	Code Selector	ECN
Bit:	0–5	6–7

There are a couple of RFCs that define what the code selector maps to. The goal of the DSCP is to interpret the selector as a somewhat arbitrary code, mapping into a specific quality of service type.

RFC 2597 defines the concept of *Assured Forwarding* (AF), the purpose of which is to allow a service provider to accept markings of packets and apply a certain amount of guaranteed bandwidth, as well as allowing more bandwidth to be given. Each class is named *AFxx*, where the first *x* is a number from one to four, representing the class of traffic, and the second x is a number from one to three, representing the drop probability from low to high (see Table 4.15).

Table 4.15: Assured Forwarding DSCP Values

Drop Probability	Class 1	Class 2	Class 3	Class 4
Low	AF11 = 10	AF21 = 18	AF31 = 26	AF41 = 34
Medium	AF12 = 12	AF22 = 20	AF 32 = 28	AF42 = 36
High	AF13 = 14	AF23 = 22	AF33 = 30	AF43 = 38

The network administrator is expected to assign meanings to the four classes, in terms of assured, set-aside bandwidth that these codes can eat into. The drop probabilities are meant to be sent by the traffic originator to make sure that, if resources are getting exhausted, some packets get more protection than others.

A different concept is defined in RFC 2598. *Expedited Forwarding* (EF) sets up a specific codepoint, 46, to allow packets to be marked as belonging to a "virtual lease line," a high-performing point-to-point measure of quality of service. (There is a wrinkle with this DSCP code as it applies to Wi-Fi: All EF tagged packets get transmitted in the class of service designated for video because of the way the EF tag is coded.)

In total, there are 21 commonly seen DSCPs: the twelve AFs, the EF codepoint, and the eight original precedence values, now known default and CS1 to CS7.

Nothing in DSCP or differentiated services defines just what the qualities of the differentiated services are to be. This is the advantage of differentiated services: the differentiation is up to the administrator, and can grow as the network grows.

4.2.3 Quality-of-Service Mechanisms and Provisioning

There are a few common ways for quality of service to be provided in networks, using enterprise-grade wireline infrastructure. The concepts all stem around handling the packets differently when it comes to queuing. Why? Most wireline networks can handle a fairly large amount of traffic, because the wireline technologies, such as Gigabit Ethernet, have enough throughput to make congestion be less of an issue. However, certain protocols are designed to take up as much bandwidth as they can—to specifically expand into the space that you give them. It will always be important on voice mobility networks to keep the voice traffic protected from these applications, especially if they cause changes in delay. Moreover, network congestion can cause loss rates to become problematic. All of the problems happen to the packets not as they are on the wire, but as they back up in queues within the choke points of the network, the routers or switches that connect the links together. What happens in those queues makes the difference.

Thankfully, using the packet classification capabilities from differentiated services, enterprise-grade wireline infrastructure can be used to both police flows that get out of hand and give the ones that are being squeezed out the help they need. These techniques go under the broad category of *queuing disciplines*, as they provide the discipline that is used to maintain order in the queues.

The idea is to take what was once one monolithic queue for the chokepoint, and to create possibly different queues, each queue leading to the same eventual chokepoint. As traffic heads towards the bottoms of the queues, an element called a *scheduler* chooses from which queues to take packets, and then provides those packets for transmission.

We'll take queuing disciplines and scheduling together for this discussion.

4.2.3.1 FIFO

The simplest behavior is to do no particularly new behavior at all. First-in, first-out (FIFO) queuing refers to using the one queue that is there, and to putting packets in with the same order in which they arrived, and pulling them out the same way. This sort of queuing is precisely what causes congestion and variable delays.

For the purposes of voice, the longer the queue gets, the longer the potential maximum delay the queue can cause the voice packet to suffer. The alternative is not much better: if the queue gets longer than it can handle, the packets will be dropped.

4.2.3.2 Classification

The first step is to determine whether there is any structure in the packets that can be used to differentiate them. Enterprise-grade classification techniques can use a wide, rich array of properties about the individual packet, including the sender, receiver, size, DSCP value, ports, applications, and routes. These can all be applied in a *stateless* manner, meaning that the router or switch need look at each packet only in isolation. An additional option exists for some routers and switches with a lot of memory and processing ability. They can use flow state to create *stateful* classification, in which previous packets that are related to the current one dictate the behavior. This distinction is identical to that used in firewalling. Once packets are classified, they can be placed into queues by their classes. These queues can be administratively created, or they can be created on the fly based on the class divisions, ensuring that packets from each class stay in separate queues.

Class-based queuing (CBQ) is an extension of this basic concept. Instead of having one level of discrimination, the concept can be extended to a hierarchy of queues, all set up by the administrator. This hierarchy can be powerful in preventing flows and users from stepping on each other, and for shaping the bursts and behavior of the traffic. *Traffic shaping* is a highly important function for variable bitrate, expansive applications, to prevent them from overwhelming other applications that may not deserve the highest prioritization, but still need to be metered.

Once the packets are classified into sibling queues, the schedulers need to be selected, to determine how to get the packets out of the queues.

4.2.3.3 Round-Robin

The simplest scheduler is the *round-robin* scheduler. As the name suggests, the round-robin scheduler takes packets in turn from each queue, wrapping around when it hits the last one. Queues with empty packets get skipped over, but otherwise, everyone gets a shot.

Round robin is good for creating packet fairness, were every class gets an equal shot at sending a packet. However, if some of the classes should have a higher priority than the other, then round robin will not suffice.

4.2.3.4 Strict Prioritization

Strict prioritization is a very simple scheduler. Classes are ordered, strictly, from highest to lowest. The scheduler always starts with the highest queue. If there are no packets in the highest-priority queue, it checks the one with the next highest priority. This continues until the scheduler finds a packet, which it then sends.

By draining the highest-priority queue before moving onto the others, strict prioritization ensures that the traffic with the highest prioritization moves right to the head of the line. Even if the lower-priority queues are heavily backed up and congested, if the highest-priority queue is empty and a highest-priority packet comes in, it will move right past the long lines and be sent first.

Strict prioritization is often good enough for voice, especially when the issue is preventing data from competing with voice. However, for elastic or variable applications where one should get more resources than the other, but not too much more, strict prioritization will not suffice either.

4.2.3.5 Weighted Fair Queuing

To provide a sense of both prioritization, of which strict prioritization may provide too much, and fairness, of which round robin may provide too little, there is the notion of *fair queuing*. In fair queuing, the goal is to provide a fair bitrate to each of the classes. Round robin provides a fair packet rate, which is the same only if the packets are all the same size. On top of fair queuing, however, the bitrate should be adjustable so that higher-quality flows get more throughput, without exhausting all the throughput available. This is the concept of *weighted fair queuing* (WFQ).

The idea behind WFQ is that each queue gets a relative weight. That relative weight is used to adjust the data rate that the queue gets. The amount of traffic that the queue gets is always based on how many other queues are active and for how long; the goal is not to tightly control throughputs or to ensure that no one queue gets ahead of the other, but that queues with equal amounts to send get their weight's worth of relative throughput.

The scheduler's goal is to give the appearance that each queue with a byte in it has a byte taken out fairly (as if, say, by round robin, though order does not matter). This gives rise to thinking about packets flowing through the queues like fluids. The output requires a given data rate, or velocity, and each of the packets are extruded through their queues a little at a time, in equal amounts. The first packet out, then, would be the one whose last byte gets drawn out first—that is, the one that finishes first. The problem, of course, is that packets are packets, not bytes, and cannot be drawn out in this manner.

What can be done is that the scheduler can do the math that simulates the bit-by-bit extraction, and make sure to dequeue packets, then, in that proportion. The scheduler calculates the expected time the packet at the end of each queue would get drawn out, in units of *virtual time*, that don't depend on real time but still flow forward. This gives the precise order of the packets that should come out. As a packet comes out, the new packet's virtual end time is calculated, and so on. This technique ensures that packets flow out in the order they should.

The weightings come into play by adjusting the velocity, in virtual time, that a queue extrudes its packets. Higher-weighted queues extrude packets more quickly, and thus those packets finish more quickly in virtual time, and hit the wire sooner.

It is important to observe that WFQ is a *work-conserving* process. Work conservation means that the scheduler never delays sending traffic. If there is a packet to send, in any queue, then at least one packet will be sent. At no time will a work-conserving process refuse to send traffic, or delay sending traffic, in hopes of getting a more even throughput. Work conservation is important for not wasting network resources for the sake of "quality."

4.2.3.6 Traffic Shaping

Traffic shaping is more severe than fair queuing. Whereas fair queuing is concerned with fairness, traffic shaping is concerned with ensuring that a precise rate of traffic is met by a given class.

Traffic shaping is usually performed through the use of some form of token bucket, first mentioned in the context of RSVP (Section 4.2.1). To recap, the idea of a token bucket is that virtual tokens, corresponding to permission to send bytes, are deposited into the virtual bucket corresponding to the queue at a fixed rate. This rate is the goal at which traffic should be sent. The token bucket then requires that a packet from the queue have enough tokens before it can be let past. This requirement ensures a constant bit rate to the flow.

Token buckets are general ways of metering the flow of traffic. Using them to shape traffic, by holding up packets until there are enough tokens for them, is clearly not work-conserving, as the hold up will happen regardless of whether the line will go idle because of it. On the other hand, token buckets have a bucket depth for a reason. If traffic does happen to go idle for a while in the queue that owns the tokens, the queue is allowed to save up its backlog of tokens for when it might need it. Once the traffic resumes, it can use up all of its saved tokens without waiting. This allows for the average traffic rate to be more manageable, even if the incoming flow is not perfectly regular.

Traffic shaping holds an important place in keeping variable flows in check, so that they do not exceed specific *service-level agreements* (SLAs), which often specify a minimum available bandwidth. The goal of an SLA is to give a fat pipe that is shared among users the appearance that it really is a dedicated thin pipe for that one user. This is reminiscent of the reason we embarked on this journey, to make packet networks seem more like dedicated circuits. For voice, a constant, inelastic traffic, traffic shaping does not hold much interest in itself for what we need. However, traffic shaping does highlight one advantage of packet-based networks. They are flexible enough to provide circuit-like throughput guarantees for some services when needed while providing expandable prioritization for other services, all on the same wire.

4.2.3.7 Policing

Policing is the other side of the coin of scheduling and queuing discipline. Instead of deciding to hold onto the packet in a queue until it has met its criteria, classes are watched for the same criteria and their packets are dropped when they exceed it.

The point of policing is that it does not require building up the long lines of delayed packets as queuing would. Instead, the policer can just observe and drop packets that go over the mark. Policing is a lot less forgiving than queuing, but it requires fewer resources in the network.

Token buckets are often used for policing. With token bucket policing, when a packet comes by that does not have enough tokens, it is simply dropped. Packets never delay in this model.

Policing is a tough tactic to get right, because it works necessarily by dropping packets that could have been queued or sent. For voice networks, where the goal is to prevent data from interfering with voice, policing is useful only for preventing runaway or hijacked voice streams, being high priority, from taking over the network. Prioritization is a better method to keep data from affecting voice quality.

4.2.3.8 Random Early Detection

Along with policing comes the idea of how to drop a packet when the queue is filling. Congestion, for data, is a major issue, and as data backs up, it can cause major problems for any traffic that shares the link with it.

The concept behind *random early detection* (RED) is that congestion can be signaled to TCP, or any other elastic and responsive traffic protocol, before the congestion gets so bad that it caused unfair loss. Congestion causes that unfair loss by affecting whichever random flow whose packet happens to be the one too many for the queue and gets dropped first. As such a flow loses packets, it slows down, and other flows expand to fit their place. To bring back less broken symmetry between the flows, random early detect uses a sliding scale of random drop probabilities to keep the backup at bay. When the queue is nearly empty, nothing is dropped. As the queue fills, however, RED kicks in by increasing its drop probability. This slow but steady increase starts backing the flows off before the queue gets too full, but the fact that it stops dropping when the queue empties gives pressure when the queues are filling and permissiveness when there is plenty of room.

On top of RED is a concept called *weighted random early detection* (WRED). WRED uses weights, based on the classifications we have seen already, to alter the drop probabilities. Using the classifications for voice allows administrators to avoid having WRED kick in for voice, which is inelastic and will not respond to being dropped, if the administrator has no ability to place voice in a separate queue or route. For data, more critical data connections, such as TCP-based SIP needed in calls, can be given a higher probability by avoiding a higher drop probability, while allowing normal data to be slowed down.

The problem with RED is the problem with policing. Packets that may have been needed to prevent the queue from going idle even though there are resources for them, causing lost work and wasted resources.

4.2.3.9 Explicit Congestion Notification

Instead of using RED, routers have the option of marking the packets, rather than dropping them. TCP endpoints that know to read for the congestion-marked packets will consider it as if the packet had, somehow, been lost, and will back off or slow down, but without causing the packet's data to disappear. This increases the performance of the network and improves efficiency, though, needless to say, it does nothing if the endpoints are not aware of the congestion notification scheme.

On TCP, *explicit congestion notification* (ECN) works by the TCP endpoints negotiating that they support this protocol. Both sides need to support it, because the only way the sender can know if an intervening router has marked a packet is for the receiver to echo that fact back to the sender over TCP itself. Once a flow is established, the sender sets the ECN bit, bit 6, in the DSCP usage of the TOS field in IP (Section 4.2.2, Table 4.14). This lets routers know that the packet supports ECN. When a router uses RED to decide that the packet should be dropped early, but notices that the packet is marked for ECN support and the router supports ECN itself, it will not drop the packet. Instead, it will set the seventh bit in the ECN header, the *CE* or *Congestion Experienced* bit, marking that the packet should be handled as if it were to have been dropped.

The TCP receiver notices that the packet has been marked, and so needs to echo this fact back in the acknowledgment. The receiver sets the *ECE*, or *ECN-echo* bit in the TCP flags field in the acknowledgement. (Section 4.1.1.4, Table 4.10). The sender gets the acknowledgement, and uses this flag to cut its congestion window in half, as if the original packet were lost.

CHAPTER 5

Introduction to Wi-Fi

5.0 Introduction

This chapter provides an introduction into wireless local area networking based on Wi-Fi, also known by its more formal standard name of IEEE 802.11. The goal of the chapter is to provide a solid background on Wi-Fi technology, looking at what needs to be done to ensure that wireless local area networks operate well as both a data network and a crucial leg of voice mobility solutions. This chapter is aimed for readers with all degrees of familiarity with wireless networking. Although not a reference on all things Wi-Fi, the chapter starts with the basics of Wi-Fi before diving into what makes voice unique over this particular type of network.

5.1 The Advantages of Wi-Fi

Until now, we've looked at why voice is interesting and what makes it work over a network, but we haven't yet examined the technologies that truly make voice mobile. The advantage of mobile voice, when working properly, is that the elements of the underlying network fade away, and user sees only a familiar phone, in a mobile package. Of course, this requires cutting the cord, allowing users to make or receive calls from anywhere. So that we can understand how and why an unwired network is able to make the elements of the network disappear to the user, so to speak, we need to dive deeper and understand what the unwired network is made of.

Wi-Fi, the wireless local area networking technology based on the work from the standards branch of the Institute of Electrical and Electronics Engineers, uses the IEEE 802.11 standard to allow portable mobile devices to connect to each other over the air, transmitting IP-based data as if they were connected directly with a cable.

But being wireless alone does not explain why Wi-Fi has become the primary wireless technology for both consumer- and enterprise-owned networks. Wi-Fi technology has a number of advantages that make it the obvious choice for wireless data, and for many circumstances, for mobile voice as well.

doi:10.1016/B978-1-85617-508-1.00001-3.

What is the Difference between Wi-Fi and IEEE 802.11?

Almost everywhere, the term *Wi-Fi* is now used to refer to the networking technology based on the IEEE 802.11 standard. There are subtle differences, however, between the two terms.

The term *Wi-Fi* is a trademark of the Wi-Fi Alliance, a nonprofit industry organization made up of nearly all of the equipment providers manufacturing IEEE 802.11–based devices: chipset vendors, consumer and enterprise access point vendors, computer manufacturers, and so on. The Wi-Fi Alliance exists for two reasons: to promote the use of Wi-Fi certified technology throughout the industry and within the press, and to ensure that wireless devices based on 802.11 work together. The term *Wi-Fi*, and the accompanying logo, can be used only for products that have passed the Wi-Fi Alliance's certification programs.

We'll discuss the Wi-Fi Alliance more later, and where the Wi-Fi Alliance's certification programs diverge from the IEEE 802.11 standard. In the meantime, remember that 802.11 and Wi-Fi mean *almost* the same thing.

Example of the Wi-Fi Alliance Certification Logo.
Note: the logo is the trademark of the Wi-Fi Alliance and is shown here for example purposes only.

5.1.1 Unlicensed Spectrum

Generally, the ability to transmit radio signals over the air is tightly regulated. Government bodies, such as the U.S. Federal Communications Commission (FCC), determine what technologies can be used to transmit over the air and who is allowed to operate those technologies (see Figure 5.1). They do this latter part by issuing licenses, usually for money, to organizations interested in transmitting wirelessly. These licenses, which are often hard to obtain, are required in part to prevent multiple network operators from interfering with each other.

The advantage of Wi-Fi, over other wireless technologies such as WiMAX (which we will cover in Chapter 7), is that no licenses are needed to set up and operate a Wi-Fi network. All that it takes to become a network operator is to buy the equipment and plug it in.

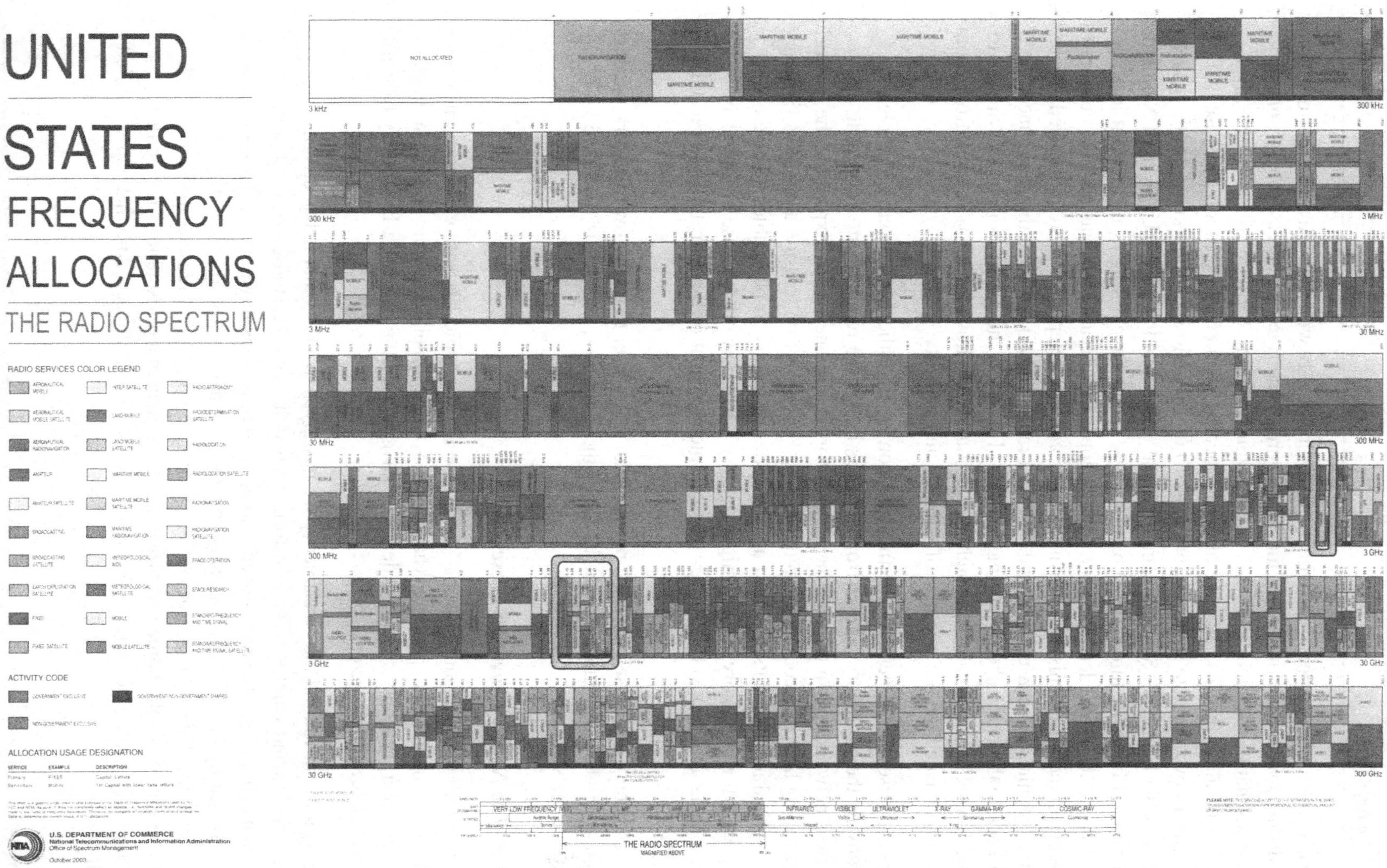

Figure 5.1: The United States Spectrum Allocation. Wi-Fi operates in the circled bands

Clearly, the array of allocations within the spectrum is bewildering. And network operators for licensed wireless technologies must be aware of the rules for at least the part of the spectrum that their technology works in, to avoid violating the terms of the license. But, thankfully, all of this is taken care of automatically when 802.11 technology is used. Wi-Fi operates in two separate stretches (or "bands") of the radio spectrum, known in the United States as the Industrial, Scientific, and Medical (ISM) bands, and the Unlicensed National Information Infrastructure (U-NII) bands. These bands have a long history, and it is no coincidence that voice lead the way. Many people first became familiar with the concept of unlicensed radio transmissions when 900MHz cordless telephones were introduced. These phones require no licenses, but have a limited range and do only one thing—connect the call back to the one and only one base station. However, the power from using wireless to avoid having to snake cables throughout the house and allowing callers to walk from room to room revealed the real promise of wireless and mobility.

For enterprises, the benefits of the freedom from using unlicensed spectrum are clear. Removing the regulatory hurdles from wireless brings the requirements for setting up wireless networks down to the same level as for wireline networks. Expanding the network, or changing how it is configured, requires no permission from outside authorities (ignoring the physical requirements such as building codes necessary to pull cables). There is no concern that a regulatory agency might reject a Wi-Fi network because of too many neighboring allocations. Enterprises gain complete control of their air, to deploy it how they see fit.

Because being unlicensed gave the potential for every user to be her own network operator, wireless networking settled into the hands of the consumer, and that is where we will continue the story.

5.1.2 The Nearly Universal Presence

Even though the focus of this book—and of so many people—is with enterprise and large-scale deployments, in explaining what makes Wi-Fi compelling, we must not lose track of the consumer, and how consumer demands have pushed the entire Wi-Fi industry forward, inevitably benefiting the enterprise.

The major contribution the consumer space has given Wi-Fi is that is has driven people to demand wireless. Three historic events changed the landscape of mobility and connectivity: the Internet moved into the home; laptops replaced desktops and were being issued by corporate IT for usage everywhere; and darkly roasted coffee came onto the scene. Or rather, for the last one, people began to find reasons to want to work and live outside of the home and office. All three demanded a simpler solution than having to drag oversized telephone cables around with each user. And that gap was filled with Wi-Fi.

Wi-Fi is now in many places that mobile users are expected to show up in. In the home, it is difficult now to find a consumer-level gateway that does not include wireless. Just as television once was the centerpiece of the living room, but contention over control of the remote and the drop in prices lead televisions to spring up in nearly every room of the house, the Internet has migrated from being connected to one prized home computer in the living room to being spread throughout the house by Wi-Fi. In the enterprise, the advantages of unwiring the network edge has lead to IT organizations peppering the office with access points. And on the road, hotels, airports, cafes, and even sporting arenas have outfitted with Wi-Fi, to try to encourage their customers to get back with their online selves as often as possible, and maybe make each one be a little more "sticky" in the meanwhile.

What this means for voice mobility is that the cycle of demand drives the technology to get ever better. Consumers' demand and expectations "pull" advanced wireless into the home, just as enterprises "push" laptops onto their employees, encouraging them to be used outside the office, therefore increasing the number of hours employees think and do their work far beyond the amount of time each employee spends in the office.

And with this cycle of demand also comes maturity of the underlying technology. Wi-Fi has gone through a number of iterations, getting faster, more powerful, and less prone to mistakes. Now, it is nearly impossible to find laptops without wireless built in. It is even an option on many desktop systems, not considered to be traditionally mobile, yet eager to be joined in on the wireless bandwagon to help company's save on cabling costs.

5.1.3 Devices

Wi-Fi was initially thought of as a data network only. Partially, this was because of an attempt to avoid the bad image that cordless phones also projected, as users were far too used to static and interference on cordless phones. But mostly, the original iterations of Wi-Fi occurred when Wi-Fi itself was struggling to find a place, and allowing users to check email or surf the Web while moving from room to room seemed to be enough of an application to motivate the fledgling industry.

But when mobile data networking took off, and people became addicted to remote email over the cellular network, the seeds were sown for device vendors to want to integrate Wi-Fi into their mobile devices. And because those devices are primarily phones, the connection of mobility to voice over Wi-Fi was natural.

Broadly, there are two categories of voice mobility devices that use Wi-Fi as a connection method. The first are Wi-Fi-only devices. These devices are often dedicated for a specific application in mind. For example, Vocera Communications makes a Wi-Fi-based communicator that is often used in hospitals to allow doctors and nurses to communicate with each other using voice recognition, rather than a keypad, to determine whom to call.

This device looks and acts more like a Star Trek communicator than a phone, but is an excellent example of voice mobility within a campus. Polycom, through its SpectraLink division, Cisco, and Ascom all make handsets that look more like a traditional mobile phone. In all of these cases, single-mode networking—using just Wi-Fi, in these examples, as the only means of connectivity—makes sense for the environment and the application.

The second type is made of mixed-mode, or integrated devices. These devices are mobile phones, made to be used with the cellular network as well as Wi-Fi. Nearly every mobile handset manufacturer is selling or is planning on selling such a device, including Research in Motion, Nokia, Samsung, and Apple with its iPhone. These devices can be made to place voice calls directly over the Wi-Fi network, rather than the cellular network, thus unlocking the entire fixed-mobile convergence (FMC) industry.

In both cases, the push from Wi-Fi networks originally designed for data allows for voice to become a leading, if not the dominating, purpose for many networks, as the maturity and variety of Wi-Fi-enabled voice devices make voice mobility over Wi-Fi possible.

5.2 The Basics of Wi-Fi

Wireline technologies are almost entirely focused on the notion of the cable. On one end lies the network, and on the other lies the client device. Starting with the original wireless telephone system, where everything—including identity—is determined merely by which port the cable connects to, the wireline technologies have only partially moved towards mobility and the concepts of link independence.

However, Wi-Fi has no cables to begin with, and so something else is needed to define the relationship between a client and the network. Wi-Fi is built upon the notion of two types of wireless devices: the *access point* and the *client*. Both use the same types of radios, but take on different roles.

5.2.1 Access Points

The access point (abbreviated as "AP"; see Figure 5.2) serves as the base station. The concept is common, from cordless phones to the large wireless carriers: the access point is what provides the "network," and the clients connect to it to gain access. Each Wi-Fi radio, whether it be in the access point or the client, is designed to send its wireless signals across a limited range, far enough to be useful but not so far as to violate the limits set by the regulations and to grossly exceed the bounds of the building the network is deployed within. This range is in the order of 100 feet, though. To set apart which device connects to the network, the access point must take on a role as some sort of master.

An access point often looks like a small brick, but with antennas and an Ethernet cable. The Ethernet cable provides the connection to the wired network, and, if power over Ethernet

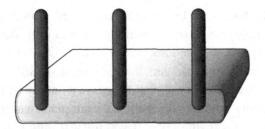

Figure 5.2: A typical Access Point

(PoE) is in use, the access point receives its power over the same cable. Access points are normally independent physical devices. Commonly, they are placed along walls, or above or below a false ceiling, to provide the maximal amount of wireless coverage with the least amount of physical impediments to the signals (see Figure 5.3). How exactly those locations are determined will be addressed later in this chapter.

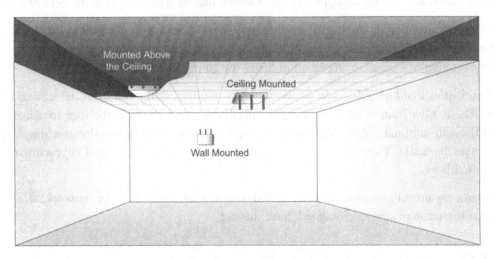

Figure 5.3: Typical Access Point mounting locations

Access points make their networks known by sending frequent wireless transmissions, known as *beacons*. These beacons describe to the client devices what capabilities the access point has, and most importantly, what network the access point is providing access to. The way the network is designated is by an arbitrary text string provided by the administrator, known as a *service set identifier* (SSID). This text string is sent in the beacons, and other transmissions, to the clients, which then provide a list of SSIDs seen to the user. Thus, when the user brings up a list of the networks that his or her laptop sees and can connect to, the list contains the SSIDs of the access points.

Because the SSID is the only way users can select which network they wants to connect to, we need to look into it a bit deeper. There are very few technical restrictions on the SSIDs except for the length, which must be less than 32 characters. However, the SSID needs to be meaningful to the user, or else he or she will not connect to it. Because SSIDs are supposed to name the network that the user is connecting to, rather than the individual access point, multiple access points can and do share the same SSID. That being said, there is nothing stopping someone else from giving an access point the SSID that belongs to your network. There is no security in the SSID itself. Eavesdroppers can trivially discover what the SSID is that your network is using (even if you use a feature known as *SSID hiding* or *SSID broadcast suppression*) and use it to either gain entry into your network or spoof your network and try to fraudulently get your clients to connect to them instead. In fact, there is nothing that prevents SSIDs from being used for nearly any purpose at all. Most of what applies to SSIDs are in the form of best practices, of which the important ones are:

- The SSID should be meaningful to the user: "employees" and "guest" are good examples of meaningful names. They may be based on the role of the user, the device the user has (such as "voice" for phones), or any other words that help the user find the network.

- When the installation shares the air with neighboring networks from other organizations, the SSID should also include text to highlight to the user what the right network is; "xyz-employees" is an example of an SSID for an organization named XYZ.

- The SSID should be able to be easily typed by the user. Although most devices show SSIDs in a list from what already are being broadcasted, allowing the user to select the SSID with minimal effort, there are many occasions on which when the user may need to type the SSID. This is especially true for mobile devices, with small keyboards or limited keys.

- Again, do not rely on obscurity of the SSID to restrict access to your network. Use real security mechanisms, as described later, instead.

5.2.2 Clients

A client is the typical end-user device. Unlike access points, which are strategically placed for coverage, clients are almost always mobile (or potentially so).

Wi-Fi clients can be general networking interface devices, such as those in laptops, or can be part of a purpose-built mobile voice handset. Either way, these clients appear to the network as endpoints, just as Ethernet devices do.

From the user's perspective, however, Wi-Fi clients add an extra complication. Unlike with wireline connections, where the user is assigned a port or cable and has the expectation that everything will work once the cable is plugged in and the process has settled down (which, for administrators, generally means that Dynamic Host Configuration Protocol (DHCP) automatic IP address discovery has completed), wireless connections have no one cable to solve all problems. The user must be involved in the connection process, even when the reason for connection or disconnection is not readily apparent. As mentioned previously, the user must learn about SSIDs. When a wireless interface is enabled, the user is normally interrupted with a list of the available networks to connect to. Knowing the right answer to this question requires an unfortunate amount of sophistication from the user, not because the user does not understand the technology, but because they usually do understand the power of mobility, and have learned to strategically hunt out wireless networks for casual email access. This is clearly evidenced by the pervasive nature of the "Free Public WiFi" ad hoc (Independent Basic Service Set, or IBSS) SSID that tends to be on so many laptops.

Ultimately, the user is responsible for knowing what the appropriate network is to connect to at any given location. Most devices do remember previous connections—including authentication credentials, in many cases—and can make the connection appear to be automatic. However, because of that caching, installations that run multiple SSIDs are often forced to deal with users not knowing exactly which network they are connected to.

Once the connection is established, the interface comes up much as a plugged-in Ethernet link does. Any automatic services, such as DHCP or Universal Plug and Play (UPnP), that run on interface startup will get kicked off, and the users will be able to communicate as if they had plugged directly into the network.

The last wrinkle comes, however, with mobility. Once the user leaves the coverage range of the one access point that it is on, the client will perform its list gathering activity (scanning) again. If it can find an SSID that it already has in its list—especially if the SSID is the same as the one the client was already associated to—the client will try to hand over to the new access point without user intervention. However, if the handoff does not succeed, or there are no more known networks in range, the client will disconnect and either warn the user with a popup or just break the connection without warning. This can come as quite a shock to the user, and can lend negative impressions about the network.

5.2.3 The IEEE 802.11 Protocol

Now that we have the basic roles established, let's look at the protocol itself.

5.2.3.1 Frame Formats

Because it belongs to the IEEE 802 family of standards, 802.11 integrates tightly into existing Ethernet networks. Wi-Fi transmissions, like their wired Ethernet brethren, are contained in what are known as *frames*. In the IEEE 802 context (including the 802.3 Ethernet series as well as 802.11), a frame is one continuous transmission of data. For 802.11, as with Ethernet, these frames usually carry a payload of 1500 bytes or less. This payload can contain one IP packet. Also as with Ethernet, 802.11 uses the 48-bit Ethernet MAC address to identify every device in the network. In fact, wireline Ethernet and 802.11

share the same pool of addresses. This is because an access point functions as a *bridge* between the wireless and wireline networking, relaying frames, and so the addresses must be unique across both technologies.

However, there are differences between Ethernet and Wi-Fi frames, the latter taking into account the wireless nature of the protocol. Unlike Ethernet frames, which generally come in only one type, 802.11 frames come in three types: *control frames*, *management frames*, and *data frames*. The first two frames are intimately involved in the underlying protocol, keeping the connection up and running. These are needed because the connection cannot be defined by a cable. Multiple devices share the same air, and the access point and clients need to keep what data is destined to them separate from the other transmissions. The data frames, not surprisingly, are the closest to the Ethernet frames, and carry data payloads. Because of the three types of frames—and the number of subtypes, mentioned shortly—all 802.11 frames have to have additional header fields.

In contrast, Ethernet headers contain three fields only. The first field is the 48-bit destination MAC address, which either names a specific device on the network or a multicast group address (including the standard broadcast address of FF:FF:FF:FF:FF:FF). The second field is the 48-bit address of the sender. The final field is the two-byte protocol field (or *EtherType*), designating whether the payload is IP (08:00), ARP (08:08), IPv6 (86:DD), or another type. After the header comes the payload—padded if the overall frame size is shorter than 64 bytes. At the very end is a four-byte CRC32 checksum as the FCS (see Table 5.1).

Table 5.1: Ethernet frame format

Destination	Source	EtherType	Frame Body	FCS
6 bytes	6 bytes	2 bytes	*n* bytes	4 bytes

To this base, 802.11 adds a third address, known as the *basic service set identifier* (BSSID). This address is the MAC address of the wireless network itself, and is needed because, unlike with Ethernet which can relay a frame from device to device, across multiple links in a switched network without confusion, 802.11 has only one link. In fact, every 802.11 transmission can be thought of as being primarily from one wireless device to another, and is addressed that way. Therefore, the usual destination and source addresses become the wireless device addresses, and the remaining address then serves to identify what the final destination of the frame is, whether that destination be on the wired or wireless links. To preserve the destination-then-source ordering of Ethernet, 802.11 places its three addresses in the order of receiving wireless device, then the sending wireless device, followed by a third address that makes sense only in context. Unfortunately, this means that the BSSID may be either the second or third address, depending on the role of the transmitter (access point or client). Table 5.8 shows all of the possible mappings.

In addition to the address, 802.11 adds the type of the frame, in what is known as the *frame control* field. This specifies the version of the frame (always 0), the type and subtype of the frame, and other flags used in the protocol. Also added is the *duration* field, which is used to help avoid certain types of interference from other wireless devices on the same channel. A *sequence control* field is also added, to help detect retransmissions (more below). The *QoS control* field identifies the quality of service parameters for the frame. Overall, the frame structure is as in Table 5.2, with the details continuing in Tables 5.3 through 5.7. Not all fields are present in every frame. One thing to keep in mind is that there is no EtherType field for 802.11 headers. Instead, 802.11 relies on the *Subnetwork Access Protocol* (SNAP) to place the type inside the data payload. This SNAP header is usually added and removed by the 802.11 device automatically, and will not show up on the wireline Ethernet networks.

> NOTE: the order of bits is backwards from IP networking—even though the byte ordering is identical. That means that bit 0 is the highest order bit in any field, but fields are always represented in the order they are sent over the air, without reordering.

Table 5.2: 802.11 Frame format

Frame Control	Duration/ID	Address 1	Address 2	Address 3	Sequence Control	QoS Control	Frame Body	FCS
1 byte	2 bytes	6 bytes	6 bytes	6 bytes	2 bytes	*2 bytes*	*n* bytes	4 bytes

Table 5.3: The frame control field

	Protocol Version	Type	Subtype	To DS	From DS	More Fragments	Retry	Power Management	More Data	Protected Frame	Order
Bit:	0–1	2–3	4–7	8	9	10	11	12	13	14	15

Table 5.4: The sequence control field

	Fragment Number	Sequence Number
Bit:	0–3	4–15

Table 5.5: The WMM quality-of-service control field

	TID/Access Class	End of Service Period	Acknowledgement Policy	Reserved	TXOP Limit or other state
Bit:	0–3	4	5–6	7	8–15

Table 5.6: The type and subtype values*

Type	Subtype	Meaning
00 – Management	0000 (0)	Association Request
00 – Management	0001 (1)	Association Response
00 – Management	0010 (2)	Reassociation Request
00 – Management	0011 (3)	Reassociation Response
00 – Management	0100 (4)	Probe Request
00 – Management	0101 (5)	Probe Response
00 – Management	1000 (8)	Beacon
00 – Management	1001 (9)	*ATIM*
00 – Management	1010 (10)	Disassociation
00 – Management	1011 (11)	Authentication
00 – Management	1100 (12)	Deauthentication
00 – Management	1101 (13)	Action
01 – Control	1000 (8)	Block Acknowledgement Request
01 – Control	1001 (9)	Block Acknowledgement
01 – Control	1010 (10)	Power Save Poll (PS-Poll)
01 – Control	1011 (11)	Request to Send (RTS)
01 – Control	1100 (12)	Clear to Send (CTS)
01 – Control	1101 (13)	Acknowledgement
01 – Control	1110 (14)	*CF-End*
01 – Control	1111 (15)	*CF-End + CF-Ack*
10 – Data	0000 (0)	Data
10 – Data	0001 (1)	*Data + CF-Ack*
10 – Data	0010 (2)	*Data + CF-Poll*
10 – Data	0011 (3)	*Data + CF-Ack + CF-Poll*
10 – Data	0100 (4)	Null (no data)
10 – Data	0101 (5)	*CF-Ack (no data)*
10 – Data	0110 (6)	*CF-Poll (no data)*
10 – Data	0111 (7)	*CF-Ack + CF-Poll (no data)*
10 – Data	1000 (8)	QoS Data
10 – Data	1001 (9)	*QoS Data + CF-Ack*
10 – Data	1010 (10)	*QoS Data + CF-Poll*
10 – Data	1011 (11)	*QoS Data + CF-Ack + CF-Poll*
10 – Data	1100 (12)	*QoS Null (no data)*
10 – Data	1110 (14)	*QoS CF-Poll (no data)*
10 – Data	1111 (15)	*QoS CF-Ack + CF-Poll (no data)*
all others		*Reserved*

* Italicized types are rarely seen.

Table 5.7: The subnetwork access protocol (SNAP) format for 802.11

AA	AA	03	00:00:00	EtherType	Payload
1 byte	1 byte	1 byte	3 bytes	2 bytes	*n* bytes

Frames may be sent from the client to the access point, or from the access point to the client. To help sort out the direction, the frame control field includes the *To DS* and *From DS* flags. To DS is sent on data frames from a client to the access point, and the From DS flag is sent on data frames from the access point to a client. Management frames do not have either of those bits set, although they too can be to or from the client. Given the direction of the frame, the address orders map to the actual devices in a certain way. Table 5.8 shows that mapping.

Table 5.8 Address fields for each frame type

Frame	To DS	From DS	Address 1	Address 2	Address 3
Data/QoS Data	1	0	Access Point (BSSID = RA)	Client (SA = TA)	Destination (DA)
Data/QoS Data	0	1	Client (DA = RA)	Access Point (BSSID = TA)	Source (SA)
Management	0	0	AP/Client (DA = RA)	Client/AP (SA = TA)	Access Point (BSSID)
RTS	0	0	Receiver (RA)	Transmitter (TA)	—
CTS	0	0	Address of transmitter of RTS (RA)	—	—
Ack	0	0	Address of transmitter of original frame (RA)	—	—
Block Ack	0	0	Receiver (RA)	Transmitter (TA)	—
Block Ack Request	0	0	Receiver (RA)	Transmitter (TA)	—
PS Poll	0	0	Access Point (BSSID)	Client (TA)	—

Again, notice how the first address is always that of the radio the transmitter is sending the frame to. For frames where either the access point or client can send the frame, the terms *transmitter address* (TA) and *receiver address* (RA) are used to identify the actual wireless devices sending and receiving. However, the terms *source address* (SA) and *destination address* (DA) are also used, and those terms may apply to wired addresses as well as wireless. Table 5.8 uses the term "client" or "access point" when it is always known who holds the address.

5.2.3.2 The Shared Medium

The 802.11 protocol is built around the concepts of Carrier Sense Multiple Access with Collision Avoidance (CSMA/CA). Basically, unlike with properly working switched Ethernet, wireless transmissions might not always get through to the receiver. In fact, they often do not. That can be because of interference on the channel, or because the devices are out of range for the speed they are transmitting at, or because of *collisions*. A collision is when two devices are transmitting at the same time, and the intended receiver hears both transmissions, cannot tell the difference between one transmission and another, and ends up receiving garbage. Unlike with wired Ethernet, in which collisions were all but eliminated with the introduction of the switch, and were detectable even when not eliminated, collisions in Wi-Fi cannot be detected while the device is transmitting. This means the following two things.

Almost all data and management frame transmissions require the receiver to send a frame back to the sender immediately in response, called an *acknowledgment*.

Transmitters must receive before they transmit, making sure that they do not transmit if another device is already on the air (*carrier sense*).

These two rules do not eliminate collisions. In fact, the particular implementation of 802.11's collision rules can lead to more collisions than wired Ethernet saw with hubs. However, they do allow for retransmissions. When a frame is sent that requires an acknowledgment but one does not come back, the sender may retransmit the same frame a short time later.

To keep track of retransmissions, and to avoid a receiver getting duplicates, the sender populates the sequence control field with a sequence number. This number, which may be unique for each sender-receiver pair (or sender-receiver-TID for QoS frames), remains the same for each retransmission of a frame, but increases across frames. The receiver can use the sequence number to filter out duplicates. The number is only 12 bits, taking on values between 0 and 4095, and thus wraps when it hits the end, with rules that prevent wrapping from confusing the receiver in almost every case. Additionally, the frame control field has a retry bit, which is set on most retransmissions. This is used to prevent duplicates from being sent to the device. Why can a receiver get duplicates when the sender stops retransmitting when it receives an acknowledgment? Because, in wireless, it is never clear whether the receiver lost the original frame, or whether it got the original frame but the sender lost the acknowledgment coming back to it. In the latter case, the receiver will see the same frame multiple times.

In 802.11, every unicast data and management frame requires an acknowledgment, except in rare cases involving special quality-of-service settings in the QoS control field. Additionally, there are a few control frames that require immediate responses. One such set of frames is used in the *request to send/clear to send* (RTS/CTS) protocol. RTS and CTS frames are sometimes used to ensure that the receiver is not busy or out of range without sending the entire data or management frame first. RTS and CTS frames are, by nature, short (though not as short as RTS/CTS transmissions are in other, non-802.11 technologies). RTS and CTS frames are sent back to back. The sender of the data or management frame will first send an RTS. If the receiver gets the RTS, it will immediately respond with a CTS. The sender will hold off after the RTS until it gets a CTS. If it does, it will respond immediately with the data or management frame, after which the receiver will respond with the acknowledgement. The order of transmissions is as follows:

```
Sender:    RTS      Data/Management

Receiver:     CTS                Ack
                   Time  →
```

In Section 5.4.3, I will cover the specifics of the complexities of how devices can know when they are safe to transmit. However, as a preview, I will cover the concept of *virtual carrier sense*, distributed as a *network allocation vector* (NAV). Not only do devices know that the air is busy because they are observing the air for transmissions before they themselves transmit—the carrier sense principle mentioned previously—but they also rely on virtual carrier sense. Virtual carrier sense allows one device to prevent another device from transmitting even if it cannot hear that the network is busy itself. There are a number of reasons why a device may want to do this, and a number of consequences that are specific to voice mobility in this, but the concept is a part of 802.11. In nearly every frame there is a *duration* field. This duration field carries the number of microseconds after the end of the frame that any device that hears the frame, *except* for the intended recipient of the frame, must remain quiet. This is an interesting concept, and is something uniquely wireless. Remember that any device in range, on the same channel, and with a compatible Wi-Fi radio can hear another device transmit, regardless of whether the transmission is destined for the overhearing device. For example, if a client in your network is speaking to your access point, devices from neighboring access points, as well as other clients for the given access point, may be able to hear the transmission. The duration field in that transmission is directed to them. That means that if device A sends to device B, devices A and B ignore the duration field, but all other devices (C, D, and so on) that can hear it are required to respect the duration field and remain quiet after the transmission has ceased, for as long as the duration field says.

Carrier sense and collisions are leading causes of overhead, wasted airtime, and poor voice quality on Wi-Fi, but nevertheless are a necessary part of the protocol. Strategies that Wi-Fi equipment can use to reduce or eliminate the problem in practice will be discussed in the remainder of this chapter and the next.

5.2.3.3 Connections and How Data Flows

Clients must connect to access points before they can exchange data. This negotiated connection establishes state on both the client and access point, and opens up the back-end network for the client to use; thus, it takes the place of the cable itself in Ethernet.

Wi-Fi, unlike cellular technologies, leaves the client in charge of ensuring that a connection is always established, and choosing which access point to connect to as the device roams around. As the moving party, the client is responsible for searching out the available SSIDs, matching them up with BSSIDs (access point MAC addresses), and then establishing the connection.

This first phase, in which the client gets the list of available networks, is called *scanning*. Clients hop from channel to channel, looking for beacons that advertise the networks they are interested in. On some channels, the clients are allowed to skip the time necessary to wait for a beacon—beacons usually come in 100 millisecond intervals, a lifetime for wireless devices—and can send a frame called a *probe request* instead. A probe request carries the capabilities of the client, and asks the access points that hear it whether they provide a given SSID. There are two particulars to look for here. Probe requests are usually broadcast, meaning that the BSSID and destination address (both have to be identical for probe requests) are the broadcast Ethernet address FF:FF:FF:FF:FF:FF. This means that any access point that hears it should respond. Furthermore, the SSID named in the probe request can be empty. If one is given, only access points that serve the SSID should respond. If one is not given, however, every access point should respond, and it should name the SSID that it provides. (Note that it is possible for Probe Requests to be sent to a specific access point, by giving its BSSID as the destination address. In those cases, an acknowledgement is required, and no other access point should respond. This is done somewhat rarely, and is mostly to check to see whether a previously known access point is still available.) For probe requests as a broadcast, no immediate acknowledgement control frame is required; rather, a *probe response* is expected. Probe responses carry almost identical information to beacons, but are dedicated to the client asking for it with a probe request.

While scanning, the client establishes a *scanning table,* mapping BSSIDs to SSID, channel, capabilities, available capacity, and similar things. Once the client is finished scanning, if it does not recognize any of the SSIDs obtained, it may present the user with the list of SSIDs and ask the user to choose an SSID. Otherwise, it filters the list by the SSID it is searching for, and then chooses what it deems to be the best access point to provide service for the SSID. How scanning works is rather complicated, and not any of the standards specify how a client chooses. Clients may choose an access point based on the strength of the signal it receives, or remembered information about the access point, or the access point's capabilities, or through complicated and obscure formulas known perhaps only to the developer of the scanning software. (We will visit the scanning process in detail in Section 6.2.2). However, once a client has made up its mind, so to speak, it proceeds to the next step.

The first frame that the client sends to an access point that lets the access point know that it has been chosen is an *Authentication* frame. The name is misleading. In the early days of 802.11, the only security available was what was known as *Wired Equivalent Privacy* (WEP). This regrettable protocol used RC4—a stream cipher—in improper ways to encrypt traffic. As an option, the same protocol could be used in the Authentication frames to authorize devices. However, the protocol was shown to be vulnerable to a specific and somewhat easy set of attacks, and has been officially deprecated by IEEE. Therefore, although it exists, and will be mentioned in the security section later, please do not use it. Given all that, the Authentication frame, which cannot be skipped, serves only to establish the first phase of the connection. The client sends an Authentication frame to the access point, and the access point will respond with a similar Authentication frame to the client, thus starting off the connection. Interestingly, however, the Authentication frame is making a comeback. The 802.11r protocol (Section 6.2.5) uses the Authentication frame to carry modern AES-based key negotiation for faster handoffs. This is not exactly authentication, but the use of the frame for more than just a pointless handshake is good to see.

The Authentication frame exchanges set the base. After those two frames, the client will send an Association Request. The Association Request carries the capabilities of the client, as well as potential requests for network resources. (Although the Probe Request also carried capabilities, the access point is not required to remember them from Probe Requests.) The access point receives the Association Request and decides whether it should accept the client. Access points are allowed to reject the client's association for any reason, but usually do so because they are full or going out of service. In either case, the access point will respond with an Association Response, which carries a status code mentioning whether the request was successful or was denied. If the status code is the one for success, the client knows it is connected. If it is one of the failure codes, the client can try another access point. Of course, if the client cannot succeed with any other access point, it may come back and try again. Unfortunately, the meaning of the failure codes is not exact, and clients may or may not use them for any particular meaning beyond that of simple failure.

If the association was a success, the Association Response from the access point will also include what is known as an *association ID* (AID). This is a number from 1 to 2007 that the access point uses for the power saving notification protocol, and happens to be unique among all clients associated to that particular access point at a given time. (Access points cannot generally have anywhere near 2007 clients associated at a given time with reasonable quality. The number they can have associated is often one to two orders of magnitude less.)

Once the association is established, the client and access point can exchange data frames. However, the story is not complete. If the access point is not using security, then the connection is established and the client and access point can exchange data just as if they were connected by a wire. However, if the access point is requiring security, then a new process will need to be started. This will be described in Section 5.6.

When the data connection is successfully established, a few things happen. The access point notifies the wired network that the client is available. Furthermore, the access point (or infrastructure behind it) establishes the client as an entry in the layer-2 forwarding table for the network. This means that frames directed to it from the wired network will properly end up being repeated by the access point for the client over the air.

Broadcasts to the wireless network go out as data frames with the same multicast or broadcast Ethernet destination address as if they were wired. Because they are multicast, no client is allowed to send an acknowledgement. They are never retransmitted, and thus clients may miss them. Broadcasts are inherently less reliable on Wi-Fi than unicast transmissions.

Upstream transmissions, however, are unique. All data transmissions from a client look like unicast transmissions to the access point. That means that, whether the destination (DA) be multicast or unicast, the receiver address (RA = BSSID) is the access point, and so the transmission is treated by 802.11 as unicast, with retries.

Client-to-client communication, then, has an interesting problem. All transmissions from a client must go to the access point. Clients are not allowed to speak to each other directly. Thus, the access point must always repeat the information from one client to another. The disadvantage of this is that it wastes airtime. However, the advantages are manifold: the access point is the arbiter of power save state, allowing it to buffer frames for sleeping clients, and it usually can reach both clients even when they are at opposite sides of the cell, where each client might not be able to reach the other. The standard designates the access point as a repeater, then, to allow this to work well. More importantly, one client has no idea where the other client is. It may be on this access point, but it may instead be on another access point, where it looks like a wired Ethernet device as far as 802.11 is concerned. To allow for mobility, the standard requires the client to only track and speak to one device: the access point. But that necessitates this hairpinning. There are provisions in the standard and activities in IEEE to attempt to remove hairpinning, but by and large these efforts are not aimed at enterprise uses, and do not go very well with mobility.

5.2.4 Infrastructure Architectures

Because an access point cannot cover an entire building, Wi-Fi mobility networks require the concept of an infrastructure. The wireless infrastructure takes into account the entire set of access points, and whatever management tools are used.

Because large campuses can have thousands of access points, there is a natural tendency for these access points to be deployed at the same time, and often to be from one vendor. The different styles of architectures reflect the different vendors' approaches to creating a network that provides the necessary functions. The ways in which wireless architectures are described often varies from source to source and from time to time. Here, we will lay out the common terms and explain how they relate to the different architectures.

There are two different ways to look at the different architectures. One way is to focus on how the wireline architecture of the wireless network is structured. It may seem odd to look at how wireless architectures work over the wire, and it is, but, for historical reasons, this is still a common framework. The wireline categorizations look as follows:

- Standalone

- Controller-based

- Controllerless

- Directly connected

The second way is to focus on the over-the-air behavior of the network. This breaks down as follows:

- Static Microcell

- Dynamic or Adaptive Microcell

- Layered

- Virtualized

We'll explore all categories. Keep in mind that some wireline categories currently only exist with certain over-the-air categories, and that will be pointed out when appropriate.

Table 5.9 lists many of the vendors, both major and minor, and what architectures are possible, as of the time of writing, with each.

Table 5.9: Selection of vendors and architectures

Vendor	Wireline Architecture	Over-the-Air Architecture
Aerohive	Controllerless	Dynamic Microcell
Aruba Networks	Controller	Dynamic Microcell
Belden/Trapeze	Controller	Dynamic Microcell
Cisco Systems	Standalone (IOS), Controller (LWAPP/CAPWAP)	Static Microcell, Dynamic Microcell
Extricom	Directly Connected	Layered
Hewlett-Packard/Colubris	Controller	Dynamic Microcell
Meru Networks	Controller	Layered, Virtualized
Motorola/Symbol	Controller	Dynamic Microcell
Xirrus	Standalone	Static Microcell, Dynamic Microcell

Before we get too far into this discussion, I should take a moment to reveal my own biases. Having been one of the inventors of two of the architectures embraced by the industry (layered and virtualized) and being heavily involved in the various standards bodies for the industry and having authored or contributed to a number of the techniques necessary for voice over Wi-Fi, I have a strong interest in certain problems in wireless that may have a solution in the techniques I helped create. But I mention those problems and solutions in sincere belief that the problems of voice mobility that you may experience, and of wireless networking in general, can be solved. In any event, I will not pull any punches, and will address strategies for every architecture you may happen to run across in modern networks.

5.2.4.1 Wireline: Standalone or "Fat"

Historically, this was the first wireline architecture for wireless. In a standalone AP network, each access point is entirely independent of the others.

In the consumer space, only standalone access points are sold today. However, the first enterprise-grade access points also fell into this style. Each access point has its own management system—whether simple or complex, web-based or command line interface (CLI). The access points each maintain their own configurations, connect to outside services (especially Remote Authentication Dial In User Service, or RADIUS) on their own, and generally have no cooperation with any neighboring access point, even from the same vendor.

Most important for mobility, each access point is its own bridge, connecting to the wired network immediately at its Ethernet port, without any tunneling. This means that the access point offers very few or no mobility services. If two access points are connected in different subnets, then the client is required to get a new IP address after a handoff, usually resulting in a dropped call. To avoid this effect, administrators are forced to distribute the subnet to every access point—for multiple-Virtual LAN (VLAN) networks, this means that each access point must be trunked back, across the access and distribution layers of the wired network.

These access points can be managed using centralized network management tools, and vendors that offer them incorporate Wi-Fi-specific functionality to attempt to mitigate the complexity of managing thousands of individual access points. The management tool may be software, installed on a server, or it may come as an appliance.

Most standalone access points are limited to the typical one or two radios. However, one manufacturer makes multiple-radio standalone access points, which they call *arrays*. This technology uses sectorization to reduce the coverage pattern for each radio, allowing over a dozen radios to be packed into the larger version. The goal here is for density: if one radio can support a certain number of clients, then 12 radios should support 12 times that amount, all with one cable pull. Understandably, wireless arrays are significantly larger than other access point types.

5.2.4.2 Wireline: Controller-Based or "Thin"

To overcome the triple concern of lack of mobility services, per-access-point RADIUS connections, and individual management with standalone access points, the wireless *controller* was introduced around 2002. The management, security, and wireline-bridging functions of 802.11 are removed from the access point and relocated to this separate appliance, called the controller. This controller looks and functions, to some extent, like a router, collecting traffic destined to or coming from the wireless network and exchanging it across one or two wired ports. Controller-based access points are left with only two nonvolatile configuration pieces: the IP address or name of the controller, and the IP address or DHCP settings it is to use when it boots up. Controller-based access points cannot operate on their own. When they power up, they seek out the controller and establish a connection, where they download their configuration into memory. In order to manage or monitor the access point, the administrator must go to the web interface or CLI of the controller.

Controllers are usually high-end data processing platforms, although every vendor offers low-end models for small office deployments. These devices have more in common with routers than with computer appliances, as they are built to tunnel data quickly.

The management advantage of controller-based architectures is that the statistics and properties of the network can be seen and altered in aggregate. Furthermore, software versioning is taken care of automatically, as the controller upgrades access points to the same version that it is running. The appearance to the administrator is that the access point is somehow "thin," or lightweight. The reality, of course, is that the controller-based access point is built of the same hardware as a standalone access point, which explains why some vendors offer the option to run either standalone or controller-based software on a given model.

Security is also performed centrally. RADIUS transactions are required by the wireless security protocols, and RADIUS needs to have the IP address and password of each device that is allowed to use it for authentication services. With controller-based architectures, there is only one IP to know—that of the controller. Also, because the controller performs the RADIUS authentications, it can cache them as needed, aiding in handoffs. There is variation within the architecture for where encryption is performed. One vendor performs the 802.11 encryption operations on the controller itself; others retain that functionality in the access points.

But most notably for voice mobility, the controller-based architectures implement a kind of transparent "Mobile IP." Data is tunneled from the access point to the controller and vice versa. This allows access points to provide services for networks that they themselves are not placed in. The advantages are readily apparent. A campus with dozens of buildings, each building with its own subnet, can install controller-based access points and yet provide a completely different set of subnets to the wireless devices. A campus-wide, flat voice subnet can be established and dedicated to voice mobility devices, without having to push the subnet throughout the campus, eliminating the need for concern about VLANs, inter-subnet handoffs, and call drops. Moreover, the tunneling used does not involve Mobile IP itself, but rather is an integrated part of the system. There are no additional steps that administrators must take to take advantage of the overlay network that the tunneling provides.

Controller-based architectures can still allow for some traffic to be bridged locally, rather than tunneled. However, this is not recommended for campus deployments—especially not for voice—because it brings up the same mobility concerns as with standalone access point deployments.

The controller-based wireline architecture currently has the most diversity with the over-the-air architectures.

5.2.4.3 Wireline: Controllerless

Controllerless access points are not standalone access points. Although this architecture does not use a controller, the access points are aware of each other and communicate, including setting up tunnels for mobility. Some of the controller functionality remains in a dedicated management appliance, but the data path function of the controller is distributed out in the access points.

This is a relatively new architecture, and not widely adopted by the vendors. The advantage claimed by this architecture is the savings of cost of the controller. In order for mobility to work, this means that access points have to take over the role of tunnel endpoints. For networks where the voice mobility subnet is never pushed out to the access layer, the controllerless access point model introduces added complexity, to ensure that enough access points are present in the voice mobility subnet to act as home agents for the voice network, and thus many access points may be required to take the place of one controller. Therefore, controllerless access point architectures lend themselves best to networks that are inherently flat or well distributed already, and where traffic patterns do not concentrate.

5.2.4.4 Wireline: Directly Connected

Directly connected architectures take the concept of centralizing to its logical limit. Instead of a controller that has a limited number of ports, this architecture offers a device that has one physical Ethernet port per access point and looks like a switch. Each access point is connected directly, using one Ethernet cable or with two cables tied together with a special booster.

Direct connection allows even more of the 802.11 functions to be centralized, which vendors may use to provide differing services. On the flip side, requiring a direct, layer-1 connection to the access point inherently limits the size of the network controlled by the appliance, and forces the appliance to be placed at the physical edge of the network.

Currently, the one vendor who offers a directly connected wireline architecture uses it to provide a layered over-the-air architecture.

5.2.4.5 Over-the-Air: Static Microcell

Static microcell over-the-air architectures usually require the administrator or a planning tool to generate the radio frequency (RF) parameters—channel selection and transmit power, in this case—for the access points. The most basic implementations just require the user to select a channel and power level. Of course, the system may have some defaults, and may even attempt to make some initial scanning to chose "better" channels. Nevertheless, once a choice is made, the choice does not change unless the administrator selects a new value or uploads a new RF plan.

This does introduce the concept of *RF planning*, which will be addressed in the section on RF (Section 5.3). The key to the static (and the subsequent dynamic) microcell architectures is the dedication of the available Wi-Fi channels to avoiding neighboring access point interference, thus resulting in an alternating pattern of channel assignments, where the closest neighbors always have different channels. For static systems, the installer is required to know how to do this by sight, or by using the RF planning tools. Furthermore, because these architectures also require reducing power levels significantly to avoid interference from second-order (further away) neighbors, and lower power levels translates into less range and smaller cell sizes, these architectures are also known as *microcell*.

Standalone access points are the most obvious candidates for static over-the-air architectures, because there is no system changing channels or power levels on the network. However, all of the wireline architectures can be made to behave statically, though how to do so may not be obvious and setting the network in that mode may not be recommended.

The advantage of the static architecture is that the RF plan is consistent, thus allowing for a more predictable coverage. The disadvantage is that the network does not react to changes in its environment, such as persistent noise or neighboring network interference.

5.2.4.6 Over-the-Air: Dynamic or Adaptive Microcell

Dynamic microcell over-the-air architectures take a different approach than static architectures. The goal of dynamic architectures is to use what is known as *radio resource management* (RRM; some vendors use similar terms) to adaptively configure the channels, power levels, and other settings of the access points.

The reason for transitioning from a stable network to one that is constantly in flux is to attempt to avoid some of the problems inherent in larger 802.11 networks, mentioned in the following sections. The key observation is that *radio resources* exist and need to be monitored somehow. Broadly, radio resources can be thought of as wireless network capacity, and they are reduced by interference, density, and mobility of wireless clients. The following sections, especially "RF Primer" and "Radio Basics," will shed light on the specifics of what impacts these radio resources.

Dynamic architectures attempt to handle the problem by constantly measuring the various fluctuations in load, density, and neighboring traffic, and then making minute-by-minute adjustments in response. The main tools in the dynamic architecture's arsenal are, as before, choosing channel settings and transmit power levels.

Dynamic architectures end up creating an alternating assignment of channels, in which every access point attempts to chose a different channel from its neighbors and a power level low enough to avoid providing too much duplicated coverage.

The advantages of dynamic radio resource management is that the network is able to avoid situations where static networks completely fail—for example, dynamic networks can continue to operate (albeit with reduced capacity) when a microwave oven is turned on, whereas static networks may succumb completely in the area around the interference. The main disadvantage, however, is that the network and its associated coverage patterns are unpredictably changing, often by the minute. This leads to a necessary tradeoff between the disease and the cure. Thus, dynamic systems provide the expert administrator with the ability to go in and turn down the aggressiveness of the adaptation, providing a choice between a more static network or more dynamic network, allowing the administrator to choose which benefits and downsides are best suited for the given deployment. You will find that many voice mobility networks have disabled many of the adaptive features of their networks to ensure a more consistent coverage.

Additionally, the smaller and changing cell sizes, along with the wide array of channels that end up being used, leads to issues with handoff that directly affect voice mobility. To help mitigate these problems, network assistance protocols can be used to increase the amount of information that clients, who decide when to hand off and where to hand off to, have at their disposal. Section 6.2.6 explores the network assistance aspects of the microcell architectures in more detail.

5.2.4.7 Over-the-Air: Layered

Layered architectures take a different approach than microcell architectures, static or dynamic. Recognizing the problems of radio resource limitations fundamental to Wi-Fi, as well as the added problem of instability produced by the dynamic architecture, the layered architecture changes the purpose of using multiple channels. Whereas dynamic architectures end up alternating channels between access points to address the problem of neighbors, layered architectures are able to solve the problem through coordination between the access points. Thus, they are able to reuse the same channel between neighboring access points. These architectures start by creating one *channel layer*, completely covering the network with just one channel. This is the most basic coverage configuration. To grow the network, the freed up channels can be used to create additional channel layers. Figure 5.4 shows the difference in channel usage between microcell architectures and layered architectures.

For channel layering to make sense, the architecture needed to resolve neighborhood problem head on. To do so, the wireline architecture needs to involve a tighter RF coordination between the access points. Currently, the two methods to achieve this are a coordinated extension to the controller wireline architecture, or to use a direct connection of access points to the appliance.

The advantage of layering is that it provides the stability to the network that was lost in the dynamic architecture, while avoiding the problems of noise that plague static architectures. An added advantage of layering is that any individual channel layer can act as one campus-wide cell, or BSSID, as far as the mobile device is concerned, without loss of the capacities of the individual access points. Thus, handoffs between access points are eliminated, providing a direct benefit for voice mobility.

5.2.4.8 Over-the-Air: Virtualized

The virtualized architecture builds upon the layering architecture, but introduces the notion of complete wireless network virtualization. Wireless LAN (WLAN) virtualization involves creating a unique virtual wireless network (a BSSID) for every mobile device. This allows the network to be partitioned for each client, providing each client with its own set of 802.11 autonegotiated features and parameters.

It's important to note that the per-device containment provided by virtualization differs from the per-device rules and access control enforcement provided by the other architectures. Containment addresses the over-the-air behavior of the client directly, using the standard to enforce the segmentation and the tight resource bounds. The client's cooperation is not needed or expected. Access control, on the other hand, is fundamentally a cooperative scheme, and clients can choose not to participate in the optional protocols required to make bidirectional access control work. Even downstream policy enforcement cannot stop a client from transmitting what it wants to upstream.

However, virtualized Wi-Fi partitions are able to maintain the per-device containment, by transferring control of the network resources from the client to the network, and then using Wi-Fi mechanisms from the network side to ensure that client behavior is limited to the resources that the client is allocated.

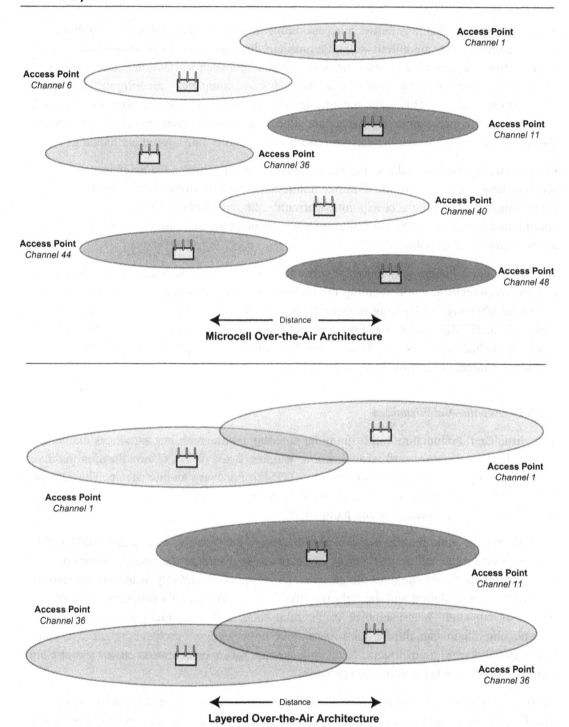

Figure 5.4: Comparison between Microcell and Channel Layering for the Same Area of Coverage

---

Architectures and 802.11 Functions

In 802.11, the concept of an "access point" is defined to carry one BSSID and one SSID over the air to a set of clients. The access point definition includes every function necessary to make the access point a bridge to wireline Ethernet, including encryption, decryption, connection management, medium access control, and timing functions.

However, this concept is only a concept, and the architectures in the market today differ by how they divide the functions of the 802.11 access point across the actual equipment deployed in the network. In general, every architecture ensures that multiple 802.11 access point concepts can be created and operated in each physical access point, thus allowing for multiple BSSIDs—and more importantly, multiple SSIDs—per access point. This starts by having multiple radios within an access point, but is most useful by allowing multiple SSIDs per radio.

This is the point of departure for the architectures. Controller-based architectures move parts of the 802.11 access point out of each physical access point and operates them, instead, in the controller, thus sharing those parts across all of the access points. This does not violate the standard, however, because the standard was designed to allow for all kinds of mappings of logical 802.11 entities to physical devices.

Ultimately, the best way to choose which 802.11 functions should be centralized—and thus, which type of architecture to invest in when creating a voice mobility network—is to choose based on how well the features meet your needs, and not on architectural principles alone.

Section 6.2.7 explores the network control aspects of the layered and virtualized architectures in more detail.

5.3 RF Primer

Understanding how Wi-Fi fits into voice mobility requires knowing how the radios work. It is tempting to want to regard Wi-Fi, because of its convenience, in the same way as wired: connect, and it just *works*, barring some rare cabling problem. However, Wi-Fi has a large number of different elements that come together to allow the wireless to work and provide high throughput, and the consequences from how some of those elements work need to be understood In this way, one of the major distinctions between voice mobility and simple data networking is that those concerned with voice mobility must become familiar with the finer details.

5.3.1 Channels

One Wi-Fi radio does not occupy the entire unlicensed spectrum, unlike frequency-hopping technologies such as Bluetooth. 802.11 divides up the spectrum into a number of different *channels*. Channels are named with whole numbers, assigned by a formula to specific center frequencies for the channels. The idea behind small number of discreet channels is to carve up the spectrum, helping pack in as many devices as possible and avoiding requiring clients to have to tune in across a wide range of frequencies, the way that analog car radios must.

The channel numbers are somewhat arbitrary, and are arranged to let you know what band they occupy. Different 802.11 radio types allow for different channel selections.

The two key properties that define how the 802.11 radio uses the spectrum are its center frequency and bandwidth. The center frequency is the one the radio uses to determine where to look for the transmissions. This concept is similar to car radios: FM channel 97.3 means that the radio tunes its center frequency to 97.3MHz. Unfortunately, Wi-Fi channels do not convert as neatly to their center frequencies. Because of this, many people and tools will either interchangeably use the center frequency or the channel number to describe the channel. Wi-Fi uses center frequencies that are always in the gigahertz range. The bandwidth tells which other frequencies are occupied by a transmission. 802.11 radios used for mobility primarily have 20MHz bandwidth, except for 802.11n radios, which can also use 40MHz bandwidths. The channel and bandwidth together show which part of the spectrum the radio occupies. Although the different 802.11 radio types may fill the carved-out part of the spectrum differently, the amount that is carved out is roughly the same for the same bandwidth. Figure 5.5 sketches the general concept.

Table 5.10 lists the channels and what radio types can use them.

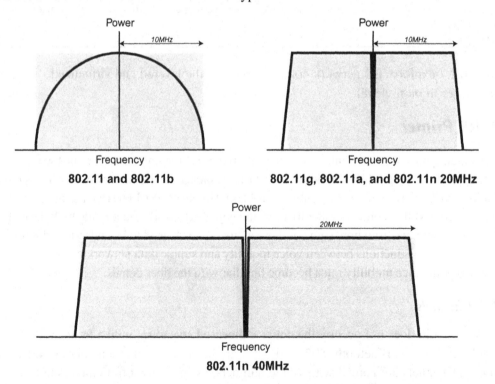

Figure 5.5: Shape of 802.11 Frequency Occupation

Table 5.10: 802.11 Channels

Channel	Frequency	US Band	11b, 11g	11a	11n	Notes	
1	2.412GHz	ISM 2.4	✓		✓	Nonoverlapping	High power: 1 W maximum.
2	2.417GHz		✓		✓		
3	2.422GHz		✓		✓		
4	2.427GHz		✓		✓		
5	2.432GHz		✓		✓		
6	2.437GHz		✓		✓	Nonoverlapping	
7	2.442GHz		✓		✓		
8	2.447GHz		✓		✓		
9	2.452GHz		✓		✓		
10	2.457GHz		✓		✓		
11	2.462GHz		✓		✓	Nonoverlapping	
12	2.467GHz	—	✓		✓	Europe, Japan, Australia. No U.S. or Canada	
13	2.472GHz		✓		✓		
14	2.484GHz		11b only			Japan only. Channel 14 does not follow the channel to frequency formula.	
36	5.18GHz	U-NII 2 Lower		✓	✓	Indoor use only. Low power: 40 mW maximum	
40	5.20GHz			✓	✓		
44	5.22GHz			✓	✓		
48	5.24GHz			✓	✓		
52	5.26GHz	U-NII 2 Upper		✓	✓	Non-DFS for equipment before July 2007	Radar detection and dynamic frequency selection (DFS) required
56	5.28GHz			✓	✓		
60	5.30GHz			✓	✓		
64	5.32GHz			✓	✓		
100	5.50GHz	U-NII 2 Extended		✓	✓		
104	5.52GHz			✓	✓		
108	5.54GHz			✓	✓		
112	5.56GHz			✓	✓		
116	5.58GHz			✓	✓		
120	5.60GHz			✓	✓	U.S., Europe, and Japan. No Canada, because of weather radar.	
124	5.62GHz			✓	✓		
128	5.64GHz			✓	✓		
132	5.66GHz			✓	✓		
136	5.68GHz			✓	✓		
140	5.70GHz			✓	✓		
149	5.745GHz	U-NII 3		✓	✓	U.S, Canada and Europe. No Japan	High power
153	5.765GHz			✓	✓		
157	5.785GHz			✓	✓		
161	5.805GHz			✓	✓		
165	5.825GHz	ISM 5.8		✓	✓	U.S., Canada and Europe. No Japan.	High power

The formula for the channels to frequencies is $2.407GHz + 0.5GHz * channel$ for the 2.4GHz band, and the simpler to remember $5GHz + 0.5GHz * channel$ for the 5GHz band. The only channels that are in the 2.4GHz band are channels 1–14. Everything else is in the 5GHz band. Therefore, channel 36 is 5.18GHz, and channel 100 is 5.50GHz.

The total number of channels is large, but many factors reduce the number that can be practically used. First to note is that the 2.4GHz band, where 802.11b and 802.11g run, only has three nonoverlapping channels (four in Japan) to choose from. Unfortunately, the eleven channel numbers available in the United States gives the false impression of 11 independent channels, and to this day there exist some Wi-Fi deployments that mistakenly use all 11 channels, causing an RF nightmare. To avoid overlapping channels, adjacent channel selections need to be four channel numbers apart. Therefore, channels 1 and 5 do not overlap. In the 2.4GHz band, custom usually spreads the channels out even a bit further, and using only channels 1, 6, and 11 is recommended. The authors of the standard recognized the problem the overlap causes, and, for the 5GHz band, disallowed overlap by preventing devices from using the intermediate channels. Therefore, no channels in the 5GHz band overlap, when it comes to 20MHz channels. (40MHz channels are an exception, and will be discussed in the section on 802.11n, Section 5.5.3.)

The unlicensed spectrum was originally designed for, and still is allocated to, other uses besides Wi-Fi. The 2.4GHz band was created to allow, in part, for microwave ovens to emit radio noise as they operate, as it is impractical to completely block their radio emissions. Because that noise prevented being able to provide the protections from interference that licensed bands have, the regulatory agencies allowed inventors to experiment providing other services in this band. And so began 802.11. The 5GHz band is, in theory, more set aside from radiation. Except for the top 5.8GHz ISM band, the 5GHz range was designed for communications devices. However, interference still exists. One primary source, and the one important from a regulatory point of view, is radar. Radars operate in the same 5GHz band. Because the radars are given priority, Wi-Fi devices in much of the 5GHz band are required to either be used indoors only, or to detect when a radar is present and shut down or change channels. This last ability is known as *dynamic frequency selection* (DFS). This is not a feature or benefit, per se, but a requirement from the various governments. DFS complicates the handoff process significantly, and will be further explored in Section 6.2.2.

5.3.2 Radio Basics

Radio technology can seem rather complicated. Wi-Fi transmitters emit power at a given strength, but receivers have to be able to correctly decode those signals at power levels up to a *billion* times weaker, just for normal wireless networks to work. This incredible swing in power levels requires quite a bit of smarts in both the definitions of the signals and in the designs of the radios themselves. It is impossible to describe every important concept in the world of radios, but we will cover those concepts that have the most impact for voice mobility.

5.3.2.1 Power and Multipath

Power, or *signal strength*, is the amount of energy the radio pushes into the air over time. Power is measured in watts. However, for wireless, a watt is a large value, and the receiver usually gets signals many millions of times weaker, once the radio waves reach it. Therefore, Wi-Fi devices tend to refer to milliwatts, or *mW*. To cover the wide range, the logarithmic *decibel* is also used. Therefore, power can also be measured in *dBm*, or decibels of milliwatts. (More formally, it can be called *dBmW*; however, the W for watt is usually left off.) *0 dBm* equals *1 mW*. The conversion formula is $P_{dBm} = 10 \times \log_{10} P_{mW}$, where P_{dBm} is the power in dBm and P_{mW} is the power in milliwatts. It is handy to remember that a 3dB change in power is roughly equal to a two-fold change in power. Note that differences between two dBm values are cited in dB, not in dBm. Therefore, the difference between an 18dBm and a 21dBm transmission power is referred to as 3dB. The reason is because the difference of any logarithmic value is really a ratio of two nonlogarithmic values, and thus 21dBm −18dBm is really stating that the ratio of the two power levels is roughly 2, or that 21dBm is 2 times stronger than 18dBm. (The units of milliwatts cancel in the ratio.)

The wide difference in power levels stems from the *inverse square law*. In *free space*, space with nothing to interfere or interact with the radio waves, the power of radio signals falls off with the square of the distance, as the power is spread evenly across space. (The reason why spreading power evenly causes the exponent to be 2 is that the area of a sphere grows as the square of the distance. Picture the same amount of radiation smeared over the surface of a sphere growing from the antenna.) This means that short amounts of distance can cause the signal to fall off rapidly. As it turns out, when thinking about wireless, the inverse square law should be thought of with a few modifications.

First, radio waves do not pass through all materials equally. The denser the material, the more the radio waves are weakened. This process is called *attenuation*, and how much an object attenuates is measured in the difference in power level the object causes over free space path loss. A vacuum, and for the most part, dry air, does not attenuate radio waves, and thus can be thought of as having 0dB of attenuation. A wooded wall might attenuate radio waves a bit, and so might have 3dB of attenuation. A cement wall, being thicker, might have 10dB of attenuation. The denser, thicker, or more conductive (metallic) the material is, the higher the attenuation Figure 5.6 shows an example of signal falloff. Certain objects are especially bad for Wi-Fi. Wire mesh, such as chicken wire lathing used in certain plaster walls, metal paneling and ductwork, and water-containing materials tend to cause high amounts of attenuation, as they absorb the radio waves better than dry and less dense materials. This shows up in deployments where old buildings, plumbing, elevator cores, and stairwells tend to disproportionately dampen the power of radio signals.

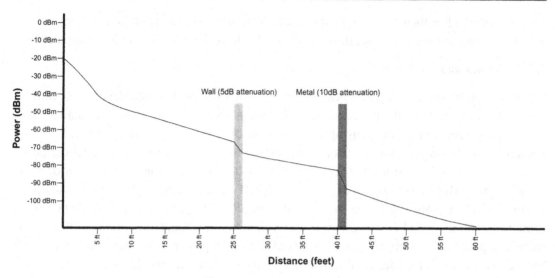

Figure 5.6: How Signals Weaken: Distance and Attenuation

Second, radio signals do not travel as far at higher frequencies than at lower ones. The higher the frequency, the more quickly signals fall off, and the more strongly they interact with and are stopped by objects. This can be understood rather easily. AM channels on radios tend to travel for hundreds of miles—long after you have left the city. That is because they are very low-frequency—in the kilohertz range. AM radio waves pass through walls, buildings, and mountains with relative ease. On the other hand, visible light only goes through glass. That is because light is a very high-frequency radio signal—400 terahertz or higher. In Wi-Fi, this is most noticeable in the difference between 2.4GHz (802.11bgn, referring to how 802.11n radios in the 2.4GHz band must support 802.11g and 802.11b as well) and 5GHz (802.11an), where the 5GHz radio waves may be stopped many feet before the 2.4GHz radio waves are, depending on the environment.

Third, radio waves reflect, and these reflections get mixed into the radio waves that are being transmitted, amplifying or weakening the signal. This effect, known as *multipath*, is caused by the constructive or destructive interference of radio waves, most often caused by reflections from multiple surfaces (see Figure 5.7). The name "multipath" comes from how the signal at any point in space can be thought of as the sum of radio waves bouncing as light rays along multiple paths. Multipath is a real problem with Wi-Fi, and is difficult to account for. That is because multipath doesn't just increase or decrease the strength of the signal. Radio waves travel at a fixed speed, the speed of light, and therefore those components of a signal that are reflected from objects further away are also slightly more delayed. The speed of light is quick—light travels at around 30 cm per nanosecond—but nanoseconds are in the range of times that Wi-Fi signals change to carry information. Therefore, the effect of multipath is add delayed echoes of the signal to the original signal, increasing the distortion of the signal as well. In environments where multipath does not

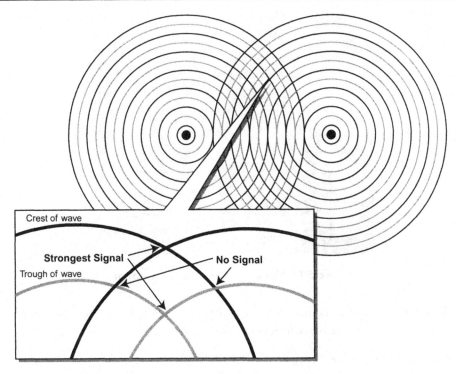

Figure 5.7: Two Transmitters and How the Signals Combine or Cancel

substantially distort the signals, the greater effect is that attenuation, and is actually modeled by altering the inverse square law's square parameter. The *path loss exponent* (PLE) is usually adjusted upwards, from the value 2, to accommodate the effects of the environment in weakening the signal over distance. Office spaces often have path loss exponents around 2.3 to 2.5. Some environments, on the other hand, such as long metallic hallways or airplanes, have path loss exponents less than 2. No, the laws of physics are not being violated—even the inverse square law is still absolutely correct with a PLE of 2 and only 2—but rather, the reflections from the long hallway make the environment serve as a *waveguide*, and those reflections constructively interfere, or amplify the signal, more than free space would. (The overall energy radiated out remains the same, just concentrated in the areas with PLE less than 2 and necessarily lessened in others.)

All this means that the strength of Wi-Fi signals depends heavily on the environment. The 802.11 standard has a diagram that is telling.

Figure 5.8, present in the IEEE 802.11 standard, shows the signal strengths throughout a one-room office. The darker the color, the weaker the signal, and the lighter the color, the stronger the signal. It is clear to see that the reflections cause intense ripple patterns across the room. These ripples are spaced a wavelength apart, and the difference from peak to trough is 50dB, or over one hundred thousand times the power.

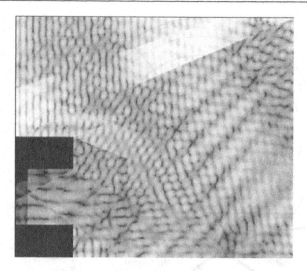

Figure 5.8: Signal Combination in a Room

The problem with the ripples is that they are nearly impossible to predict. This has an impact on the performance of the wireless network.

5.3.2.2 Noise and Interference

Radio noise occurs in Wi-Fi networks from three main sources. Each source contributes to the *noise floor*, which represents the background signal power level that will always be present. Signals that can be received must be a certain power level above the noise floor, in order for the signal to be distinct from the din. All of the major theory behind radio reception and information transfer relate the capacity to send information, and the amount of information that can be sent, to the ratio of the signal power level to this noise floor.

The first source of noise is the most basic and least interesting to voice mobility networks once it is understood. *Thermal noise* is the noise produced in every RF frequency because of the temperature of the world and all of the things in it. Thermal noise is a basic property of the fact that radio waves are light, and light is radiated because of the temperature of objects. The phenomenon that causes a piece of metal to become red when heated also causes the thermal background radiation within the Wi-Fi bands. The noise is not only coming from the nonradiating components in the room. The radios themselves—the antennas, the circuit paths, and the amplifiers and active components—all themselves are injecting thermal noise in the radio circuits. This is most important for the receiver's circuits, as this noise is what can drown out the weak signal coming in. In Wi-Fi networks, the noise floor tends to get no lower than around −100dBm, measured by Wi-Fi devices. This sort of noise is expected to be random and predictable, but predictable in how it is random. Once the thermal noise level is learned, it should not change by more than a couple of dB as time goes on, and that variation is as much the result of temperature variations as the Wi-Fi equipment heats up.

The second source of noise comes from non-Wi-Fi devices. Microwave ovens, cordless phones, Bluetooth devices, and industrial machinery all inject noise into the Wi-Fi bands. Devices are allowed, by regulatory rules, to inject this noise into the ISM bands that Wi-Fi uses, as long as the power level does not exceed certain thresholds. But intentionally generated communications signals tend to be of greater concern, as coexistence between different technologies in the unlicensed spectrum is mandatory, and Wi-Fi is not the only technology. The two ISM bands, the 2.4GHz and the 5.8GHz, are where these devices should be expected. The 5.8GHz band can especially have surprising sources of noise, because there are outdoor applications for the 5.8GHz band, using high-power but low-beam-width transmitters. A neighbor could easily set up such a system and interfere with a Wi-Fi network already in place.

The third, and most important source of noise is *self noise*, or noise generated by Wi-Fi from its operation. This noise always comes from neighboring access points' devices, which generate enough power that their signal energy reaches some of all of the devices in the given cell. In the best case, those interfering signals may reach the devices as weak, undecipherable power, contributing to an effective rise in the noise floor, but not necessarily resulting in a direct change into the operation of the Wi-Fi protocol. This sort of noise is always present in Wi-Fi networks that see more than a trivial amount of use. The more densely deployed the network is, as a mixture of access points and the clients that use them, the higher the noise floor will rise because of the contribution of this power. The noise floor can rise as high as −80dBm in many networks when being used, and even higher when the network comes under stress. In networks where the density is reasonably elevated, the effect of the other devices is stronger than noise, as it directly affects the Wi-Fi protocol by causing the devices to detect the distant transmissions and defer transmission. Even when this does not happen, the bursty on-and-off nature of Wi-Fi can mean that transmissions in progress can experience bit errors as the interference disrupts the radios themselves, without being detected as noise or energy outright.

5.3.3 RF Planning

RF planning is designed to address the two problems of multicellular networks. The first problem is to ensure that the coverage levels within the network are high enough that the expected data rates, based on the minimum required signal to noise ration, can be achieved at every useful square foot of the building or campus environment. The second problem is to avoid the intercell interference which results from multiple devices transmitting on the air without mitigation.

Proper RF planning is an expensive, time-consuming process. The basics of RF planning are for the installers to predict what the signal propagation properties will be in the expected environment. This sort of activity always requires using sophisticated RF prediction tools. RF prediction tools operate by requiring the operator to designate the locations and RF properties—attenuation, mostly—of each physical element in the building, the furniture, the walls, the floors, and the heavy machinery. Clearly a laborious process, the operator must copy in the location of these elements one at a time. Some tools are intelligent enough to take CAD drawings or floor-plan maps and estimate where the walls are, but an operator is required to verify that the guesses are not far from reality. RF planning tools then use RF calculations, based on electromagnetic principles, to determine how much the signal is diminished or attenuated by the environment. The planning tools need to know the transmit power capabilities and antenna gains of all of the access points that will be deployed in the network.

RF planning can be used this way to assist in determining where access points ought to be located, to maximize coverage given the particular SNR requirements. Because RF planning uses exact equations to predict the effects of the environment, it can be only as good as the information it is given. Operators must enter the exact RF and physical properties of the building to have a high likelihood of getting an accurate answer. For this reason, RF planning suffers from the *garbage-in-garbage-out* problem. If the operator has uncertainty about the makeup of the materials in the building, then the results of the RF plan share the same uncertainty.

Furthermore, RF planning cannot predict the effects of multipath. Multipath is more crucial than ever in wireless networking, because the latest Wi-Fi radios take advantage of that multipath to provide services and increase the data rate. Not being able to predict multipath places a burden on RF planning exercises, and requires RF planners to look for the worst-case scenarios.

Using RF planning tools to determine what power levels or channel settings each access point takes, then, is not likely to be a successful proposition as the network usage increases. Unfortunately, Wi-Fi self noise is a problem that does not show itself until the network is being heavily used, at which point it shows with vigor. Until then, as the network is just getting going, self noise will not be present at high levels and will not occupy 100% of the airtime. Thus, network administrators will see early successes with almost any positioning of Wi-Fi equipment, and can gain a false sense of security. (It is important to note that this is a property of trying to predict how RF propagates. Tools or infrastructure that constantly monitor and self-tune suffer the same problems, but with the added wrinkle that the self-tuning is disruptive, and yet will be triggered when the noise increases and the network needs to be disrupted the least.)

The one place where RF planning shows strength is in determining a rough approximation of the number and position of access points that are needed to cover a building. This does not require the sort of accuracy as complete RF plan, and tends to work well because of the fact that Wi-Fi networks are planned for a much higher minimum SNR than is necessary to cover the building. That higher SNR is required, however, to establish a solid data rate, and so what appears to be padding or overprovisioning from a coverage point of view can be lost capacity from a data rate point of view. Nonetheless, determining the rough number of access points needed for large deployments is a task that can do with some automation, and RF planning tools used only to plan for coverage (and not for interference), can be reasonably effective—even more so if the infrastructure that is deployed is able to tolerate the co-channel interference that is generated.

5.4 Wi-Fi's Approach to Wireless

The designers of 802.11 took into account the RF properties to create a technology that could transmit in the face of the obstacles in RF environments. Over time, they developed multiple differing, but related, radio technologies that built upon each one's previous 802.11 radio types and modern RF design principles to continue to improve the speed, range, and resiliency of the transmissions.

Let's look at the principles behind 802.11 transmissions at the physical layer.

5.4.1 Data Rates

A data rate, in 802.11, is the rate of transmission, in megabits per second (Mbps) of the 802.11 header and body. The 802.11 MAC header, the body, and the checksum (but not the physical layer header) are transmitted at the same data rate within each frame.

A data rate represents a particular encoding scheme, or way of sending bits over the air. Each data rate can be thought of as coming from its own *modem*, designed just for that data rate. An 802.11 radio, then, can be thought of has having a number of different modems to chose from, one for each data rate. (In practice, modern radios use digital signal processing to do the modulation and demodulation, and therefore the choice of a modem is just the choice of an algorithm in microcode on the radio or software used to design the radio itself.)

Each data rate has its own tradeoff. The lowest data rates are very slow, but are designed with the highest robustness in mind, thus allowing the signal to be correctly received even if the channel is noisy or if the signal is weak or distorted. These data rates are very inefficient, in both time and spectrum. Packets sent at the lowest data rates can cause network disruption, as they occupy the air for many milliseconds at a time. Although one millisecond sounds like a short amount of time, if each packet were, say, ten milliseconds long, then the highest throughput an access point could get would be less than 1.2Mbps for 1500-byte packets.

The higher data rates trade robustness for speed, allowing them to achieve hundreds of megabits per second. The description of the 802.11 radio types will walk through the principles involved in packing more data in. Occasionally, someone may mention that this effect is related to *Shannon's Law*. Shannon's Law states that the maximum amount of information that can be transmitted in a channel increases logarithmically with the signal-to-noise ratio. The stronger the signal is than the noise floor, the faster the radio can transmit bits. Lower data rates do not take advantage of high SNRs as well as higher data rates do. As data rates go higher, the radios become increasingly optimistic about the channel conditions, trying to pack more bits by making use of the higher fidelity that is possible. That higher fidelity is held to a smaller distance from the radio, and so higher data rates travel less far. (But note that 802.11 uses a concept to ensure that every device within the longest range knows of a transmission, no matter what the data rate is.) Think of it as saying that the amount of available "space" in a channel is determined by the SNR. More SNR means that more bits can be packed, by reducing the "space" between bits. Of course, the smaller the "space" between bits, the harder it becomes to tell the bits apart.

Data Rates and Throughput

Data rates in 802.11 refer to how fast the bits of the frame are transmitted over the air. Numbers as high as 300Mbps exist for the latest 11n devices; Section 5.5 explains each radio type. However, there is a significant gap between the data rate and the highest possible throughput that an application can see.

The main reason for this are that there is a tremendous amount of overhead in 802.11. Because each frame is preceded by a low-data rate header (the preamble), as well as mandatory random waiting times (the backoff), much of the airtime is spent in negotiating which device can transmit. This limits one-way traffic—such as UDP streams—to a significantly lower throughput than the data rate the frames are going at. The peak throughput varies significantly, depending on the vendors and products involved, but good rules of thumb are:

- 802.11b: 11Mbps data rate → around 8Mbps UDP throughput
- 802.11a/g: 54Mbps data rate → around 35Mbps UDP throughput
- 802.11n: 300Mbps data rate → around 250Mbps UDP throughput

Furthermore, 802.11 is a *half-duplex* network, meaning that upstream and downstream traffic compete for the same airtime. Thus, TCP traffic, which must have one upstream packet (also called acknowledgments) for every two downstream data packets, has an even lower throughput, such as:

- 802.11b: 11Mbps data rate → around 6Mbps TCP throughput
- 802.11a/g: 54Mbps data rate → around 28Mbps TCP throughput
- 802.11n: 300Mbps data rate → around 190Mbps TCP throughput

5.4.2 Preambles

Because 802.11 allows transmitters to choose from among multiple data rates, a receiver has to have a way of knowing what the data rate a given frame is being transmitted at. This information is conveyed within the *preamble* (see Figure 5.9).

The preamble is sent in the first few microseconds of transmission for 802.11, and announces to all receivers that a valid 802.11 transmission is under way. The preamble depends on the radio type, but generally follows the principle of having a fixed, well-known pattern, followed by frame-specific information, then followed by the actual frame. The fixed pattern at the beginning lets the receiver train its radio to the incoming transmission. Without it, the radio might not be able to be trained to the signal until it is too late, thus missing the beginning of the frame. The training is required to allow the receiver to know where the divisions between bits are, as well as to adjust its filters to get the best version of the signal, with minimum distortion. The frame-specific information that is included with the preamble (or literally, the Physical Layer Convergence Procedure (PLCP) following the preamble, although the distinction is unnecessary for our purposes) names two very important properties of the frame: the data rate the frame will be sent at, and how long the frame will be.

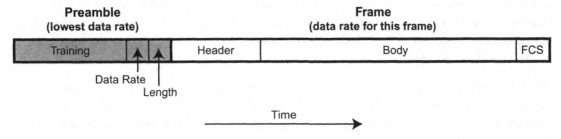

Figure 5.9: 802.11 Preambles Illustrated

All preambles are sent at the lowest rate the radio type supports. This ensures that no matter what the data rate of the packet, every radio that would be interfered with by the transmission will know a transmission is coming and how long the transmission will last. It also tells the receiver what data rate it should be looking for when the actual frame begins. All devices within range of the transmitter will hear the preamble, the length field, and the data rate. This range is fixed—because the preamble is sent at the lowest data rate in every case, the range is fixed to be that of the lowest data rate. Note that there is no way to change the data rate at which the preamble is sent. The standard intentionally defines it to be a fixed value—1Mbps for 802.11b, and 6Mbps for everything else.

When a radio hears a preamble with a given data rate mentioned, it will attempt to enable its modem to listen for that data rate only, until the length of the frame, as mentioned in the preamble, has concluded. If the receiver is in range of the transmitter, the modem will be able to properly detect the frame. If, however, the receiver is out of range, the receiver will hear garbage. The garbage will not pass the checksum (also garbage), and so will be discarded.

To prevent radios from interpreting noise as a preamble, and locking to the wrong data rate for a possibly very long length, the frame-specific information has its own checksum bit or bits, depending on the radio type. Only on rare occasions will the checksum bit fail and cause a false reception; thus, there is no concern for real deployments.

In summary, a receiver then works by first setting its radio to the lowest common denominator: the lowest data rate for the radio. If the fixed sequence of a preamble comes in, followed by the data rate and length, then the radio moves its modem up to the data rate of the frame and tries to gather the number of bits it calculates will be sent, from the length given. Once the amount of time necessary for the length of the frame has concluded, the radio then resets back to the lowest data rate and starts attempting to receive again.

5.4.3 Clear Channel Assessment and Details on Carrier Sense

Now that we've covered the preamble, you can begin to understand what the term *carrier sense* would mean in wireless.

The term *clear channel assessment* (CCA) represents how a radio determines if the air is clear or occupied. Informally, this is referred to as *carrier sense*. As mentioned previously, transmitters are required to listen before they transmit, to determine whether someone else is also speaking, and thus to help avoid collisions.

When listening, the receiver has a number of tools to help discover if a transmission is under way. The most basic concept is that of *energy detection*. A radio can figure out whether there is energy in the channel by using a power meter. This power meter is usually the one responsible for determining the power level, often stated as the Receive Signal Strength Indication (RSSI) of a real signal. When applied to an unoccupied channel, the power meter will detect the noise floor, often around −95dBm, depending on the environment. However, when a transmission is starting, the power meter will detect the signal being sent, and the power level measured will jump—let's say, to -70dBm for this example. That difference of 25dB can be used by the radio to clue in that it should attempt to turn on its modem and seek out the preamble. This allows the radio to have its modem off until real signals come by.

Energy detection can be used as a form of carrier sense to trigger the CCA. When done that way, non-802.11 noise that crosses a certain threshold, determined by the radio, will show up as an occupied channel for as long as the noise is present. This allows the radio to avoid transmitting into a channel at the same time as interference is present. In the 2.4GHz band, microwave ovens can often trigger the energy detection thresholds on radios, causing the radios to stop transmitting at that time.

On the other hand, energy detection for CCA has its limitations. If the noise coming in is something that would not interfere with the transmission, but does trip the energy detection threshold, then airtime is being wasted. Therefore, the carrier acquisition portion of CCA comes into play. Radios know to look for specific bit patterns in a transmission, such as the preamble. When they detect these bit patterns, they can assert CCA as well. Or, more importantly, when they detect some energy in the channel but cannot detect these bit patterns, they can conclude that there is no legitimate 802.11 signal and suppress CCA.

5.4.4 Capture

Radios are very sensitive, and can usually sense far-off transmissions. Here, sensing does not mean receiving the data within the frame correctly, but rather, hearing the preamble or the energy of the signal and deferring its own transmission for the proper time. Often, however, multiple signals come in at overlapping times, and at widely different power levels. The receiver locked into a distant signal will have that signal drowned out when a higher power signal comes in. That need not result in the loss of both signals. If the higher power signal is powerful enough, the receiver might be able to lock onto that higher signal. This concept is called *capturing*, as the receiver is captured by the more powerful transmission.

Capture works when the higher signal is more powerful than the lower signal by at least the SNR required for the data rate of the signal. In this sense, the receiver can look at the original, lower-powered signal as noise, rather than as a valid 802.11 signal.

The capture effect may be an important phenomenon within dense wireless deployments. Devices are most likely to allow for capture when they are receiving signals that are so weak that only CCA has been enabled, and the receiver is not making headway in receiving the distant signal. In those cases, the receiver may have not locked onto a signal, and so might be able to lock onto the stronger signal when it overlaps. For radios where the distant signal is being detected and demodulated correctly enough that the radio has locked into the distant signal, meaning that it has trained to the signal and is receiving what it believes to be in-range information (even if corrupt), it is unclear whether the radio will unlock and look for a new signal on an energy jump. Some radios will, and others won't, and it is difficult to know, from the packaging and documentation, whether a given radio will capture effectively.

5.4.5 Desensitization and Noise Immunity

One way of addressing the problem with CCA is for the access point to support *noise immunity* or *desensitization*. Noise immunity increases the amount of energy required before the radio defers, by making the radio less sensitive to weaker signals. This has the effect of reducing the access point's receive range, or the range that the radio can hear other transmissions within. But it also has the effect of increasing the strength of noise required to block the radio from sending.

This is useful, for example, when there are nearby non-802.11 noise sources, such as those from cordless phone systems or microwave ovens. If the noise source's signals reach the radio at −85dBm, and the radio used energy detection sensitive to −90dBm, then the radio would be unable to transmit. However, desensitizing the radio by adjusting the minimum energy level required to sense carrier to above −85dBm, such as −80dBm, would allow the radio to transmit.

Noise immunity settings are not the solution to every problem with noise, however. Desensitizing the radio reduces the receive range of the radio, increasing the risk of creating hard-to-detect coverage holes and causing interference to other 802.11 devices because of hidden nodes.

5.4.6 Hidden Nodes

Carrier sense lets the transmitter know if the channel near itself is clear. However, for one transmitter's wireless signal to be successfully received, the channel around the receiver must be clear—the transmitter's channel doesn't matter. The receiver's channel must be clear to prevent interference from multiple signals at the same time. However, the transmitter can successfully transmit with another signal in the air, because the two signals will pass through each other without harming the transmitter's signal.

So why does 802.11 require the transmitter to listen before sending? There is no way for the receiver to inform the transmitter of its channel conditions without itself transmitting. In networks that are physically very small—well under the range of Wi-Fi transmissions—the transmitter's own carrier sensing can be a good proxy for the receiver's state. Clearly, if the transmitter and receiver are immediately next to each other, the transmitter and receiver pretty much see the same channel. But as they separate, they experience different channel conditions. Far enough away, and the transmitter has no ability to sense if a third device is transmitting to or by the receiver at the same time. This is called the *hidden node problem.*

Figure 5.10 shows two transmitters and a receiver in between the two. The receiver can hear each transmitter equally, and if both transmitters are sending at the same time, the receiver will not be able to make out the two different signals and will receive interference only. Each transmitter will perform carrier sense to ensure that the channel around it is clear, but it won't matter, because the other transmitter is out of range. Hidden node problems generally appear this way, where the interfering transmitters are on the other side of the receiver, away from the transmitter in question.

802.11 uses RTS/CTS as a partial solution. As mentioned when discussing the 802.11 protocol itself, a transmitter will first send an RTS, requesting from the receiver a clear channel for the entire length of the transmission. By itself, the RTS does not do anything for

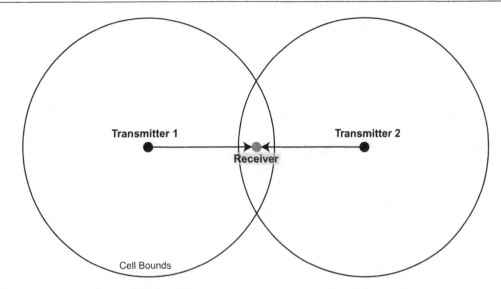

Figure 5.10: Hidden Nodes: The receiver can hear both transmitters equally, but neither transmitter can hear the other.

the transmitter or receiver, because the data frame that should have been sent would have the same effect, of silencing all other devices around the sender. However, what matters is what the receiver does. The CTS it sends will silence the devices on the far side from the sender, using the duration value and virtual carrier sense to cause those devices to not send, even though they cannot detect the following real data frame (see Figure 5.11).

This is only a partial solution, as the RTSs themselves can get lost because of hidden nodes. The advantage of the RTS, however, is that it is usually somewhat shorter than the data frame or frames following. For the RTS/CTS protocol to be the most effective against hidden nodes, the RTS and CTS must go out at the lowest data rate. However, many devices send the RTSs at far higher rates. This is done mostly to just take advantage of RTSs determining whether the receiver is in range, and not to avoid hidden nodes.

Furthermore, the RTS/CTS protocol has a very high overhead, as many data packets could be sent in the time it takes for an RTS/CTS transmission to complete.

5.4.7 Exposed Nodes

Unfortunately, the opposite problem to hidden nodes also occurs. Many times, especially in dense networks, the protocol or carrier sense causes devices that could transmit successfully to defer instead. This is called the *exposed node problem*, because nodes are exposed to being blocked when they need not be.

Because there are two carrier sense types—virtual and physical—there are two types of exposed node problems. The first type happens when a device sets its virtual carrier sense based on an RTS but does not hear the subsequent data frame. 802.11 has a partial

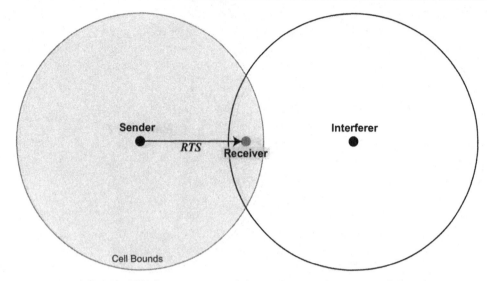

First, the RTS allocates airtime and prevents devices in the sender's cell from transmitting--including the receiver.

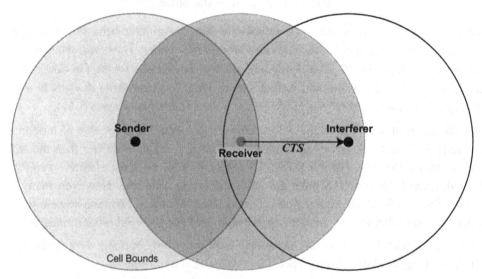

Them, the CTS allocates airtime and prevents devices in the receiver's cell from transmitting--including the interferer.

Now the sender can transmit the data free of interference.

Figure 5.11: RTS/CTS for Hidden Nodes: The CTS silences the interfering devices.

provision for this, called *NAV resetting*. If a device hears an RTS, sets its carrier sense, but fails to sense the data frame that should have come after the CTS, it may reset its virtual carrier sense and begin transmitting. Unfortunately, this solution addresses only the RTS case, and RTSs are somewhat rare.

The second type, more damaging to dense voice networks, is with physical carrier sense. Here, a far away device triggers the sender to not transmit, as the sender picks up and obeys the far away device's preamble fields or signal energy. However, in many cases, the faraway device is sending to a device even farther away, and there is no risk of interfering with that signal if the transmitter were to transmit. Figure 5.12 illustrates this idea.

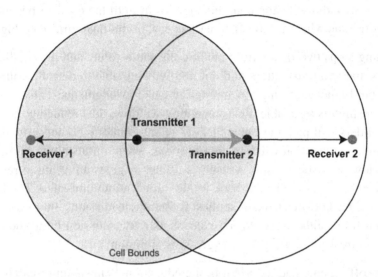

Figure 5.12: The Exposed Node Problem. Transmitter 1 inadvertently prevents Transmitter 2 from transmitting to Receiver 2, even though there would be no interference.

5.4.8 Collisions, Backoffs, and Retries

Multiple radios that are in range of each other and have data to transmit need to take turns. However, the particular flavor of 802.11 that is used in Wi-Fi devices does not provide for any collaboration between devices to ensure that two devices do take turns. Rather, a probabilistic scheme is used, to allow for radios to know nothing about each other at the most primitive level and yet be able to transmit.

This process is known as *backing off,* as is the basis of *Carrier Sense Multiple Access with Collision Avoidance,* or CSMA-CA. The process is somewhat involved, and is the subject of quite a bit of research, but the fundamentals are simple. Each radio that has something to send waits until the channel is free. If they then transmitted immediately, then if any two radios had data to transmit, they would transmit simultaneously, causing a collision, and a receiver would only pick up interference. Carrier sense before transmission helps avoid a radio transmitting only when another radio has been transmitting for a while. If two radios do decide to transmit at roughly the same time—within a few microseconds—then it would be impossible for the two to detect each other.

To partially avoid the collisions, each radio plays a particular well-scripted game. They each pick a random nonnegative integer less than a value known as the *contention window* (CW), a small power of 2. This value will tell the radio the number of *slots*, or fixed microsecond delays, that the radio must wait before they can transmit. The goal of the random selection is that, hopefully, each transmitter will pick a different value, and thus avoid collisions. When a radio is in the process of backing off, and another radio begins to transmit during a slot, the backing-off radio will stop counting slots, wait until the channel becomes free again, and then resume where it left off. That lets each radio take turns (see Figure 5.13).

However, nothing stops two radios from picking the same value, and thus colliding. When a collision occurs, the two transmitters find out not by being able to detect a collision as Ethernet does, but by not receiving the appropriate acknowledgments. This causes the unsuccessful transmitters to double their contention window, thus reducing the likelihood that the two colliders will pick the same backoff again. Backoffs do not grow unbounded: there is a maximum contention window. Furthermore, when a transmitter with an inflated contention window does successfully transmit a frame, or gives up trying to retransmit a frame, it resets its contention window back to the initial, minimum value. The key is to remember that the backoff mechanism applies to the retransmissions only for any one given frame. Once that frame either succeeds or exceeds its retransmission limit, the backoff state is forgotten and refreshed with the most aggressive minimums.

The slotted backoff scheme had its origin in the educational Hawaiian research network scheme known as *Slotted ALOHA*, an early network that addressed the problem of figuring out which of multiple devices should talk without using coordination such as that which token-based networks use. This scheme became the foundation of all contention-based network schemes, including Ethernet and Wi-Fi.

However, the way contention is implemented in 802.11 has a number of negative consequences. The denser and busier the network, the more likely that two radios will collide. For example, with a contention window of four, if five stations each have data, then a collision is assured. The idea of doubling contention windows is to exponentially grow the window, reducing the chance of collisions accordingly by making it large enough to handle the density. This would allow for the backoffs to adapt to the density and business of the network. However, once a radio either succeeds or fails miserably, it resets its contention window, forgetting all adaptation effects and increasing the chance of collisions dramatically.

Furthermore, there is a direct interplay between rate adaptation—where radios drop their data rates when there is loss, assuming that the loss is because the receiver is out of range and the transmitter's choice of data rate is too aggressive—and contention avoidance. Normally, most devices do not want to transmit data at the same time. However, the busier the channel is, the more likely that devices that get data to send at different times are forced to wait for the same opening, increasing the contention. As contention goes up, collisions go up, and rate

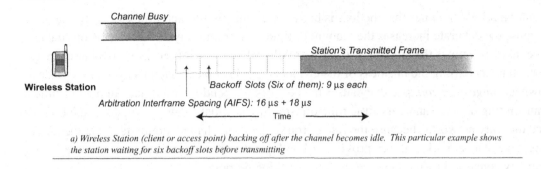

a) Wireless Station (client or access point) backing off after the channel becomes idle. This particular example shows the station waiting for six backoff slots before transmitting

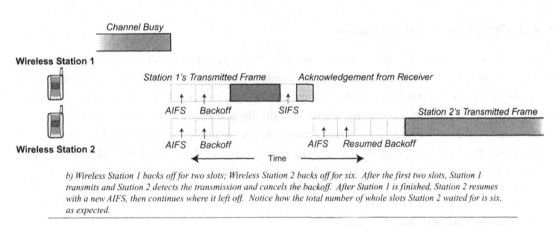

b) Wireless Station 1 backs off for two slots; Wireless Station 2 backs off for six. After the first two slots, Station 1 transmits and Station 2 detects the transmission and cancels the backoff. After Station 1 is finished, Station 2 resumes with a new AIFS, then continues where it left off. Notice how the total number of whole slots Station 2 waited for is six, as expected.

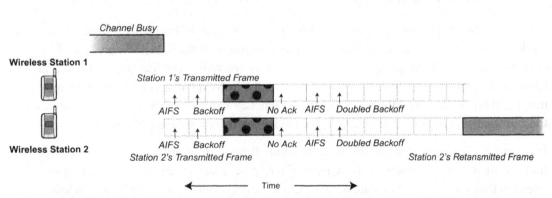

b) Bpth Wireless Stations back off for two slots and then transmit. Therefore, both transmissions collide. Because Wi-Fi uses Collision Avoidance, and cannot detect collisions directly, the only indication to the stations is the lack of an Acknowledgement. After the time passes that an Acknowledgement should have been received by, both stations double their contention window, pick new random backoffs (which may be a larger number of slots than before, and likely will not be equal again) and start over.

Figure 5.13: The backoff procedure for two radios.

adaptation falsely assumes that the loss is because of range issues and drops the data rate. Dropping the data rate increases the amount of time each frame stays on air—a 1Mbps data frame takes 300 times the amount of time a 300Mbps data frame of the same number of bytes takes—thus increasing the business of the channel. This becomes a vicious cycle, in a process known as *congestion collapse* that causes the network to spend an inordinate amount of time retransmitting old data and very little time transmitting new data. This is a major issue for voice mobility networks, because the rate of traffic does not change, no matter what the air is doing, and so a network that was provisioned with plenty of room left over can become extremely congested by passing over a very short tipping point.

5.4.9 Challenges to Voice Mobility

You may ask why 802.11 has as many seemingly negative aspects as have been mentioned. It is important to remember that 802.11 was designed to reduce the cost and complexity of the radio as much as possible, and this simplicity has allowed Wi-Fi to blossom as much as it has. However, because the design was not originally geared at large, dense, and mobile deployments, many challenges exist for voice mobility deployments over Wi-Fi. This chapter and the next will address the ways in which voice mobility deployments can be successfully produced over Wi-Fi networks, and will explore where each type of deployment works and where it may be limited.

5.5 The Wi-Fi Radio Types

Let's now look at the types of radios. Radio types, in Wi-Fi, are designated by the amendment letter for the part of the standard that added the description of the specific radio type. The original 802.11 standard, produced in 1999, has only one radio type, but was so limited in bandwidth that it was not broadly used commercially. Subsequent amendments have provided new radio types, starting with 802.11b, which launched wireless LANs into the forefront, and this constant stream of amendments has kept Wi-Fi at the cutting edge of wireless.

It's worthwhile to spend a little bit of time going through the mechanics of how these radios transmit their signals. Voice mobility networks require a keener sense of how wireless operates than convenience or data networks, and having some insight on how wireless transmitters operate can provide significant benefit in planning for and understanding the network. Feel free to skim this section and use it as a reference.

5.5.1 802.11b

802.11b is the first radio that was added to the original 802.11 standard. 802.11b builds on the original technology used for 802.11 radios, and because of this, we will refer to 802.11b as including the original 802.11. 802.11b has four data rates, and is designed to be able to transmit at data rates as high as 11Mbps. See Table 5.11.

Table 5.11: The 802.11b data rates

Speed	Preamble Type
1Mbps	Long
2Mbps	Long/Short
5.5Mbps	Long/Short
11Mbps	Long/Short

The lowest two data rates—1Mbps and 2Mbps—belong to the original 802.11-1999 standard, and were what broke the ground into the concept of a wireless "Ethernet." They happen to be referred to here as 802.11b rates in keeping with current industry practices.

802.11 signals are radio waves, and the goal of any digital radio is to encode a binary signal on those radio waves in a way that maximizes the chance of the signal being received while simultaneously minimizing the complexity of the receiver. The original 802.11-1999 signals are of a type known as *Direct Sequence Spread Spectrum* (DSSS).

The starting point for the 802.11b signals is known as the *carrier*. The carrier has the center frequency of the signal; channel 11's carrier frequency is at 2.462GHz. By itself, a carrier is a pure tone—a sine wave—at that frequency. These carriers do not possess any information, but they do carry energy in a distinct form that the receivers can use to identify a real signal, rather than just noise. The radio's job is to add the information in on top of the carrier.

What makes a radio technology distinct is precisely how it encodes that information onto the carrier. The manner of encoding is known as a *modulation* technique. Two well-known ones are *amplitude modulation* (AM) and *frequency modulation* (FM). Both are used in car radios, and so should be familiar places to start.

Amplitude modulation works by varying the amplitude, not surprisingly, of the carrier to convey the information. The receiver locks onto the carrier, and then uses the change in amplitude to determine what the original signal was. For a car radio that uses AM (Figure 5.14), the original signal will vary at the frequencies that are audible—usually up to 3000 Hz. The carrier, however, varies much faster, at the 1000 kilohertz range. When the original signal moves up in amplitude, so does the carrier; thus, when plotted with amplitude over time, the carrier looks like it has the original signal as an envelope, limiting the bounds it can oscillate in. Receivers can be incredibly simple, by just filtering out the high-frequency portion of the signal—using basic circuits to blend out oscillations around the carrier frequency but allowing the audible frequencies through—the original signal can be recovered.

Frequency modulation, on the other hand (Figure 5.15), works by varying the center frequency based on the data. Instead of varying the amplitude, the carrier's frequency is offset by the value of the data signal. The receiver notices as the carrier drifts one way or

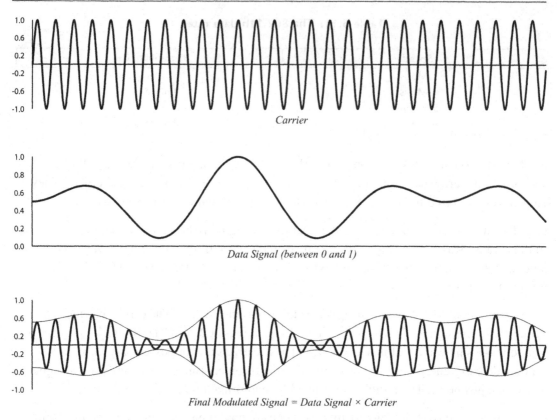

Figure 5.14: Analog Amplitude Modulation

the other from the center frequency, and uses that drift to derive the original data. For car radios, which are usually analog, that drift becomes the signal that is sent to the speakers.

There is a third type, *phase modulation* (PM or ΦM, as Φ is the custom mathematical symbol for phase, measured in degrees from 0 to 360), which works by altering the phase of the signal, or the offset of the carrier in time. Figure 5.16 shows phase modulation, and how it compares to frequency modulation. Note how the phase modulated signal is offset—in this example of simple of phase modulation, the modulated appears to be reflected vertically, the result of its phase having been altered by 180 degrees—when the data signal is non-zero, whereas the frequency modulated signal clearly has been sped up or slowed down for non-zero data.

Because the phase of a signal is the offset from the carrier in degrees, and the amplitude is just a distance the peak of the carrier is offset from the origin, an amplitude and phase modulated signal can be represented using polar coordinates. In the descriptions here I will try to avoid unnecessary mathematics, but the concept that the type of modulation applied to a carrier can be represented by polar coordinates is powerful for those interested in

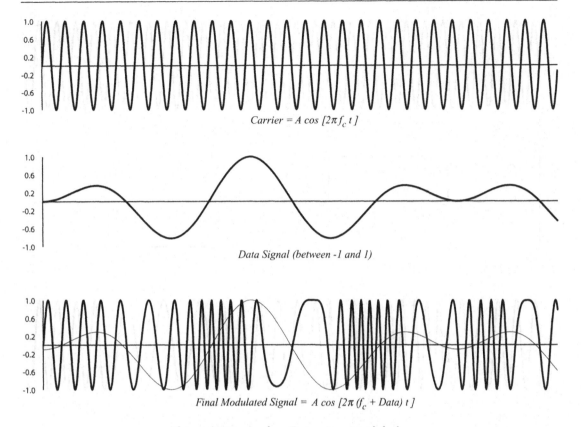

Figure 5.15: Analog Frequency Modulation

understanding just what radios do to the signal to be able to transmit the data. The polar coordinate representation is one of the two common methods for representing complex numbers, and the designs of radios are always though of as involving complex math. (A complex number is type of number that can represents the square roots of negative numbers, something not allowed in real numbers.) For those interested in the details, the appendix at the end of this chapter explores the mathematics.

The term "baseband" refers to the signal converted into a function that, when multiplied by the carrier, will produce the ultimate function. Therefore, the baseband signal can be thought of as the encoded signal, but before any particular channel or center frequency has been chosen. This means that the baseband signal is identical for every channel. Because the baseband encoding is the most complicated part of the radio—it does the business of giving the signal its flavor, if you will, the part of a radio that encodes or modulates the signal is also called the baseband.

In digital systems, which 802.11 belongs to, phase modulation is known by the term *phase shift keying* (PSK), representing that the phase is shifted abruptly because of the digital

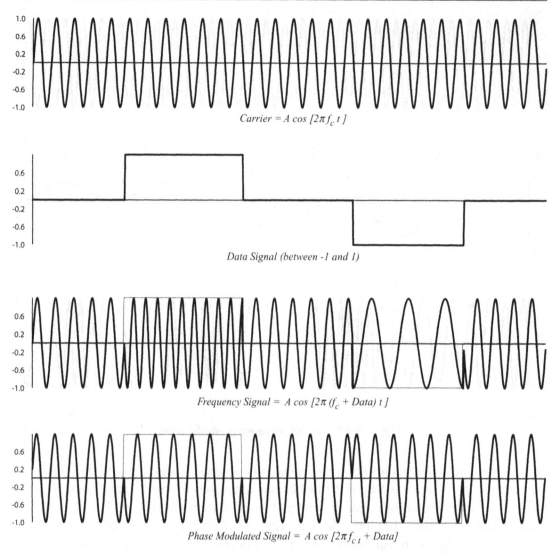

Carrier = A cos [2π f_c t]

Data Signal (between -1 and 1)

Frequency Signal = A cos [2π (f_c + Data) t]

Phase Modulated Signal = A cos [2π f_{c t} + Data]

NOTE: Because the phase modulation maps the incoming signal into the range of 180
degrees, plus or minus, the encoded signal can only differentiate between a zero and
non-zero value. -1 and 1 become -180 degrees and 180 degrees, which are equal.

Figure 5.16: Frequency versus Phase Modulation

nature of the incoming signal, and the shift is "keyed" by the digital data. Understanding
these terms is not necessary to understanding Wi-Fi, but it can offer some insight into why
some radios do better in certain environments than others.

5.5.1.1 1Mbps

For 1Mbps, 802.11 uses what is called *binary phase shift keying* (BPSK). This simply
means that the phase is shifted forward by 180 degrees for a 1 bit in the data, and left

unmodified for a 0 bit in the data. Figure 5.17 shows what is called the *constellation* for this representation. Don't let it be intimidating: this is just the polar coordinate system mentioned previously, with each X representing the modulation of the carrier represented by each bit. Notice that the distance from the origin *A* is always 1: BPSK does not use amplitude modulation at all.

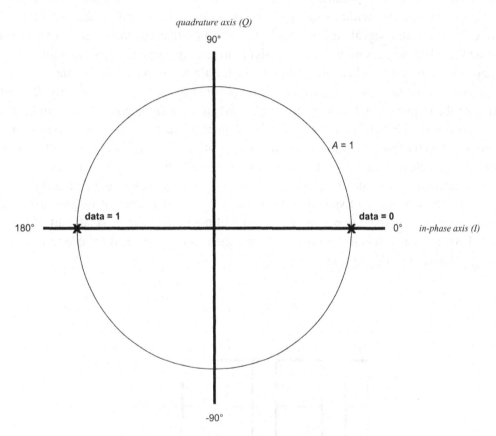

Figure 5.17: Binary Phase Shift Keying

The modulating signal, however, is not just the stream of bits that make up the frame. Additional work goes in to prepare the incoming bitstream for transmission so that it will better survive the sometimes nasty conditions of the RF channel. Each incoming data bit is mapped to 11 transmission bits. These outgoing 11 bits are called, together, one *symbol*, and each of the eleven bits is individually called a *chip*. The process of mapping bits to chips is known *spreading*, the term deriving from the effect the mapping has of spreading out the frequency range the signal takes on. (Or, one can think of the process as spreading the information contained in the original data bit over the various chip bits of the symbol.) This expanding of one data bit to 11 chip "bits" is done to protect the signal from various types of interference, and the mapping is very specific sequence known as a *Barker sequence*. The sequence used happens to be

+1, −1, +1, +1, −1, +1, +1, +1, −1, −1, −1

where +1 represents the 1 bit and −1 represents the 0 bit. This sequence might seem to be random, but it was chosen because it possesses an important real-world property. As the signal encoded with this Barker sequence reflects off surfaces, it will be delayed a bit relative to the original signal, and each delayed reflection that hits the receiver will appear to be added together, causing the received signal to appear blurred out as the sequences overlap. The important mathematical property of the Barker sequence is that the *autocorrelation*, or the sum of the element-wise product of the sequence and any delayed version of the sequence, will have a strong peak when no delay is present and will be close to zero otherwise. This allows the receiver to use math to hunt out the main signal—the part that comes directly from the transmitter and not off of reflected surfaces—and filter out the echoes to get a cleaner signal. Furthermore, the increasing of the rate by a factor of 11 provides redundancy in time, which causes spreading in frequency to the final 20MHz bandwidth. The advantage of spreading the signal over a wider range of frequencies is that many interference sources are narrow-band and will then affect only a small part of the encoded signal, rather than the entire signal, and the receiver may still be able to recover the original data. See Figure 5.18.

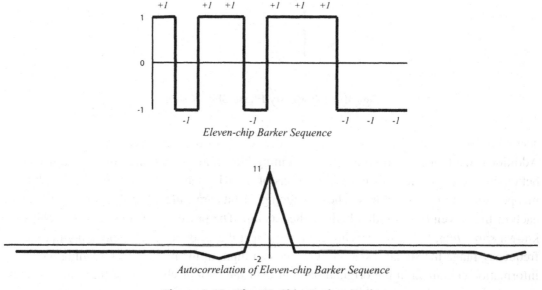

Eleven-chip Barker Sequence

Autocorrelation of Eleven-chip Barker Sequence

Figure 5.18: The 11-Chip Barker Code

Two more points: the data bits are differentially encoded, meaning that if two bits to send are 11, the chip sequence is the Barker sequence, followed by the Barker sequence multiplied by −1. The differential encoding is used to make it unnecessary for the receiver to know whether 0 degrees is a 0 bit and 180 degrees is a 1 bit, or vice versa. Also, the data bits go through a process known as *scrambling*, which ensures that data that is made of predominantly one value—such as a string of zeros—has enough variation over the air to prevent the signal from not changing enough and drifting to one side or the other of the center frequency, something which makes the signal harder to transmit or receive. The process is called scrambling because the bits are modified with a pseudorandom sequence—there is no security aspect in this, however. The scrambled bits end up becoming the ones that are then spread out into the 11-chip symbols.

5.5.1.2 2Mbps

The 2Mbps data rate is very similar to the 1Mbps data rate. The main difference is that the data bits are taken pairwise (after scrambling), and encoded so that two-bit pairs map to 11-chip symbols. This doubles the data rate: the two bits are packed in by encoding both bits into one symbol, with the symbol taking the same time as a 1Mbps one. Looking at the constellation (Figure 5.19), the four possible values of the two bits—00, 01, 10, and 11—are equally spaced around the circle—at degrees 0, 90, 180, and 270 (-90). In this sense, the two bits are packed in by splitting "space", meaning room in the constellation, rather than sending one after the other by splitting "time".

The downside to packing in two bits per symbol is that the distance between the constellation points on the symbol has gone down. This is important, because noise shows up at the receiver as an uncertainly about whether a given symbol belongs as one of the—in this case, four Xs. If there are only two possible symbols, as with the 1Mbps rate, then the receiver's value would have had to have drifted very far to cause a mistake. With four possible symbols, however, the received symbol need not drift as far to cause an error. This is what was being referred to earlier when discussing how lower data rates are more robust, whereas higher data rates require increased fidelity. The more noise, the worse the higher data rates do, and the more mistakes they make. In 802.11, one mistake in choosing the symbol causes the entire frame, good bits and bad, to be dropped as an error (Figure 5.20).

Therefore, the 2Mbps data rate requires both a higher absolute signal strength and a higher SNR to be received, and thus will not carry with quite the same range as 1Mbps rates will.

5.5.1.3 5.5Mbps and 11Mbps

The 5.5Mbps data rate was the first one introduced in the 802.11b amendment itself. Recognizing that 2Mbps was not fast, compared with the predominant 10Mbps wireline speeds from 10BASE-T Ethernet, the IEEE introduced the 5.5Mbps and 11Mbps data rates.

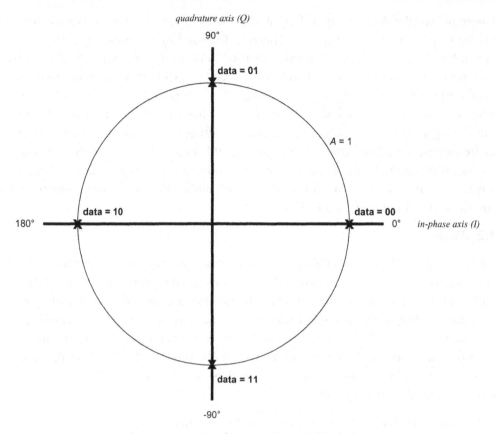

Figure 5.19: Quadrature Phase Shift Keying

To get to the faster data rate, IEEE decided not to continue splitting space by adding more points to the constellation. Instead, they looked at the 11 chips, and decided to pack more bits in by changing the data bits throughout the 11-chip sequence, thus splitting time. Instead of assigning one or two bits for the entire 11-chip sequence, both 5.5Mbps and 11Mbps use an eight-chip sequence (still sent out at 11 megachips/second), and then map either four or eight bits to each eight-chip symbol. These eight-chip symbols are QPSK encoded, and the (complex) value of each of the eight chips is determined by four bit-pairs that are input. The particular formula used to determine the eight chip values from the eight input bits is designed to increase the reliability of the signal, and is a type of *complementary code keying* (CCK).

The 11Mbps rate maps the eight data bits per symbol to the four bit-pairs directly. The 5.5Mbps data rate is produced by using only four data bits to get the eight-bit input to the formula by using a specified mapping up front, adding greater redundancy.

Because of the greater redundancy, 5.5Mbps has a longer range and less SNR requirements than 11Mbps.

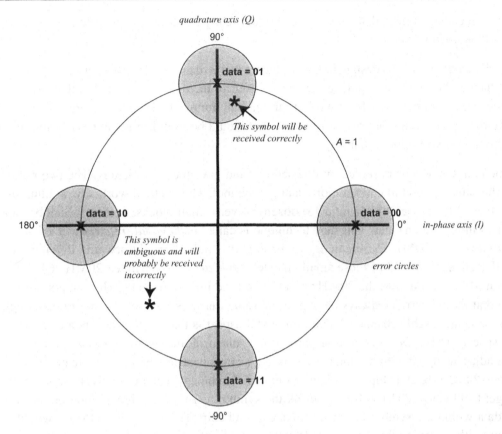

Figure 5.20: Symbol Error

5.5.1.4 Short and Long Preambles

802.11b allows the transmitter to choose from either a short or long preamble for the 2, 5.5, and 11Mbps data rates. The intent of allowing a short preamble, being almost exactly half as short in time as the long preamble will be to free up some additional airtime, at the risk of slightly decreasing the chances that there will be enough time for a radio to lock onto the signal. In the early days of 802.11b-only networking, this statement may have been more true. Now, however, most of the distinction is lost, and short preambles should always be recommended, unless old legacy devices are in use. Most of the issues with short preambles now are from those old devices not accepting short preambles.

5.5.2 802.11a and 802.11g

802.11a and 802.11g expanded the data rates and network efficiency significantly, leading to a top data rate of 54Mbps. 802.11a—for the 5GHz band—and 802.11g—for the 2.4GHz band—are nearly identical except for the band in use and some additional rules. Their introduction was a turning point in Wi-Fi, allowing the designers to use a newer, different way of encoding signals than before that would greatly enhance the speed of the network, as well as the robustness.

The challenge to 802.11 designers was to increase the data rates by packing more bits into the channel, but without requiring the signal itself to fluctuate in time so fast that delay effects prevented the radio from working in real environments. In fact, if there was a way to make the signal even more robust to delay, that would be useful in getting Wi-Fi into more environments with more reliability.

In the end, the designers settled on one scheme and gave them two letters. The two radios use the same method to encode more data per symbol. This method works by dividing up the channel from one 20MHz block to dozens of very small blocks, each of which becomes its own miniature channel. This new technique is called *orthogonal frequency division multiplexing* (OFDM). The frequency division part is rather straightforward: dividing the 20MHz channel into 48 of those small channels (*subcarriers*), separated 20MHz/64 = 5/16th MHz apart allows the 20MHz band to be used more efficiently. (It is important to note that the subcarriers always belong to the same transmission, and are not to be thought of as separate, usable channels.) The orthogonal aspect is that the subcarriers are encoded and spaced so that the sidelobes, or parts of the subcarrier that would ordinarily interfere with adjacent signals) do not interfere with adjacent subcarriers here. Why create more carriers? 802.11b, at 1Mbps, has 1-microsecond symbols, which are relatively short in time. To get to 11Mbps, 802.11b had to shrink the symbol to 73% the original duration and vary the data within the symbol for each 90-nanosecond chip. Trying this for 54Mbps would require either varying the data every 18 nanoseconds, or making major sacrifices on the redundancy. Either way, the signal would not be able to travel through most environments very well, and Wi-Fi would not be able to transmit at high speeds past a few feet. However, by spreading the data out across 48 subcarriers, the data rate on each subcarrier is dropped back to something similar to the 1Mbps 802.11b data rate—1.125Mbps. Bumping up the modulation to six scrambled bits per symbol then allows the symbol to be held without fluctuation for longer and reduces the bandwidth of the subcarrier to its final value, resulting in a growth from 1 microsecond symbols to 4 microseconds. By packing in more bits per Hz, the signal can become more resilient over time.

This advance was quite exciting for its time, and is still one of the two major underpinnings for modern 802.11n radios.

Table 5.12 shows the data rates available to 802.11ag. There are now eight data rates unique to 802.11ag.

The two additions to 802.11b for each subcarrier are the growth of the constellations and the use of an *error correction code* built into the radio. The first part, the constellation expansion, builds upon the concepts used to expand from 1Mbps to 2Mbps to pack more

Table 5.12: The 802.11ag data rates

Speed	Modulation	Bits per Symbol per Subcarrier	Coding Rate
6Mbps	BPSK	1	½
9Mbps	BPSK	1	¾
12Mbps	QPSK	2	½
18Mbps	QPSK	2	¾
24Mbps	16-QAM	4	½
36Mbps	16-QAM	4	¾
48Mbps	64-QAM	6	⅔
56Mbps	64-QAM	6	¾

bits in. The constellation points are no longer on a circle; rather, they are spaced evenly on a square grid, with 64-QAM being more dense in both amplitudes and phases. Figure 5.21 shows the two constellations—this time, the labels for which bits go to which constellation point are left off for clarity.

As you may notice, the use of four different modulation schemes—BPSK, QPSK, 16-QAM, and 64-QAM—lead to four different numbers of coded bits per symbol, but there are eight, not four, data rates. To allow for two data rates for the same modulation type, 802.11ag introduces the error correction code. Error correction codes are methods of expanding digital data into something with more bits than the original in such a way that some of those expanded bits can be lost without causing a loss in the original data. In a pure sense, this is the opposite goal from data compression, where redundant bits are removed, rather than added. The extra bits are wasteful if there are no errors in the data, but if there are errors in the data, the code will protect the data from loss. Redundancy in wireless, as we have seen, is a good thing to have. We've seen spreading sequences already, and error correction codes can be thought of as a way to spread the bits as well, but over time rather than over frequencies. (Of course, frequency and time are two sides of the same coin.)

The simplest error correction codes are just to repeat bits of the data. A bit stream of 101101 can be doubled, to produce 101101101101. This stream protects against any one bit being lost. Let's say that the stream arrives at the receiver as 101x01101101, where x is the lost bit. The receiver can clearly recover the lost bit, by looking at the copy that makes up the second half of the transmission. Doubling protects the stream from up to 50% loss, if the right 50% is lost, but can fail at two losses if the two bits lost are from the same source. Because the ability to correct for errors depends not only on the number of bits lost, but the positions of the lost bits, doubling makes for a very poor error-correction code in practice. To protect against exactly one bit loss in the most efficient manner possible, just sum the

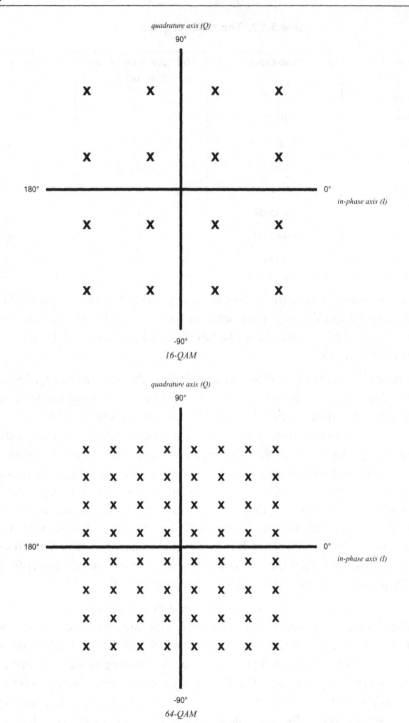

Figure 5.21: 16-QAM and 64-QAM

source bits and record whether that sum is even or odd as one additional bit. If one bit is lost, the sum of the surviving bits can be compared in evenness to the sum bit, therefore recovering the original data. This is called a *parity check code*, and conceptually introduces the concept of using arithmetic on some or all of the source bits to produce the extra bits.

802.11ag uses a *convolutional encoder* to expand the source bits, in this case producing twice the number of bits but, unlike the doubler, doing so more intelligently to avoid some loss patterns of equal bit losses being worse than others. This expansion still produces only one data rate, at a $\frac{1}{2}$-coding rate per modulation. To produce the other coding rates, 802.11ag just uses the property that the error-correcting code can tolerate loss, and goes ahead and starts tossing bits to get to a higher coding rate. This process, called *puncturing*, sounds inefficient by reducing redundancy that was just added (why not just not add as many extra bits in the first place?), but ends up saving on the complexity of the coding hardware, the radio.

The overall picture, then, for 802.11ag encoding is for the data bits to be scrambled (as with 802.11b), then expanded with the error-correcting code, then split among the subcarriers, and then modulated.

Because of the larger number of things going on in the signal, there is more risk of losing data if the receiver's timing goes off a bit from the sender's. To compensate, timing is maintained by the addition of four *pilot* subcarriers. These carry a known signal pattern—like the preamble—but do so for the length of the transmission. If the receiver's clock speeds up or slows down relative to the sender's, then the constellation would essentially rotate, and as the constellation points are now closer together than with 802.11b, the rotation would cause the receiver's bits to jumble. The pilot subcarriers' known pattern lets the receiver adjust as needed.

5.5.2.1 Preambles, Slots, and Optimizatoins

The preambles of 802.11ag are at 6Mbps, the lowest 802.11ag data rate. They are also significantly shorter than the short 802.11b preambles: 40 microseconds for 802.11ag. Improvements in signal processing technology since 802.11b came along allowed the designers to not need to provide as much synchronization time.

Furthermore, the radios can stop receiving and start transmitting more quickly for 802.11ag than they were expected to be able to do for 802.11b. This allows 802.11ag to use shorter slots than 802.11b, allowing for less wasted time. For 802.11g, short slots are an option, and are determined based on the presence of 802.11b clients, to prevent 802.11g clients from using a different contention scheme. 802.11a always assumes the faster slots.

5.5.2.2 802.11b Protection

802.11a has no legacy clients to deal with. However, 802.11g is in the 2.4GHz band, and has to avoid destroying 802.11b performance when the two devices are present together.

This destruction would occur because 802.11b radios use carrier detection (see Section 5.4.3) to determine whether the channel is clear before transmitting, and that means that the 802.11b devices are looking for 802.11b transmissions. 802.11g transmissions, however, look nothing like 802.11b, and so 802.11b radios would end up only seeing 802.11g as some sort of foreign interference.

To prevent this from disrupting any 802.11b device's traffic, 802.11g introduces the notion of *CTS-to-self protection*. Because 802.11b clients can only see 802.11b traffic, a way to stop them from transmitting when an 802.11g transmission will start is for the 802.11g device to send an 802.11b (legacy) CTS message first. This CTS message, sent not as a part of an RTS/CTS transaction (see Section 5.4.6) but from the 802.11g sender to, nominally, itself, sets the virtual carrier sense for all devices that can hear it. The CTS frame has a Duration field—or the length of time to quiet the other stations—long enough to let it finish the 802.11g transmission that will follow.

CTS-to-self protection is automatically turned on for any AP that has 802.11b legacy clients associated to it, or for access points who overhear neighboring access points that have 802.11b clients assigned to them. This CTS-to-self message can be incredibly inefficient, and has the potential to disrupt voice mobility networks, as will be mentioned later in this chapter.

802.11g data rates, even though they are identical to 802.11a data rates, may go by the additional term *ERP* in product literature. The term, short for *Extended Rate PHY*, is typical of the language used in the 802.11 standard (worse abuses will come up in the next section), but is good to know for when it occasionally slips into product documents for users. When you see ERP, think 802.11g, and when you see non-ERP, think 802.11b-only legacy devices. Neither term is correct for 802.11a devices, which are just known as 802.11a.

5.5.3 802.11n

54Mbps seemed like a lot at the time, but enterprise wireline networks operate at 100Mbps or more. To allow for even higher data rates, IEEE has embarked upon the 802.11n standard. 802.11n revolutionizes Wi-Fi by adding another radio breakthrough, as well as a long list of additional enhancements and optimizations.

The most important addition 802.11n brings is the use of a technology called *multiple-in, multiple-out* (MIMO). MIMO does something that seems counterintuitive—almost magical—to those used to thinking about how two radios transmitting at once cause collisions and destroy wireless networks. MIMO transmits multiple signals at once, on the same channel, at the same time, and at the same power levels. However, MIMO is not magic, just math, and is able to greatly increase the speed of the network.

MIMO works by requiring each device to have multiple antennas. These antennas are not terribly far apart—a few inches at the most—but they need to be present. MIMO then splits the data across those antennas, sending out multiple *spatial streams*. These streams go over the air at the same time, and the receiver uses multiple antennas to pick up this transmission, applies some math to separate back out the streams, and then recombines the data.

We will go through how this multiple simultaneous transmissions work in a moment. However, because MIMO has some general rules that the products using them need to follow, let us start with those rules, and the state of the technology.

802.11n is a very new standard, and, for 2008 and most of 2009, was not yet finished. However, every major Wi-Fi vendor was selling 802.11n-based products. How can this happen? As it turns out, major parts of 802.11n were complete enough for vendors to build products to from 802.11n Draft 2.0. The features that were complete enough, and were also interesting enough to encourage users to purchase products based on them, were written down by the Wi-Fi Alliance into an industry certification program, also known as 802.11n Draft 2.0. This program specifies a rigorous set of interoperability tests, to ensure that vendors that pass it have built their devices to the same specification (as in, they didn't make major errors).

The existence of the certification program should bring you comfort in knowing that 802.11n devices will work together. For 802.11n Draft 2.0 products, the Wi-Fi Alliance, which uses its role to ensure that devices interoperate, happened to do their work a bit earlier than IEEE. As it has turned out, however, the Draft 2.0 feature set and program are essentially the same as those for the final standard. This makes sense, because the vendors that figure prominently in IEEE and the Wi-Fi Alliance had a tremendous incentive to ensure that the final standard was only minimally changed from the draft.

What features do 802.11n devices commonly provide? Table 5.13 lists them.

The main feature is the ability to provide 300Mbps data rates for clients. This is achieved by using two spatial streams and double-wide 40MHz channels. (The standard defines up to four spatial streams, and some—but not most—devices can accommodate three streams, yet the overwhelming majority to date can use only two.) Furthermore, the WMM and aggregation optimizations go a long way towards closing the gap between the data rate and the actual highest application throughput.

Ignoring the MIMO aspect, 802.11n is similar to 802.11ag. It too is OFDM, with exactly the same subcarrier setup as 802.11ag, except for increasing the number of data subcarriers by four to 52 (and thus slightly increasing the frequency width). Additionally, the symbol is usually still 4 microseconds; however, there is an optional mode known as *short guard interval* that shaves 400 nanoseconds off the symbol's length, getting a slightly higher kick in throughput. There are still eight modulation and coding rates that establish the data rates,

Table 5.13: 802.11n Common features

802.11n Features	Meaning
Two spatial streams	Doubles performance over non-MIMO
Aggregation	Allows for very high efficiency
WMM	Quality-of-service is mandatory (see Section 6.0.1)
WPA2	High-grade encryption support is mandatory (see Section 5.6.1.1)
40MHz wide channels	Doubles performance over 20MHz
300Mbps	Top data rate
MRC/Receive Beamforming	Longer range in some cases

except that one BPSK mode is removed and another 64-QAM mode, with an even higher coding rate, is added. These eight rates, however, are multiplied into a much higher number of rates, based on the channel width, the number of spatial streams, and the guard interval. All together, the notion of data rates being signaled in the products by Mbps has been abandoned, and replacing it is the concept of the *modulation and coding scheme* (MCS), based on a small number representing what the actual parameters are in a table. This is similar to how simple channel numbers represent much more complicated frequencies.

Table 5.14 contains the common two-stream set for 802.11n devices and encompasses 60 different data rate options, each with its own slightly different SNR requirement, channel width, or robustness.

There are additional features that 802.11n has as options, which are not commonly implemented but could be of great benefit for voice mobility. Among these features are *space-time block codes* (STBC), transmit beamforming, and extended power save capabilities. However, because these features have not yet become commonplace, their use is rather limited.

5.5.3.1 MIMO

As mentioned before, MIMO lets the devices transmit at, three times, or four times by using as many spatial streams. The general rule for MIMO, as a theory, is that the number of spatial streams usable is the lesser of the number of antennas that can be used simultaneously on the transmitter and the receiver. In theory, a 100-antenna transmitter and a 50-antenna receiver could allow for a 50 (< 100) spatial stream radio.

In practice, there are limits. 802.11n defines only four spatial streams maximum. The Wi-Fi Alliance Draft 2.0 certification tested for only two spatial streams, and most devices today remain only capable of two streams. The reasons are rather simple. More antennas require more room for antennas, and it is hard to find room to place them. Also, more antennas

Table 5.14: The 802.11n data rates

MCS	Modulation	Bits per Symbol per Subcarrier	Coding Rate	Spatial Streams	Speed			
					20MHz Long GI	20MHz Short GI	40MHz Long GI	40MHz Short GI
0	BPSK	1	½	1	6.5	7.2	13.5	15
1	QPSK	2	½		13	14.4	27	30
2	QPSK	2	¾		19.5	21.7	40.5	45
3	16-QAM	4	½		26	28.9	54	60
4	16-QAM	4	¾		39	43.3	81	90
5	64-QAM	6	⅔		52	57.8	108	120
6	64-QAM	6	¾		58.5	65	121.6	135
7	64-QAM	6	⅚		65	72.2	135	150
8	BPSK	1	½	2	13	14.4	27	30
9	QPSK	2	½		26	28.9	54	60
10	QPSK	2	¾		39	43.3	81	90
11	16-QAM	4	½		52	57.8	108	120
12	16-QAM	4	¾		78	86.7	162	180
13	64-QAM	6	⅔		104	115.6	216	240
14	64-QAM	6	¾		117	130	243	270
15	64-QAM	6	⅚		130	144.4	270	300

require more radio chains—an 802.11n radio is actually made of multiple copies of the parts needed to make an 802.11ag radio work—and those are expensive and draw power. Finally, practical considerations prevent higher numbers of simultaneous streams from working well.

Now, for the description of why MIMO works. For this example, assume that the sender and receiver each have three antennas, but follow the industry norm of using only two spatial streams. The sender divides its signal into two spatial streams, then spreads its two streams across the three antennas. Those two streams from three signals bounce around the environment, and end up as three different signals at the receiver, one for each of the receiver's antennas. Each of those three signals is some different combination of the three signals from the sender, and thus is some different combination of the two spatial streams. This is where the math sets in. The different combinations are usually very different. If you write out the two spatial streams, the effects of the sender's spreading and the channel's bouncing, and the receiver's receiving, you can produce a matrix equation, from linear algebra. Because the combinations are different for each antenna—linearly independent, in fact—the receiver can undo the effects of the channel using linear algebra and produce the original streams. For further explanation, see the appendix at the end of this chapter.

Basically, the MIMO receiver uses the preamble of the frame that is sent to discover what the effects of the channel are on the streams, and then uses that to undo those effects.

The effect of having multiple antennas when only one spatial stream is used (and a main effect as a part of MIMO) is for *beamforming*. There are two parts to beamforming: receive beamforming, and transmit beamforming. The term beamforming arose from RADAR, where stationary equipment used electronics and a large number of antennas to set up interference patterns just right to concentrate a signal in a direction or to a point, as if the antennas were mounted on a swivel and were pointed, although they are not.

Receive beamforming isn't necessarily thought of the same way as transmit beamforming, but is the major reason why 802.11n has higher range than 802.11abg. There are a number of techniques for doing what could be called receive beamforming. One term used surprisingly often by vendors in describing their products in data sheets is *maximum ratio combining* (MRC). To understand it at a high level, the receiver is twiddling with how it combines the signals received on each of its antennas to maximize the power of the final signal it received. Because of the way interference patterns and combinations work, it turns out that the twiddling it does is a unique pattern (**H**, the channel matrix, if you read the math briefing) based on the client's location. But because receive beamforming learns that pattern when it sees the preamble (same as MIMO), this is not a problem, and the end result is an apparent amplification of the signal, thus increasing range for reception. 802.11n clients with MIMO have longer range on legacy access points than legacy clients do, for that reason. Therefore, if you need to extend the range of a couple of clients, your best bet is to upgrade the clients to 802.11n. (Upgrading the access points without upgrading the clients may not increase range at all in many cases.)

Transmit beamforming is also possible. 802.11n defines two types of beamforming, known as *explicit* and *implicit beamforming*. Explicit beamforming uses the cooperation of the receiver to determine the best way of combining signals across the transmitter's antennas for forming the signal to the receiver. This cooperation requires features on the receivers that are not commonly implemented. Implicit beamforming simply requires the transmitter, assuming that the channel it sees from the receiver when that device transmits is the same as what the receiver sees from the transmitter. By assuming this reciprocity, the transmitter does not need to involve the receiver. However, it is forced into its guess, which may not be correct, and requires that the transmitter always keep tabs on the receiver's channel conditions by either sending an RTS to it before every packet, thus eliciting a responding

CTS that will help uncover the channel conditions, or by winging it and hoping the receiver doesn't move much. For this reason, some vendors are limiting their transmit beamforming support to only legacy, non-802.11n clients.

Transmit beamforming is an interesting concept for voice mobility networks based on the microcell approach of reducing transmit power levels to begin with.

5.5.3.2 Legacy Support

Legacy, in the context of 802.11n, means 802.11ag as well as 802.11b. To avoid the same problem of 802.11g possibly being interfered with by 802.11b here, with 802.11ag stepping on 802.11n, 802.11n uses a different form of protection. Instead of the wasteful CTS frames, and the protocol necessary to decide whether a CTS frame is needed, 802.11n uses a special preamble. The preamble first starts off as an 802.11ag preamble, not only at 6Mbps but signaling a 6Mbps data rate for the following data, and including a length that will encompass the entire 802.11n frame. However, as soon as the preamble is over, the 802.11n radio stops transmitting in 802.11ag, and switches over to 802.11n, where it continues with more preamble fields, including the real data rate. 802.11n devices see the entire frame. 802.11ag devices see the preamble only, but defer just as if the frame were all 802.11ag but out of range. Thus, the technologies do not interfere.

For 802.11b clients, 802.11n still uses an 802.11b CTS frame.

5.5.3.3 Aggregation

802.11 frames have to have the preamble, but there was no particular reason that one preamble couldn't cover multiple frames. 802.11n introduces this concept with frame aggregation, or A-MPDUs.

A-MPDUs are a special type of jumbo frame that contains multiple 802.11 data frames sent from and to the same wireless device. Up to four milliseconds or roughly 64,000 bytes can be packed into one of these aggregates. In almost every sense, an A-MPDU can just be thought of as a concatenation of data: every byte is sent at the same data rate that the preamble calls out, and the A-MPDU is retransmitted in full if the expected acknowledgment does not come back immediately from the receiver. However, unlike a simply larger data frame, if some of the data frames within the aggregate are received and others are not, the receiver can indicate this by using a special Block Acknowledgment. This block acknowledgment specifies precisely which of the senders' frames were received and which had errors in them. For those frames that were not received, the sender can choose to add those to later aggregates, thus not wasting time resending frames that already arrived.

The main benefit of aggregation is to reduce the overhead of the preamble and backoff for 802.11n.

5.5.3.4 Double-Wide Channels (40MHz)

Double-wide channels work by bonding together two adjacent 20MHz channels into one larger 40MHz channel. This 40MHz channel acts just like a wider 20MHz channel, but offering data rates slightly higher than twice that of the 20MHz channel. (The slight increase over twice comes from using up the gap between channels as usable bandwidth.)

There is some inconsistency in the naming of this feature in the industry. Some devices call it "double-wide," others call it "channel bonding," and others call it just "40MHz." We'll use "double-wide" and "40MHz" for this description.

Double-wide channels are named by the 20MHz channels that they occupy. There are a few nomenclatures in active use, but all are just slight variations of the basic concept. There is a *primary channel* and an *extended channel* in a 40MHz channel name. 40MHz channels can operate with 20MHz devices, but only on one 20MHz half of the double-wide channel. This one half is the primary channel. The other half is used only for the rest of the 40MHz transmissions, and is thus the extended channel. For example, a 40MHz channel selection that has a primary channel of 36 and an extended channel of 40 can be written as 36+1, (36, +1), 36U, 36H, and so on. The same 40MHz channel, but with the primary being on the other half, can be written as 40−1, (40, −1), or 40L.

It is useful to keep in mind, however, that a 40MHz frame only sees one 40MHz channel, and is not split itself into two separate 20MHz frames for each half, unlike what the terminology suggests.

There are a few considerations for double-wide channels. The first is that both 20MHz channels that are within the 40MHz should be empty. This is especially true for the extension channel, which should not have any other access points using that channel, unless a layered or virtualized over-the-air architecture is used that can support that. Certainly, no access point should be deployed such that its primary (or only) channel is 40 if 36+1 is being used anywhere around that access point.

The second consideration is that some devices cannot support 40MHz, and so will use only the primary half of the channel. This is obviously true for legacy (non-802.11n) devices, but is also true for some 802.11n devices. Some 802.11n devices are not designed to take advantage of double-wide channels, and a few early 802.11n devices had support for double-wide channels in the 2.4GHz band turned off, though newer devices from those manufacturers have solved that problem.

The third consideration is that there are only half as many 40MHz channels as there are 20MHz channels, and the direction a 40MHz channel may extend is limited. This is to avoid the overlap mentioned earlier, but ends up leading to channel waste unless channels are carefully planned. For example, channels 36 and 40 can be bonded, but channels 40 and 44 can never be, by rule. If channel 36 is being used by an existing 20MHz network, then channel 40 can never be used with a 40MHz channel in that case.

The 2.4GHz band does not have the limitation of which direction a 40MHz channel can extend, but because the bandwidth is so limited, only one 40MHz channel can be created successfully. The network or administrator can usually choose which. Keep in mind that dynamic architecture vendors recommend against using double-wide channels in the 2.4GHz band, because using it on that architecture eliminates a channel that is necessary for the alternating channel plan to work with. Layered and virtualized architectures do not have this limitation.

5.5.3.5 Coming Down the Road

802.11 is still growing. As of this writing, there is a push to expand the technology to increasing bands—such as the spectrum opened up by the end of analog television broadcasts in 2009. These technologies will not show up for a few years, however, and should not affect voice mobility networks that are being considered for deployment in the near future.

5.6 Security for 802.11

Security is a broad subject, and there is an entire chapter dedicated to the unique challenges with security for voice mobility later. But any component of voice mobility over Wi-Fi will require some use of 802.11's built-in encryption. Keep in mind that securing the wireless link is not only critical, but may be the only encryption used to prevent eavesdroppers from listening in on sensitive voice calls for many networks.

802.11 security has both a rich and somewhat checkered past. Because of the initial application of 802.11 to the home, and some critical mistakes by some of the original designers, 802.11 started out with inadequate protection for traffic. But thankfully, all Wi-Fi-certified devices today are required to support strong security mechanisms.

Nevertheless, administrators today do still need to keep in mind some of the older, less secure technologies—often because the mobile handset might not correctly support the latest security, and it may fall to you to figure out how to make an old handset work without compromising the security of the rest of the network.

A secure wireless network provides at least the following (borrowed from Chapter 8):

- *Confidentiality*: No wireless device other than the intended recipient can decrypt the message.

- *Outsider Rejection*: No wireless device other than a trusted sender can send a message correctly encrypted.

- *Authenticity and Forgery Protection*: The recipient can prove who the original composer of the message is.

- *Integrity*: The message cannot be modified by a third party without the message being detected as having been tampered with.

- *Replay Protection*: A older but valid message cannot be resent by an attacker later, thus preventing attackers from replaying old transactions.

Some of these properties are contained in how the encryption keys get established or sent from device to device, and the rest are contained in how the actual encryption or decryption operates.

5.6.1 Wi-Fi Security Technologies

Let's start with the earliest security for Wi-Fi. It is useful to read about WEP, the first and oldest method for Wi-Fi security, to understand the basic terms and concepts for wireless security in general.

5.6.1.1 WEP

WEP (Wired Equivalent Privacy) was the first attempt to secure 802.11. Unfortunately, the privacy it provided was neither equivalent to wired nor very good. Its very design does not protect against replays, meaning that an attacker can record prior valid traffic and replay it later, getting the network to repeat actions (such as charging credit cards) without detecting it. Furthermore, WEP uses for encryption RC4, an algorithm that was not designed to be used in the way WEP uses it, leading to ways of reverse-engineering and cracking the encryption without the key. Finally, WEP uses a very poor message integrity code.

All of that said, WEP is a good place to look to learn the mechanics of security in 802.11, as the later and better security additions replaced the broken pieces but did not destroy the framework.

> NOTE: It is the author's recommendation to not use WEP in existing or new networks, under any circumstances, because of the known flaws. Please consider the studying of WEP to be an academic exercise at this point, and do not allow vendors to talk you into using it.

5.6.1.1.1 Keying

WEP starts off with an encryption key, or a piece of knowledge that is known by the access point and the client but is sufficiently complicated that outsiders—attackers, that is—shouldn't be able to guess it.

There is one and may be two or more WEP keys. These keys are each either 40 bits (WEP-40) or 104-bits (WEP-104) long, and are usually created usually from text passwords, although they can be entered directly as hexadecimal numbers. Manually entered keys are called *pre-shared* keys (PSK). WEP provides very little signaling to designate that encryption is in use, and there is no way to denote whether the short or long keys are being used. If any security, at all, is used in the network, the "Privacy" flag in the network's beacons are set. Clients that want to use WEP had to associate to the network and start sending encrypted traffic. If the keys matched, the network made forward progress and the user was happy. If the keys did not match, the user would not be able to do much, but would otherwise not know what the error was. As you can see, this is not an ideal situation, and is avoided in the modern, post-WEP protocols.

There are some more complicated possibilities, which are not worth going over, except to note that the origin of the confusing 802.11 term "authentication" for the first phase of a client's connection to the network came from an old method of using WEP to verify the key before association. This security method is completely ignored by post-WEP protocols, which use a different concept to ensure that clients have the right key. Therefore, the two Authentication frames are now considered vestigial, and carry no particularly useful information in them.

5.6.1.1.2 Encryption

The encryption key is not used directly to encrypt each packet. Instead, it is concatenated with a per-packet number, called the *initialization vector* (IV), to create the key that RC4 uses to encrypt the data. The initialization vector can be any number. Transmitters would start at zero, and add one for each frame sent, until it hit the end of the three-byte sequence, where it would start over at zero again. Why have a per-packet key, when the original key was supposedly secret? To answer this, let's look at the encryption algorithm for WEP, which is based on RC4.

RC4 is a stream cipher, meaning that it is designed to protect a large quantity of flowing, uninterrupted data, with minimal overhead. It is used, for example, to protect secure web traffic (HTTPS), because web traffic goes across in a stream of HTML. RC4 is really a pseudorandom number generator, with cryptographic properties to ensure that the stream of bits that comes out is hard to reverse-engineer. When given a key, RC4 generates an infinite number of bits, all appearing to be random. These bits are then matched up, bit-by-bit, to the incoming *plaintext*, or not yet encrypted, data. Each bit of the plaintext is added to each matching bit of the RC4 stream, without carry. This is also known as taking the *exclusive or* of two bits, and the logic goes that the resulting "sum" bit is 1 if either of the incoming bits are 0, and 0 otherwise. The mathematical operation is represented by the $\oplus$ symbol, and so the four possibilities for the exclusive or are as follows: $0 \oplus 0 = 0$, $0 \oplus 1 = 1$, $1 \oplus 0 = 1$, and $1 \oplus 1 = 0$. When applied to the plaintext and RC4 together, the resulting stream looks as random as the original RC4 stream, but has the real data in it. Only a receiver with the right key can recreate the RC4 stream, do the same bitwise exclusive or to the encrypted data, and recover the original data. (The exclusive or operation has the property that any number that has any other number added twice provides the same number back: $n \oplus d \oplus d = n$. Therefore, applying the exclusive or of the RC4 stream twice to the original data, once by the encryption algorithm and once by the decryption algorithm, gets the plaintext data back.)

So far, so good. However, an attacker can use the properties of the exclusive or to recover the plaintext in certain cases, as well. If two frames come using the same per-frame key— meaning the same IV and WEP key—an eavesdropper can just add the two encrypted frames together. Both frames have the same per-frame key, so they both have the same RC4 stream, causing the exclusive or of the two encrypted frames to cancel out the identical RC4 stream and leave just the exclusive or of the two original, plaintext frames. The exclusive or of two plaintext frames isn't terribly different from having the original plaintext: the attacker can usually guess at the contents of one of the frames and make quick work discovering the contents of the other.

This isn't a flaw with RC4 itself, as it is with using any exclusive or cipher—a type of *linear cipher*, because $\oplus$ is really addition modulo 2—as they are vulnerable to bit-by-bit attacks unless other algorithms are brought in as well.

Okay, so that explains the per-frame keying and the IV, and why it is not a good solution for security. In summary, replays are allowed, and the IV wraps and key reuse reveals the original plaintext. Finally, the per-frame key doesn't include any information about the sender or receiver. Thus, an attacker can take the encrypted content from one device and inject it as if it were from another. With that, three of the problems of WEP are exposed. But the per-frame keying concept in general is sound.

5.6.1.1.3 *Integrity*

To attempt to provide integrity, WEP also introduces the *integrity check value* (ICV). This is a checksum of the decrypted data—CRC-32, specifically—that is appended to the end of the data and encrypted with it. The idea is that an attacker might want to capture an encrypted frame, make possibly trivial modifications to it (flipping bits or setting specific bits to 0 or 1), and then send it on. Why would the attacker want to do this? Most active attacks, or those that involve an attacker sending its own frames, require some sort of iterative process. The attacker takes a legitimate frame that someone else sends, makes a slight modification, and sees if that too produces a valid frame. It discovers if the frame was valid by looking for some sort of feedback—an encrypted frame in the other direction—from the receiver. As mentioned earlier, RC4 is especially vulnerable to bit flipping, because a flipped bit in the encrypted data results in the flipping of the same bit in the decrypted data. The ICV is charged with detecting when the encrypted data has been modified, because the checksum should hopefully be different for a modified frame, and the frame could be dropped for not matching its ICV.

As mentioned before, however, WEP did not get this right, either. CRC-32 is not cryptographically secure. The effect of a bit flip on the data for a CRC is known. An attacker can flip the appropriate bits in the encrypted data, and know which bits also need to be flipped in the CRC-32 ICV to arrive at another, valid CRC-32, without knowing what the original CRC-32 was. Therefore, attackers can make modifications pretty much at will and get away with it, without needing the key. But again, the concept of a per-frame *message integrity code* in general is sound.

5.6.1.1.4 Overall

WEP alters the data packet, then, by appending the ICV, then encrypting the data field, then prepending the unencrypted IV. Thus, the frame body is replaced with what is in Table 5.15.

Table 5.15: 8.02.11 Frame Body with WeP (Compare with Table 5.2)

IV	Key ID	Data	ICV
3 bytes	1 byte	$n-8$ bytes	4 bytes

The issues described are not unique to RC4, and really applies to how WEP would use any linear cipher. There are also some problems with RC4 itself that come out with the way RC4 is used in WEP, which do not come out in RC4's other applications. All in all, WEP used some of the right concepts, but a perfect storm of execution errors undermined WEP's effectiveness. Researchers and attackers started publishing what became an avalanche of writings on the vulnerability of WEP. Wi-Fi was at risk of becoming known as hopelessly broken, and drastic action was needed. Thus, the industry came together and designed 802.11i.

5.6.1.2 RSNA with 802.11i

802.11i addresses the major problems with WEP. The first problem, the inability to establish per-connection keys, and the inability to use different encryption algorithms, was fixed by a better protocol.

On top of that, 802.11i introduced two new encryption and integrity algorithms. *Wi-Fi Protected Access* (WPA), version one, was created to quickly work around the problems of WEP without requiring significant changes to the hardware that devices were built out of. WPA introduced the *Temporal Key Integrity Protocol* (TKIP), which sits on top of WEP and fixes many of the problems of WEP without requiring new hardware. TKIP was designed intentionally as a transition, or stopgap, protocol, with the hopes that devices would be quickly retired and replaced with those that supported the permanent solution, the second of the two algorithms.

Wi-Fi Protected Access version 2 (WPA2), as that permanent solution, required completely new hardware by not worrying about backwards compatibility. WPA2 uses AES to provide better security and eliminate the problems of using a linear stream cipher. A better integrity algorithm ensures that the packet has not been altered, and eliminates some of the denial-of-service weaknesses that needed to be introduced into TKIP to let it ward off some of the attacks that can't be directly stopped.

A word, first, on nomenclature. For those of you in the know, you might know that WPA has both TKIP and AES modes, 802.11i has slightly different TKIP and AES modes, and that both were harmonized in WPA2. However, practically, there really is no need to know that. For the remainder of this chapter, I will use WPA to mean TKIP as defined in WPA, WPA2 to mean AES as defined in the standard, and 802.11i to mean the framework under which WPA and WPA2 operate. This is actually industry convention—WPA and TKIP go hand in hand, and WPA2 and AES go hand in hand—so product documentation will most likely match with this use of the terms, but when there is doubt, ask your vendors whether they mean TKIP or AES.

802.11i first introduced the idea of a per-connection key negotiation. Each client that comes into the network must first associate. For WEP, which has no per-connection key, the client always used the user-entered WEP key, which is the same for every connection. But 802.11i introduces an additional step, to allow for a fresh set of per-connection keys every time, yet still based on the same master key.

Networks may still used preshared keys. These are now bumped up to be 128 bits long. For WPA or WPA2, this mode of security is known as *Personal*, because the preshared key method was intended for home use. Enterprises can also use 802.1X and a RADIUS server to negotiate a unique key per device. This mode of security is known as *Enterprise*. For example, "WPA2 Enterprise" refers to using WPA2 with 802.1X. Either way, the overall key is called the *pairwise master key* (PMK). This is the analog to the original WEP key.

Now, when the client associates, it has to run a four-message protocol, known as the *four-way handshake*, to determine what should be used as the key for the connection, known as the PTK (the *pairwise temporal key* or *pairwise transient key*). This whole concept of derived keys is known as a *key hierarchy*.

The four way handshake is made of unencrypted data frames, with Ethernet type of EAPOL (0x888E), and show up as the specific type of Extensible Authentication Protocol over LAN (EAPOL) message known as an EAPOL Key message. These four messages can be seen by wireless capture programs, and mark the opening of the data link between the client and the access point. Before the four-way handshake, clients and access points cannot exchange any data besides EAPOL frames. After the handshake, both sides can use the agreed-upon key to send data.

Message 1 of the four-way handshake is sent by the access point to the client, and signals the security settings of the access point (as contained in something called the *RSN IE*, shown in Table 5.16). The RSN IE contains the selection of encryption and integrity algorithms. The message also contains something called a *nonce*, which is a random number that the access point constructs (more on this shortly) and which will be mixed in with the PMK to produce the PTK.

Table 5.16: The security settings in the RSN IE

Element ID	Length	Version	Group Cipher Suite	Pairwise Cipher Suite Count	Pairwise Cipher Suite List	AKM Suite Count	AKM Suite List	RSN Capabilities	PMKID Count	PMKID List
1 bytes	1 byte	2 bytes	4 bytes	2 bytes	n bytes	2 bytes	m bytes	2 bytes	2 bytes	p bytes

Message 2 is sent in response, from the client to the access point, and includes the same information, but from the client: a client RSN IE, and a client nonce. Once the client has chosen its nonce, it has enough information to produce the PTK on its end. The PTK is derived from the two nonces, the addresses of the access point and client, and the PMK. At this point, it might seem like the protocol is done: the client knows enough to construct a PTK before sending Message 2, and the access point, once it gets the message, can use the same information to construct its own PTK. If the two devices share the same PMK—the master key—then they will pick the same PTK, and packets will flow. This is true, but the protocol needs to do a little bit more work to handle the case where the PMKs do not agree. To do this, the client "signs" Message 2 with a *message integrity code* (MIC). The MIC used is a cryptographic hash based on both the contents of the message and the key (PTK). Thus, the access point, once it derives its own PTK from its PMK and the nonces, can check to see whether the client's sent MIC matches what it would generate using its own PTK. If they match, then the access point knows that message 2 is not a forgery *and* the client has the right key. If they do not match, then the access point drops the message.

If Message 2 is correct, then Message 3 is sent by the access point, and is similar to Message 1 except that it too is now "signed" by the MIC. This lets the client know that the access point has the right key: at Message 2, only the access point could detect an attacker, but not the client. Also, the client can now verify that the access point is using the same security algorithms as the client—a mismatch would only occur if an attacker is injecting false RSN IEs into the network to try to get one side or both to negotiate to a weaker algorithm (say, TKIP) if a stronger algorithm (say, AES) is available. Finally, for WPA2, the client learns of the multicast key, the *group temporal key* (GTK), this way, as it is encrypted with the PTK and sent as the last part of the message.

Message 4 is a response from the client to the access point, and validates that the client got Message 3 and installed all of the correct keys.

The nonces exist to prove to each side that the other side is not replaying these messages—that is, that the other side is alive and is not an attacker. Imagine that the access point sends its nonce. An attacker trying to replay a previous, valid handshake for the same client could send an old Message 2, but the MIC on that Message 2 can never be correct, because it would always be based on the access point nonce recorded previously and was used in that previous handshake, and not the new one that the access point just created. Thus, the access point always can tell the difference between a client that is really there, and one that is just replayed from the past. The client can use its nonce to do the same thing. Also, if either side has the wrong PMK—which would happen with preshared keys if someone typed one of the keys wrong—the devices can catch it in the four-way handshake and not pretend to have a working connection.

Overall, the four-way handshake lets the two sides come together on a fresh connection key every time. The four way handshake is the same, except for some minor details such as choice of algorithm, for WPA and WPA2.

By the way, keep in mind that the four-way handshake is only designed to provide a new PTK every time based on the same PMK, to provide a fresh PTK and eliminate the problem of old or stale keys that WEP has. The four-way handshake is *not* designed to hide the PTK from attackers who have the PMK. This is an important point: if an attacker happens to know the PMK already—such as a preshared key that he or she stole or remembered—then every PTK ever generated from that PMK, in the past and in the future, can be broken with minimal effort. This is known as a lack of *forward secrecy* and is a major security flaw in preshared key networks.

In other words, you must keep the PMK secret. Do not share preshared keys, ever—even if you have stopped using that preshared key and moved to a new one long ago. If an attacker had been recording your past conversations, when the old preshared key was in use, and someone leaks the preshared key to this attacker, your old conversations are in jeopardy. This has serious consequences for voice mobility networks, and will be reiterated in Chapter 8 on security for voice mobility.

5.6.1.3 WPA and TKIP

TKIP was designed to run on WEP hardware without slowing the hardware down significantly. To do this, TKIP is a preprocessing step before WEP encryption. RC4 is still the encryption algorithm, and the WEP CRC-32 could not be eliminated. However, TKIP adds features into the selection of the per-frame key, and introduces a new MIC to sit beside the CRC-32 and provide better integrity.

Table 5.17: 8.02.11 Fram Body with WPA (Compare with Table 5.15)

Expanded IV	Data	MIC	ICV
8 bytes	*n* - 8 bytes	8 bytes	4 bytes

The first change is to expand the IV and key ID fields to eight bytes total (see Table 5.17). The expanded fields gives a six-byte IV, now called the *TKIP sequence counter* (TSC). The goal is to give plenty of room so that the TSC nearly never needs to wrap. Furthermore, if it does get close to wrapping, the client is required to renegotiate a new PTK. This prevents key reuse. Finally, the TSC is used to provide the replay protection missing in WEP. The TSC is required to go up by one for each message. Each side keeps the current TSC that it is sending with, and the one it last received successfully from the other side. If a frame comes in out of order—that is, if it is received with an old TSC—the receiver drops it. An attacker can no longer replay valid but old frames. And, of course, although it can try to invent new frames, even with higher TSCs, the receiver won't update the last good TSC unless the frame is decryptable, and it will not be because the attacker does not know the key.

The second change is to come up with a better way of producing the per-frame key. The per-frame key for TKIP uses a new algorithm that takes into account not only the now larger IV and the PTK, but the transmitter's address as well. This algorithm uses a cryptographic device known as an *S-box* to spread out the per-frame key in a more even, random-looking pattern. This helps avoid the problems with weak RC4 per-frame keys, which were specific WEP per-frame keys that caused RC4 to leak information. The result of this algorithm is a brand new per-frame key for each frame, which avoids many of the problems with WEP.

Unfortunately, the underlying encryption is still WEP, using a linear cipher vulnerable to bit flipping. To catch more of the bit flips, a new, cryptographically "better" MIC was needed. WPA uses Michael, a special MIC designed to help with TKIP without requiring excessive computation. It is not considered to be cryptographically secure in the same sense as is WPA2, but is considered to be significantly better than CRC-32, and thus can be used to build secure networks with some caveats. In this case, the designers were aware of this limitation up front, and designed Michael to be good enough to provide that transition from WEP to something more secure down the road (which became AES).

The Michael MIC is not just a function of the data of the packet. It also depends on the sender's address, the receiver's address, and the priority of the packet, as well as the PTK. Michael is designed to avoid the iterative guessing and bit flipping that WEP is vulnerable to. Furthermore, it is based on the entire frame, and not just individual fragments, and so avoids some fragmentation attacks that can be used against WEP. The result of Michael is the eight-byte MIC, which is placed at the end of the frame before it is sent for WEP encryption.

Because the designers know that Michael isn't enough, they also built in a provision for detecting when an attack is under way. Attackers try to modify frames and submit them, and see if the modified frames get mistaken as being authentic. Most of the time, they fail, and these modified frames are not decryptable. With WEP, a nondecryptable frame is silently dropped, with no harm. However, a frame with a bad MIC should never happen in a properly functioning system, and is a sign that the network is under attack. To help prevent these attacks from being successful, WPA adds the concept of *countermeasures*. If two frames with bad MICs (but good FCSs, so that we know they are not corrupted by radio effects) are received in a 60-second interval, the access point kicks all of the clients off and requires them to renegotiate new keys. This drastic step introduces a painful denial-of-service vulnerability into TKIP, but is necessary to prevent attackers from getting information easily. Of course, having countermeasures doesn't increase the robustness of the underlying algorithms, but kicking off all of the clients ensures that the attacker has to start from scratch with a new PTK.

Overall, TKIP was an acceptable bridge from WEP to WPA2. The designers rightfully recognize that TKIP is itself flawed, and is subject to a few vulnerabilities of its own. Besides the obvious denial-of-service attacks, TKIP also still allows for attacks that attempt to guess at certain parts of the particular messages and make some minor, but arbitrary, alterations to the packets successfully. Although workarounds exist for these types of attacks, TKIP will never be entirely hassle-free.

Therefore, I recommend that you migrate to WPA2 for every device on the network.

5.6.1.4 WPA2 and AES

WPA2 introduces a new encryption algorithm, using the Advanced Encryption Standard (AES). This cipher was produced to be used as a standard algorithm wherever encryption is needed.

AES is a *block cipher*, unlike RC4. A block cipher takes blocks of messages—fixed chunks of bytes—and encrypts each block, producing a new block of the same size. These are nonlinear ciphers, and so the bit-flip attacks are significantly harder. AES was specifically designed and is believed to be practically impervious to those styles of attacks. With block ciphers, each block starts off independently, a bit of a downside compared to stream ciphers. To remove that independence, WPA2 also uses what is called *Counter* mode, a simple concept where later blocks are made to depend on previous blocks.

The MIC used is also based on AES, but is used as a cryptographic hash. This use is called *cipher block chaining* (CBC), and essentially uses the same concept of making later blocks depend on earlier ones, but only outputting the last block as the result. This small block (128 bits) is dependent on every bit of the input, and so works as a signature, or hash.

The overall algorithm used is known as *Counter Mode with Cipher Block Chaining-Message Authentication Code* (CCMP).

Table 5.18: 8.02.11 Fram Body with WPA2 (Compare with Table 5.15)

PN = IV	Data	MIC
8 bytes	n–8 bytes	8 bytes

Table 5.18 shows the frame body used with WPA2. As with WPA, WPA2 has essentially the same expanded IV. Because WPA2 isn't using TKIP, the name has been changed to the *packet number* (PN), but serves the same purpose, starting at 0 and counting up. The PN is used for replay detection, as well as ensuring per-frame keying. The MIC is also eight bytes, but uses CBC-MAC rather than Michael. With new hardware, the last vestige of WEP can be dropped, and the old ICV is removed.

Because the WPA2 MIC is considered to be cryptographically strong, the designers of WPA2 eliminated the countermeasures that WPA has. It is still true that no frame should come in with an invalid MIC; however, the administrator can be alerted to deal with it in his own time, as there are not any known exploits that can be successfully mounted against WPA2 using an invalid MIC to date.

5.6.1.5 Wi-Fi Link Security: Summary

To summarize, 802.11 security is provided by three different grades of technology: the outdated and broken WEP, the transition-providing WPA, and the secure and modern WPA2.

WPA and WPA2 are both built on the same framework of 802.11i, which provides a rich protocol for 802.11 clients and access points to communicate and negotiate which over-the-air encryption and integrity algorithms should be used.

Networks start off with a master key—either a preshared key, entered as text by the user into the access point and mobile device, or generated in real time by enterprise-grade authentication systems. This master key is then used to derive a per-connection key, called the PTK. The PTK is then used to encrypt and provide integrity protection for each frame, using either TKIP for WPA or AES for WPA2.

It bears repeating that preshared keys, for all grades of 802.11 security, have problems that cause both security and management headaches. The biggest security headache is that the privacy of the entire network is based on that PSK being kept private for eternity. If a PSK is ever found out by an attacker—even if that key has been retired or changed a long time ago—then the attacker can use that key to decrypt any recordings of traffic that were taken when the PSK had been in use. Furthermore, because preshared keys are text and are common for all devices, they are easy to share and impossible to revoke. Good users can be fooled into giving the PSK away, or bad users—such as employees who have left the organization—can continue to use the preshared keys as often as they desire.

These problems are solved, however, by moving away from preshared keys to using 802.1X and EAP. Recently, some vendors have been introducing the ability to create per-user preshared keys. The advantage of having per-user keys is that one user's access can be revoked without allowing that user to compromise the rest of the network. The problem with this scheme, however, is the continued lack of forward secrecy, meaning that a user who has his password stolen can still have decrypted every packet ever sent or will send using that key. For this reason, 802.1X is still recommended, using strong EAP methods that provide forward secrecy.

5.6.2 802.1X, EAP, and Centralized Authentication

Up to now, we've discussed Wi-Fi's self contained security mechanisms. With WPA2, the encryption and integrity protection of the data messages can be considered strong. But we've only seen preshared keys, or global passwords, as the method the network authenticates the user, and preshared keys are not strong enough for many needs.

The solution is to rely on the infrastructure provided by centralized authentication using a dedicated *Authentication, Authorization, and Accounting* (AAA) server. These servers maintain a list of users, and for each user, the server holds the *authentication credentials* required by the user to access the network. When the user does attempt to access the network, the user is required to exercise a series of steps from the authentication protocol demanded by the AAA server. The server drives its end of the protocol, challenging the user, by way of a piece of software called a *supplicant* that exists on the user's device, to prove that the user has the necessary credentials. The network exists as a pipe, relaying the protocol from the AAA server to the client. Once the user has either proven that she has the right credentials—she apparently is who she says she is—the AAA server will then tell the network that the user can come in.

The entire design of RADIUS was originally centered around providing password prompts for dial-up users on old modem banks. However, with the addition of the *Extensible Authentication Protocol* (EAP) framework on top of RADIUS, and built into every modern RADIUS server, more advanced and secure authentication protocols have been constructed. See Figure 5.22.

The concept behind EAP is to provide a generic framework where the RADIUS server and the client device can communicate to negotiate the security credentials that the network administrator requires, without having to concern or modify the underlying network access technology. To accomplish this last feat, the local access network must support 802.1X.

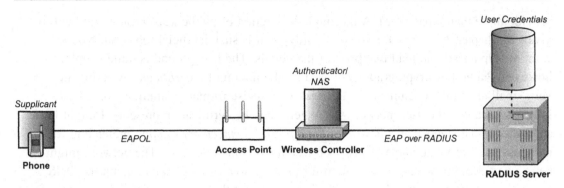

Figure 5.22: The Components of RADIUS Authentication Over Wi-Fi

5.6.2.1 What Is Authentication in 802.1X?

Let's first define exactly what authentication is, and what the technology expects out of the authentication process. We've mentioned credentials immediately preceding this section. An authentication credential is something that one party to communication has that the other parties can use to verify whether the user is really who he claims he is and is authorized to join the network.

In the preshared key case, the authentication credential is just the preshared key, a global password that every user shares. This is not very good, because every user appears identical, and there is no way for users to know that their networks are also authentic. Authentication should be a two-way street, and it is important for the clients to know that the network they are connecting to is not a fraud. With preshared keys, anyone with the key can set up a fraudulent *rogue* access point, install the key, and appear to be real to the users, just as they can arbitrarily decrypt over-the-air traffic.

Normal computer account security, such as what is provided by email servers, enterprise personal computers, and *Active Directory* (AD) networks, generally uses the notion that a user has a unique, secret password. When the user wants to access the network, or the machine, or the email account, she enters her password. If this password matches, then the user is allowed in. Otherwise, he or she is not.

(In fact, to prevent the system administrators from having access to the user's password, which the user might use in other systems and might not want to share, these systems will record a cryptographically hashed version of the password. This version, such as the MD5-hashed one mentioned in the next section, prevents anyone looking at it from knowing what the original password is, yet at the same time allows the user to type their password at any time, which leads to a new MD5-hashed string that will be identical to the one recorded by the system if and only if the passwords are identical.)

This identifies the user, but what about the network, which can't type a password to prove itself to the user? More advanced authentication methods use public key cryptography to provide more than a password. A thorough description of public key cryptography will be given in Chapter 8 security for voice mobility, as it is such a crucial topic, and you may want to skip ahead and read that part for the details. The background is quite simple, however. Public key cryptography is based on the notion of a *certificate*. A certificate is a very small electronic document, of an exact and precise format, containing some basic information about the user, network, or system that the certificate represents. I might have a certificate that states that it is written for *jepstein@somecompany.com*, pretending for a moment that that is the name of my user account at some company. The network might have a certificate that states it is written for *network.somecompany.com*, using the DNS name of the server running the network. To ensure that the contents of the certificate are not downright lies made up in the moment, each certificate is signed using another certificate, that of a *certificate authority* who both parties need to trust in advance. Finally, each certificate includes some cryptographic material: a *public key*, that is shouted out in the certificate, and a *private key*, which the owner of the certificate keeps hidden and tells no one. This private key is like a very big, randomly generated password. The difference is that the private key can be used to encrypt data that the public key can decrypt, and the public key can be used to encrypt data that the private key can decrypt. This allows the holder of the certificate to prove his or her identity by encrypting something using his or her private key. It also allows anyone else in the world to send the holder of the certificate a private message that only the holder can decrypt.

Certificates are necessary for network authentication. When the user tries to authenticate to the network, the network will prove its identity by using its private key and certificate, and the client will accept it only if the network gives the right information based on that certificate. Certificates are also useful for user authentication, because the same properties work in reverse. The EAP method known as EAP-TLS requires client certificates. Most of the other Wi-Fi-appropriate EAP methods use only server certificates, and require client passwords instead.

To recap, authentication over Wi-Fi means that the user enters a password or sends his certificate to the AAA server, which proves his identity, while the network sends its certificate to the client, whose supplicant automatically verifies the network's identity—just like how web browsers using HTTPS verify the server's identity.

It is the EAP method's job to specify whether passwords or certificates are required, how they are sent, and what other information may be required. The EAP method also is required to allow the AAA server and the client to securely agree to a master key—the PMK—which is used long after authentication to encrypt the user's data. The EAP method also must ensure that the authentication process is secure even though it is sent over an open, unencrypted network, as you will see in the following section on 802.1X.

The administrator is allowed to control quite a bit about what types of authentication methods are supported. The AAA administrator (not, you may note, the network administrator, unless this is the same person) determines the EAP methods, and thus the certificate and authentication requirements. The AAA administrator also chooses how long a user can keep network access until he or she has to reauthenticate using EAP. The network administrator controls the encryption algorithm—whether to use WPA or WPA2. Together, the two administrators can use extensions to RADIUS to also introduce network access policies based on the results of the AAA authentication.

5.6.2.2 802.1X

802.1X, also known as EAPOL, for EAP over LAN, is a basic protocol supported by enterprise-grade Wi-Fi networks, as well as modern wired Ethernet switches and other network technologies. The idea behind 802.1X is to allow the user's device to connect to the network as if the RADIUS server and advanced authentication systems did not exist, but to then block the network link for the device for all other protocols except 802.1X, until authentication is complete. The network's only requirements are twofold: prevent all data traffic from or to the client except for EAPOL (using Ethernet protocol 0x888E) from passing; and taking the EAPOL frames, removing the EAP messages embedded within, and tunneling those over the RADIUS protocol to the AAA server.

The job of the network, then, is rather simple. However, the sheer number of protocols can make the process seem complex. We'll go through the details slowly. The important thing to keep in mind is that 802.1X is purely a way of opening what acts like a direct link between the AAA server and the client device, to allow the user to be authenticated by whatever means the AAA server and client deem necessary. The protocols are all layered, allowing the highest-level security protocols to ride on increasingly more specific frames that each act as blank envelopes for its contents.

Once the AAA server and the client have successfully authenticated, the AAA server will use its RADIUS link to inform the network that the client can pass. The network will tear down its EAPOL-only firewall, allowing generic data traffic to pass. In the same message that the AAA server tells the network to allow the client (an EAP Success), it also passes the PMK—the master key that the client also has and will be used for encryption—to the network, which can then drop into the four-way handshake to derive the PTK and start the encrypted channel. This PMK exchange goes in an encrypted portion of the EAP response from the RADIUS server, and is removed when the EAP Success is forwarded over the air. The encryption is rather simple, and is based on the shared password that the RADIUS server and controller or access point have. Along with the PMK comes a session lifetime. The RADIUS server tells the controller or access point how long the authentication, and subsequent use of the keys derived from it, is valid. Once that time expires, both the access point and the client are required to erase any knowledge of the key, and the client must reauthenticate using EAP to get a new one and continue using the network.

For network administrators, it is important to keep in mind that the EAP traffic in EAPOL is *not* encrypted. Because the AAA server and the client have not agreed on the keys yet, all of the traffic between the client and the RADIUS server can be seen by passive observers. This necessarily limits the EAP methods—the specific types of authentication—that can be used. For example, in the early days of 802.1X, an EAP method known as EAP-MD5 was used, where the user typed a password (or the client used the user's computer account password), which was then hashed with the MD5 one-way cryptographic hash algorithm, and then sent across the network. Now, MD5 is flawed, but is still secure enough that an attacker would have a very hard time reverse-engineering the password from the hash of it. However, the attacker wouldn't need to do this, as he could just replay the same MD5 hashed version himself, as if he were the original user, and gain access to the network. For this reason, no modern wireless device supports EAP-MD5 for wireless authentication.

5.6.2.3 Key Caching

Because the work required establishing a PMK when 802.1X and RADIUS are used is significant, WPA2 provides for a way for the PMK to be cached for the client to use, if it should leave the access point and return before the PMK expires.

This is done using *key caching*. Key caching works because each PMK is given a label, called a *PMKID*, that represents the name of the RADIUS association and the PMK that was derived from it. The PMKID is specifically a 128-bit string, produced by the function

```
PMKID = HMAC-SHA1-128(PMK, "PMK Name" || AA || SPA),
```

where AA is the BSSID Ethernet address, SPA is the Ethernet address of the client, and HMAC-SHA1-128 is the first 128 bits of the well-known SHA1-based HMAC function for producing a cryptographic one-way signature with the PMK as the key. The double-pipes ("||") represent bitwise concatenation. The "PMK Name" ASCII string is used to prevent implementers from putting the wrong function results in the wrong places and having it work by accident.

From this, it is pretty clear to see that a client and access point can share the same PMKID only if they have the same PMK and are referring to each other.

When the client associates, it places into its Reassociation message's RSN information element (Table 5.16) the PMKID it may have remembered from a previous association to the access point. If the access point also remembers the previous association, and still has the PMK, then the access point will skip starting 802.1X and will proceed to sending the first message in the four-way handshake, basing it on the remembered PMK.

This caching behavior is not mandatory, in the sense that either side can forget about the PMK and the connection will still proceed. If the client does not request a PMKID, or the access point does not recognize or remember the PMKID, the access point will still send an

EAP Request Identity message, and the 802.1X protocol will continue as if no caching had taken place.

5.6.3 Putting It All Together: An Example

The client passes through a number of phases when associating to a Wi-Fi network that uses enterprise-grade security. To help understand how everything fits together, we will go through one example authentication, using WPA2 and the EAP method EAP-PEAP, which requires each mobile device to have a username and password. The password will be sent, securely tunneled through PEAP, to the RADIUS server, which is usually attached to a Microsoft Active Directory server.

Each message that is sent will be represented by a table, showing the relevant contents of the message. The aim is to allow the reader to follow along, when analyzing wireless packet capture traces, what the individual steps mean, when a client associates to the network. As a matter of presentation, when information that might be important is repeated in subsequent messages, it will be omitted for those messages.

Step 1: Associate with the Wi-Fi Network

The mobile device, having scanned for the SSID of the network desired—let's call it *voice* for this example—has found an access point that is advertising the *voice* SSID.

The client requests a connection with the access point by sending an 802.11 Authentication message, requesting *open authentication*, meaning that the client does not want to use WEP. See Table 5.19.

Table 5.19: 802.11 Authentication message from client to AP

Frame Control	Destination Address	Source Address	BSSID	Algorithm Number	Authentication Sequence
Authentication	AP Address	Client Address	AP Address	0 (Open System)	1

The access point accepts the open connection by responding with its own 802.11 Authentication message, to the client, simply stating that the request is a success. See Table 5.20.

Table 5.20: 802.11 Authentication message from AP to client

Frame Control	Destination Address	Source Address	BSSID	Algorithm Number	Authentication Sequence	Status Code
Authentication	Client Address	AP Address	Client Address	0 (Open System)	1	0 (Success)

The client then sends an 802.11 Association Request message to the access point, informing the access point of its Wi-Fi capabilities, supported extensions and 802.11 features (Table 5.21).

Table 5.21: 802.11 Association request message from client to AP

Frame Control	Destination Address	Source Address	BSSID	Capabilities	SSID	Information Elements
Association Request	AP Address	Client Address	AP Address	Capabilities	voice	Radio, Security, and QoS Capabilities

The access point accepts the association request and sends an 802.11 Association Response message to the client, announcing success, providing the client with the access point's capabilities and its network-wide configuration parameters.

At this point, the client cannot speak to any other access point without disconnecting or being disconnected, but it cannot send or receive any real data traffic. The client must first use EAPOL to authenticate.

Step 2: Authenticate with the AAA Server

The sends an EAPOL Start message (Table 5.22), encoded as a Wi-Fi Data frame with Ethernet protocol 0x888E, sent to the Ethernet address of the access point. This message is optional, but when sent is meant to request that the access point should start the EAP exchange.

Table 5.22: 802.11 EAPOL start message

Frame Control	Destination Address	Source Address	BSSID	EtherType	EAPOL Type
Data	AP Address	Client Address	AP Address	0x888E	Start

At around the same time, the access point will usually voluntarily send an EAPOL message with an EAP Request Identity message inside (Table 5.23), triggering the start of the authentication process. The Request Identity message is the EAP way of asking the client to announce who he or she is.

Table 5.23: 802.11 EAP request identity

Destination Address	Source Address	Ether-type	EAPOL Type	EAP Code	EAP Type	Identity
Client Address	AP Address	0x888E	0=EAP	1=Request	1=Identity	hello

The client receives the request for the identity and responds with identity to use (Table 5.24). Let's call the user "user", in the domain "LOCATION". PEAP uses a separate protocol (MSCHAPv2) for the presentation of the real username and password. The identity given in the outer protocol may or may not matter, depending on the RADIUS server. In this example, the outer identity is the same one given as the real, inner identity: "LOCATION\user".

Table 5.24: 802.11 EAP response identity

Destination Address	Source Address	Ether-type	EAPOL Type	EAP Code	EAP Type	Identity
AP Address	Client Address	0x888E	EAP	2=Response	Identity	LOCATION\ user

This response triggers the start of the PEAP protocol, tunneled over EAP, tunneled over EAPOL, carried over 802.11. The first message is from the RADIUS server, through the access point, and informs the client that PEAP is beginning (Table 5.25).

PEAP uses TLS as the outer tunnel, within which the encrypted username and password are passed. The first message in the TLS exchange is what is known as a TLS Client Hello (Table 5.26). The Client Hello passes the client's nonce, used as a part of the key derivation protocol. The client will specify a number of cipher suites, but must specify RSA public key encryption with RC4 stream encryption and either MD5 or SHA hashes.

Table 5.25: 802.11 EAP request PEAP

Destination Address	Source Address	Ether-type	EAPOL Type	EAP Code	EAP Type	Flags
Client Address	AP Address	0x888E	EAP	Request	25=PEAP	Start

Table 5.26: 802.11 PEAP client hello

Destination Address	Source Address	EAP Code	EAP Type	TLS Type	Handshake Type	Nonce	Cipher Suites
AP Address	Client Address	Response	PEAP	22=Handshake	1=Client Hello	*random*	*Many*

The server will respond with a Server Hello. The Server Hello message will specify the server's nonce, a session ID (which is usually not taken advantage of by wireless clients), one of the client's cipher suite to use for the rest of the process, and the beginning of a chain of certificates for the RADIUS server, which identifies itself as being valid. The client will usually verify that the server is signed by a valid certificate authority somewhere along the path and is allowed to serve the role it does, unless the client's administrator has explicitly disabled this check. Because certificates are much longer than the maximum EAPOL packet, the PEAP Server Hello and Certificate will be divided up over many consecutive EAPOL frames from the access point. After the certificate, the server may include a request for the client to send a certificate. This would be used by PEAP to short-circuit the inner tunnel and revert to plain TLS, if the client has a certificate. Usually, PEAP is not used with client certificates, so the client will ignore this request and trigger the password exchange. If requested, the types of certificates and distinguished names of acceptable certificate authorities, one of whom needed to have signed any client certificate given, will be provided. The message ends with a Server Hello Done. See Table 5.27.

Table 5.27: 802.11 PEAP server hello and certificate, usually split across multiple EAPOL messages

Destination Address	Source Address	EAP Code	Flags	TLS Type	Handshake Type	Session ID	Nonce	
Client Address	AP Address	Request	0x40 = More	Handshake	2=Server Hello	arbitrary	random	...

	Handshake Type	Server Certificate	Handshake Type	Certificate Types	Distinguished Names	Handshake Type
...	11=Certificate	X509 Certificate	13=Certificate Request	RSA, DSS...	Any Names	14=Server Hello Done

The client will respond to the intermediate server certificate messages with empty responses, to keep the request/response protocol going (Table 5.28).

Table 5.28: 802.11 EAP response PEAP

Destination Address	Source Address	Ether-type	EAPOL Type	EAP Code	EAP Type
AP Address	Client Address	0x888E	EAP	Response	PEAP

When the Server Hello Done message arrives at the client, the client will kick off the second, inner phase of PEAP. First, the client responds with a Certificate handshake message. If the client were going to provide a certificate, it would do so here. However, with normal PEAP, the certificate message will be empty. Following this is the Client Key Exchange. Let's assume that the server and client agreed to RSA public key encryption. The client chooses a random 48-byte premaster key, which is encrypted by the server certificate's RSA public key, and then packaged in the key field. Following this comes the Change Cipher Spec message (Table 5.29), to inform the server that all future communications will take place using encryption based on the key. Finally, the first encrypted message is introduced, which is a marker, encrypted by the key, that states that the cipher change is done.

Table 5.29: 802.11 PEAP client change cipher spec

Destination Address	Source Address	EAP Code	Flags	TLS Type	Handshake Type	Client Certificate	...
AP Address	*Client Address*	Response	*none*	Handshake	Certificate	*none*	

Handshake Type	Key	Handshake Type	Handshake Type	Encrypted Handshake
22=Client Key Exchange	*Pre-master Key*	20=Change Cipher Spec	32=Encrypted Handshake Message	Finished *(encrypted with TLS PRF)*

The server now responds with a Change Cipher Spec and Finished message (Table 5.30), to mark the switch over of the protocol completely to the inner TLS tunnel.

Table 5.30: 802.11 PEAP server change cipher spec

Destination Address	Source Address	EAP Code	TLS Type	Handshake Type	Handshake Type	Encrypted Handshake
Client Address	*AP Address*	Request	Handshake	Change Cipher Spec	Encrypted Handshake Message	Finished *(encrypted with TLS PRF)*

The client, once again, sends an empty response (Table 5.31).

Table 5.31: 802.11 EAP response PEAP

Destination Address	Source Address	Ether-type	EAPOL Type	EAP Code	EAP Type
AP Address	*Client Address*	0x888E	EAP	Response	PEAP

Now, the inner MSCHAPv2 protocol can take place. Table 5.32 will peel back the inner TLS tunnel and reveal the contents. The inner tunnel will also present an EAP exchange, but using MSCHAPv2, rather than TLS.

Table 5.32: 802.11 PEAP encrypted request identity

Destination Address	Source Address	EAP Code	TLS Type	EAP Code (encrypted with RC4)	EAP Type (encrypted)
Client Address	*AP Address*	Request	23=Application Data	Request	Identity

The first step of MSCHAPv2 is for the server to request the identity of the client.

The next step is for the client to respond, in an encrypted form, with the real identity of the user (Table 5.33). If the previous, outer response had been something arbitrary, the server will find out about the real username this way.

Table 5.33: 802.11 PEAP encrypted response identity

Destination Address	Source Address	EAP Code	TLS Type	EAP Code (encrypted)	EAP Type (encrypted)	Identity (encrypted)
Client Address	AP Address	Response	Application Data	Response	Identity	LOCATION\ user

The server then responds with a challenge (Table 5.34). The challenge is a 16-byte random string, which the client will use to prove its identity.

Table 5.34: 802.11 PEAP encrypted MSCHAPv2 challenge

Destination Address	Source Address	EAP Code	TLS Type	EAP Code (encrypted)	EAP Type (encrypted)	CHAP Code (encrypted)	Challenge (encrypted)
Client Address	AP Address	Request	Application Data	Request	MSCHAPv2	Challenge	random

The client responds to the challenge. First, it provides a 16-byte random challenge of its own. This is used, along with the server challenge, the username, and the password, to provide an NT response (Table 5.35).

Table 5.35: 802.11 PEAP encrypted MSCHAPv2 response

Destination Address	Source Address	EAP Code	TLS Type	EAP Code (encrypted)	CHAP Code (encrypted)	Peer Challenge (encrypted)	Response (encrypted)
AP Address	Client Address	Response	Application Data	Response	Response	random	NT response

Assuming the password matches, the server will respond with an MSCHAPv2 Success message (Table 5.36). The success message includes some text messages which are intended to be user printable, but really are not.

The client now responds with a success message of its own (Table 5.37).

The server sends out an EAP TLV message now, still encrypted, indicating success (Table 5.38). The exchange exists to allow extensions to PEAP to be exchanged in the encrypted

Table 5.36: 802.11 PEAP encrypted MSCHAPv2 server success

Destination Address	Source Address	EAP Code	TLS Type	EAP Code *(encrypted)*	CHAP Code *(encrypted)*	Authenticator Message *(encrypted)*	Success Message *(encrypted)*
Client Address	AP Address	Request	Application Data	Request	Success		

Table 5.37: 802.11 PEAP encrypted MSCHAPv2 client success

Destination Address	Source Address	EAP Code	TLS Type	EAP Code *(encrypted)*	CHAP Code *(encrypted)*
AP Address	Client Address	Response	Application Data	Response	Success

tunnel (such as a concept called *cryptobinding*, but we will not explore the concept further here).

Table 5.38: 802.11 PEAP encrypted MSCHAPv2 server TLV

Destination Address	Source Address	EAP Code	TLS Type	EAP Code *(encrypted)*	TLV Result *(encrypted)*
Client Address	AP Address	Request	Application Data	33=TLV	Success

The client sends out an EAP TLV message of its own, finishing up the operation within the tunnel (Table 5.39).

Table 5.39: 802.11 PEAP encrypted MSCHAPv2 server TLV

Destination Address	Source Address	EAP Code	TLS Type	EAP Code *(encrypted)*	TLV Result *(encrypted)*
AP Address	Client Address	Response	Application Data	TLV	Success

Now, the server sends the RADIUS Accept message to the authenticator. This message includes the RADIUS master key, derived from the premaster key that the client chose. This key is sent to the authenticator, where it becomes the PMK for WPA2 or the input to the PMK-R0 for 802.11r. The authenticator then generates an EAP Success message (Table 5.40), which is sent over the air to the client.

The sheer number of packets exchanged in this 802.1X step is what leads to the need for key caching for mobile clients in Wi-Fi, described in Section 6.2.4, and also eliminates the need to perform the 802.1X negotiation except on the first login of the client.

Table 5.40: 802.11 EAP success

Destination Address	Source Address	EAP Code
Client Address	AP Address	3=Success

Step 3: Perform the Four-Way Handshake

Both the authenticator and the client have the PMK. The four-way handshake derives the PTK. The first message (Table 5.41) sends the authenticator's nonce, and a copy of the access point's RSN information.

Table 5.41: 802.11 Four-way handshake message one

Destination Address	Source Address	EAPOL Type	Key Type	Flags	Nonce	RSN IE
Client Address	AP Address	Key	RSN (WPA2)	Ack	random	Same as in Beacon

The client generates the PTK, and sends the next message (Table 5.42), with its nonce and a copy of the client's RSN information, along with a MIC signature.

Table 5.42: 802.11 Four-way handshake message two

Destination Address	Source Address	EAPOL Type	Flags	Nonce	MIC	RSN IE
AP Address	Client Address	Key	MIC	random	hash	Same as in Association

The third message, also with a MIC, delivers the GTK that the authenticator is currently using for the BSS, encrypted (Table 5.43).

Table 5.43: 802.11 Four-way handshake message three

Destination Address	Source Address	EAPOL Type	Flags	MIC	GTK
Client Address	AP Address	Key	Install, Ack, MIC	hash	encrypted

Finally, the client responds with the fourth message (Table 5.44), which confirms the key installation.

Finally, the client is associated to the access point, and both sides are encrypting and decrypting traffic using the keys that came out of the 802.1X and WPA2 process.

Table 5.44: 802.11 Four-way handshake message four

Destination Address	Source Address	EAPOL Type	Flags	MIC
AP Address	*Client Address*	Key	Ack, MIC	*hash*

Appendix to Chapter 5 Wi-Fi

5A.1 Introduction

I have often been asked about the "whys" of Wi-Fi: why the 802.11 standard was designed the way it was, or why certain problems are still unsolved—even the ones people don't like to talk about—or how can a certain technique be possible. Throughout this book, I have tried to include as much information as I think would be enlightening to the reader, including insights that are not so easy to come across. Nevertheless, there is a lot of information that is out there, that may help satisfy your curiosity and help explain some of the deeper whys, but might not be necessary for understanding wireless networking. How does MIMO work? Why is one security mode that much better than the other? This book tries to answer those questions, and this appendix includes much of the reasoning for those answers.

This appendix is designed for readers who are interested in going beyond, but might not feel the need to see the exact details. Thus, although this discussion will use mathematics and necessary formulas to uncover the point, care was not taken to ensure that one can calculate with what is presented here, and the discussion will gloss over fundamental points that don't immediately lead to a better understanding. I hope this appendix will provide you with a clearer picture of the reasons *behind* the network.

5A.2 What Do Modulations Look Like?

Let's take a look at the mathematical description of the carrier. The carrier is a waveform, a function over time, where the value of the function is the positive or negative.

The basic carrier is a sine wave:

$$f(t) = \sin(2\pi f_c t) \tag{1}$$

where f_c is the carrier's center frequency, t is time, and the amplitude of the signal is 1. Sine waves, the basic function from trigonometry, oscillate every 360 degrees—or 2π radians, being an easier measure of angles than degrees—and are used as carriers because they are the natural mathematical function that fits into pretty much all of the physical and mathematical equations for oscillations. The reason is that the derivative of a sinusoid—a sine function with some phase offset—is another sinusoid.

($\frac{d}{dt}\sin(t) = \cos(t) = \sin(t - \pi/2)$.) This makes sine waves the simplest way most natural

oscillations occur. For example, a weight on a spring that bounces will bounce as a sine wave, and a taught rope that is rippled will ripple as a sine wave. In fact, frequencies for waves are defined specifically for a sine wave, and for that reason, sine waves are considered to be pure tones. All other types of oscillations are represented as the sum of multiple sine waves of different frequencies: a *Fourier transform* gets the mathematical function representing the actual oscillation into the frequencies that make it up. Pictures of signals plotted as power over frequency, such as envelopes, are showing the frequency, rather than time, representation of the signal. (Envelopes, specifically, show the maximum allowable power at each frequency.)

Modulations affect the carrier by adjusting its phase, its frequency, its amplitude (strength), or a combination of the three:

$$f(t) = A(t)\sin[2\pi(f_c + f(t))t + \phi(t)] \tag{2}$$

with amplitude modulation *A(t)*, frequency modulation *f(t)*, and phase modulation *ϕ(t)*. The pure tone, or unmodulated sine wave, starts off with a bandwidth of 0, and widens with the modulations. A rule of thumb is that the bandwidth widens to twice the frequency that the underlying signal changes at: a 1MHz modulation widens the carrier out to be usually at least 2MHz in bandwidth. Clearly, the bandwidth can be even wider, and is usually intentionally so, because spreading out the signal in its band can make it more impervious to narrow bandwidth noise (spread spectrum). The frequency of the carrier is chosen so that it falls in the right frequency band, and so that its bandwidth also falls in that band, which means that the center frequency will be much higher than the frequency of the modulating signal.

AM modulates the amplitude, and ΦM modulates the phase. Together (dropping FM, which complicates the equations and is not used in 802.11), the modulated signal becomes

$$f(t) = A(t)\sin[2\pi f_c t + \phi(t)] \tag{3}$$

In this case, the modulation can be plotted on a polar graph, because polar coordinates measure both angles and lengths, and *A(t)* becomes the length (the distance from the origin), and *ϕ(t)* becomes the angle.

Complex numbers, made of two real numbers *a* and *b* as *a + bi*, where *i* is the square root of −1, happen to represent lengths and angles perfectly for the mathematics of waves, as *a* is the value along the *x* axis and *b* is the length along the *y* axis. In their polar form, a complex number looks like

$$A(\cos\phi + i\sin\phi) = Ae^{i\phi} \tag{4}$$

where e^n is the exponential of n. The advantage of the complex representation is that one amplitude and phase modulation can be represented together as one complex number, rather than two real numbers. Let's call the modulation $s(t)$, because that amplitude and phase modulation will be what we refer to as our signal.

Even better, however, is that the sine wave itself can be represented by a complex function that is also an exponent. The complex version of a carrier can be represented most simply by the following equation

$$f(t) = \mathrm{Re}\left\{e^{2\pi i f_c t}\right\} \tag{5}$$

meaning that f(t) is the real portion (the a in $a + bi$) of the exponential function. This function is actually equal to the cosine, which is the sine offset by 90 degrees, but a constant phase difference does not matter for signals, and so we will ignore it here. Also for convenience, we will drop the Re and think of all signals as complex numbers, remembering that transmitters will only transmit power levels measured by real numbers.

Because the signal is an exponential, and the modulation is also an exponential, the mathematics for modulating the signal becomes simple multiplication. Multiplying the carrier in (5) by the modulation in (4) produces

$$f(t) = s(t)e^{2\pi i f_c t} = Ae^{i\phi}e^{2\pi i f_c t} = Ae^{2\pi i f_c t + i\phi} = Ae^{i(2\pi f_c t + \phi)} \tag{6}$$

where the amplitude modulation A and the phase modulation ϕ adds to the angle, as needed. (Compare to equation (3) to see that the part in parentheses in the last exponent matches.)

All of this basically lets us know that we can think of the modulations applied to a carrier independently of the carrier, and that those modulations can be both amplitude and phase. This is why we can think of phase-shift keying and QAM, with the constellations, without caring about the carrier. The modulations are known as the *baseband* signal, and this is why the device which converts the data bits into an encoded signal of the appropriate flavor (such as 802.11b 11Mbps) is called the baseband.

Even better, because the carrier is just multiplied onto the modulations, we can disregard the carrier's presence throughout the entire process of transmitting and receiving. So, this is the last we'll see of *f(t)*, and will instead turn our intention to the modulating signal *s(t)*.

The complex modulation function *s(t)* can be thought of as a discrete series of individual modulations, known as *symbols*, where each symbol maps to some number of bits of digital data. The complex value of the symbol at a given time is read off of the constellation chart, based on the values of the bits to be encoded. Each symbol is applied to the carrier (by multiplication), and then held for a specified amount of time, much longer than the oscillations of the carrier, to allow the receiver to have a chance to determine what the change of the carrier was. After that time, the next modulation symbol is used to modulate the underlying carrier, and so on, until the entire stream of symbols are sent. Because it is convenient to view each symbol one at a time as a sequence, we'll use $s(0)$, $s(1)$, $s(2)$, … $s(n)$, and so on, to represent symbols as a sequence, where n is a natural number referring to the correct symbol in the sequence.

5A.3 What Does the Channel Look Like?

When the transmitter transmits the modulated signal from equation (6), the signal bounces around through the environment and gets modified. When it reaches the receiver, a completely different signal (represented by a different function) is received. The hope is that the received signal can be used to recover the original stream of modulations $s(t)$.

Let's look at the effects of the channel more closely. The transmitted signal is a radio wave, and that wave bounces around and through all sorts of different objects on its way to the receiver. Every time the signal hits an object, three things can happen to it. It will be attenuated—its function can be multiplied by a number A less than 1. Its phase will be changed—its function can be multiplied by $e^{i\phi}$ with ϕ being the angle of phase change, which happens on reflections. And, after all of the bouncing, the signal will be delayed. Every different reflection of the signal takes a different path, and all of these paths come together on the receiver as a sum.

We can recognize the phase and attenuation of a signal as the $Ae^{i\phi}$ from equation (4), because a modulation is a modulation, whether the channel does it or transmitter. Thus, we will look at the effects of modulating, the channel, and demodulating together as one action, which we will still call the channel. Every reflection of the signal has its own set of A and ϕ. Now let's look at time delay. To capture the time delay of each reflection precisely, we can create a new complex-valued function $h(t)$, where we record, at each t, the sum of the As and ϕs for each reflection that is delayed t seconds by the time it hits the receiver. If there is no reflection that is delayed at t, then $h(t) = 0$ at that point. The function $h(t)$ is known as the *impulse response* of the channel, and is generally thought of to contain every important aspect of the channel's effect on the signal. Let's give one trivial example of an $h(t)$, by building one up from scratch. Picture a transmitter and a receiver, with nothing else in the entire universe. The transmitter is 100 meters from the receiver. Because the speed of light is 299,792,458 meters per second, the delay the signal will take as it goes from the transmitter to the receiver is 100/299,792,458 = approximately 333 nanoseconds. Our $h(t)$ starts off as all zero, but at $h(t = 333ns)$, we do have a signal. The phase is not changed, so the value of h at that point will just be the attenuation A, which, for the distance, assuming a 2.4GHz signal and other reasonable properties, is −80dBm. It's easy to see how other reflections of the signal can be added. The received signal, $y(t)$, is equal to the value of $h(\tau)$ multiplied by (modulating) the original signal $s(t)$, summed across all τ (as the different reflections add, just as water waves do). This sum is an integral, and so we can express this as

$$y(t) = \int h(\tau)s(t-\tau)d\tau = h(\tau)*s(t) \tag{7}$$

where * is the *convolution* operator, which is just a fancy term for doing the integral before it.

Radio designers need to know about what properties the channel can be assumed to have in order to make the math work out simply—and, as is the nature of engineering—these simplifying assumptions are not entirely correct, but are correct enough to make devices that work, and are a field of study into itself. A basic book in signal theory will go over those details. However, we can simply this entire discussion down to two simple necessary points:

1. A delay of a slowly modulated but quickly oscillating sine wave looks like a phase offset of the original signal.

2. Noise happens, and we can assume a lot about that noise.

Phase offsets are simple multiplications, rather than complex integrals, so that will let us replace equation (7) with

$$y(t) = hs(t) + n \tag{7}$$

where h is now a constant (not a function of time) equal to the sums of all of the attenuations and phase changes of the different paths the signal takes, and n is the thermal noise in the environment. (Assumption 1 is fairly severe, as it turns out, but it is good enough for this discussion, and we needed to get rid of the integral. By forcing all of the important parts of $h(t)$ to happen at $h(t = 0)$, then the convolution becomes just the multiplication. For the interested, the assumption is known as assuming *flat fading*, because removing the ability for h to vary with time removes the ability for it to vary with frequency as well. Flat, therefore, refers to the look of h on a frequency plot.)

The receiver gets this $y(t)$, and needs to do two things: subtract out the noise, and undo the effects of the channel by figuring out what h is. n is fairly obvious, because it is noise, which tends to look statistically the same everywhere except for its intensity. Now, let's get h. If the receiver knows what the first part of $s(t)$ is supposed to look like—the preamble, say—then, as long as h doesn't change across the entire transmission, then the receiver can divide off h and recover all of $s(t)$, which we will call $r(t)$ here for the received signal:

$$r(t) = \frac{y(t) - n}{h} = \frac{hs(t)}{h} = s(t) \tag{8}$$

That's reception, in a nutshell.

(Those readers with an understanding of signal theory may find that I left out almost the entire foundation for this to work, including a fair amount of absolutely necessary math to get the equations to figure through [the list is almost too long to present]. However, hopefully all readers will appreciate that some of the mystery behind wireless radios has been lifted.)

5A.4 How Can MIMO Work?

If radios originally seemed mysterious, then MIMO could truly seem magical. But understanding MIMO is nowhere out of reach. In reality, MIMO is just a clever use of linear algebra to solve equations, such as equation (8), for multiple radios simultaneously.

In a MIMO system, there are N antennas, so equation (7) has to be done for each antenna:

$$y_i(t) = h_i s_i(t) + n_i, \text{ with on } i \text{ for each antenna} \tag{9}$$

written as vectors and matrices

$$\mathbf{y}(t) = \mathbf{H}\mathbf{s}(t) + \mathbf{n} \tag{10}$$

with one dimension for each antenna. $\mathbf{H}$ now is a matrix, whose diagonal serves the original purpose of h for each antenna as if it were alone. But one antenna can hear from all of the other antennas, not just one, and that makes $\mathbf{H}$ into a matrix, with off-diagonal elements that mix the signals from different antennas. For the sake of it, let's write out the two-antenna case:

$$\begin{pmatrix} y_1(t) \\ y_2(t) \end{pmatrix} = \begin{bmatrix} h_{11} & h_{12} \\ h_{21} & h_{22} \end{bmatrix} \begin{pmatrix} s_1(t) \\ s_2(t) \end{pmatrix} + \begin{pmatrix} n_1 \\ n_2 \end{pmatrix} \tag{11}$$

or, multiplied out,

$$y_1(t) = h_{11}s_1(t) + h_{12}s_2(t) + n_1$$

$$y_2(t) = h_{22}s_2(t) + h_{21}s_1(t) + n_2 \tag{12}$$

Each receive antenna gets a different (linear!) combination of the signals for each of the transmitting antennas, plus its own noise.

The receiver's trick is to undo this mixing and solve for s_1 and s_2. **H**, being a matrix, cannot be divided, as could be done with the scalar h in (8). However, the intermixing of the antennas can be undone, if they can be determined, and if the intermixing is independent from antenna to antenna. That's because equations (10–12) are a system of linear equations, and if we start off with a known $s(t)$, we can recover the hs. If we start the sequence off with a few symbols, such as $s_1(0)$ and $s_2(0)$, that are known—say, a preamble—then the receiver, knowing **y**, **n**, and **s**(0), can try to find the values **H** = (h_{11} h_{12}, h_{22}, h_{22}) that make the equations work. Once **H** is known, the data symbols **s**(t) can come in, and those are now two unknowns across two equations. We can't divide by **H**, but in the world of matrices, we invert it, which will get us the same effect. **H** is an invertible matrix if each of the rows is linearly independent of the others. (Not having linearly independent rows produces the matrix version of dividing by zero: more sense can be made of such a thing with matrices just because they have more information, but still, there remains an infinite number of solutions, and that won't be useful for retrieving a signal.)

So, the analog of equation (8) is

$$\mathbf{r} = \mathbf{H}^{-1}(\mathbf{y} - \mathbf{n}) = \mathbf{H}^{-1}(\mathbf{Hs}) = \mathbf{s} \qquad (13)$$

That's the intuition behind MIMO. Of course, no one builds receivers this way, because they turn out to be overly simplistic and not to be very good. (It sort of reminds me of the old crystal no-power AM radio kits. They showed the concept well, but no one would sell one on the basis of their quality.)

The process of determining **H** is the important part of MIMO.

One other point. You may have noticed that, in reality, any measured **H** is going to be linearly independent, just because the probability of measuring one row to be an exact linear combination of any of the others is practically small. (The determinant of **H** would have to exactly equal 0 for it to not be invertible.) However, this observation doesn't help, because **H** has to be more than just invertible. Its rows have to be "independent" enough to allow the numbers to separate out leaving strong signals behind. There is a way of defining that.

The key is that $\mathbf{HH}^H$, the channel matrix multiplied by its conjugate transpose (this product originates from information theory, as shown in the next section), reveals information about how much information can be packed into the channel—basically, how good the SNR will be for each of the spatial streams. The reason is that the spatial streams do, in fact, interfere with each other. When the channel conditions are just right, mathematically, then the interference is not an issue. But when the channel leads to enough cross-stream interference, MIMO can shut down.

5A.5 Why So Many Hs? Some Information Theory

For those readers curious as to why there are so many instances of the letter H in the formulas, read on. $\mathbf{HH}^H$ comes from attempting to maximize the amount of information that the radio can get through the channel and into $\mathbf{y}$ by picking the best values of $\mathbf{s}$. That matters because $\mathbf{H}$ acts unevenly on each symbol. Some symbols may be relatively simple and robust, and others may be more finicky and sensitive. The definition of information is from information theory (and is purely mathematical in this context), and we are looking for trying to maximize the amount of mutual information from what the transmitter sent ($\mathbf{s}$) and what the receiver got ($\mathbf{y}$).

Intuitively, the measure of the maximum information something can carry is the amount of different values the thing can take on. We are looking at $\mathbf{y}$, and trying to see how much information can be gleaned from it. The channel is something that removes information, rather than, say, adding it. For MIMO, this removal can happen just by going out of range, or by being in range but having a spatial stream that can't get high SNR even though a non-MIMO system would have that high SNR. The latter can only come from some form of interference between the spatial streams.

To look at that, we need to look at the amount of stream-to-stream correlation that each possible value of $\mathbf{y}$ can have. This correlation, across all possible values of $\mathbf{y}$, is known as the *covariance* of $\mathbf{y}$, or cov($\mathbf{y}$), and can be represented by an N×N matrix. MIMO works best when each of the antennas appears to be as independent as possible of the others. One way of doing that would be for the covariance of one antenna's signal signal y_i with itself to be as high as possible across all covalues of $\mathbf{y}$, but for the covariance of a signal y_i with a signal y_j from another antenna to be as close to zero as possible. The reason is that the higher the covariance of an antenna's signal with itself, the more variation from the mean has been determined, and that's what information looks like—variation from the mean. When the other terms are zero, the cross-stream interference is also zero. A perfect MIMO solution would be for every antenna on the receiver to connect directly with every antenna on the transmitter by a cable. That would produce a covariance matrix with reasonable numbers in the diagonal, for every antenna with itself, and zeros for every cross-correlation between antennas.

The determinant of the covariance matrix measures the amount of interference across antennas. The closer to 0 the determinant is, the more the interantenna interference degrades the signal and reduces the amount of information available. The determinant is used here because it can be thought of as defining a measure of just how independent the rows of a matrix are. Let's look at the following. The determinant of the identity matrix **I**, a matrix where each row is orthogonal, and so the rows taken as vectors point as far away from the others as possible, is 1. The determinant of a matrix where any two of the row or columns are linearly dependant is 0. Other matrices with normalized rows fall in between those two, up to a sign. Geometrically, the determinant measures the signed volume of the parallelepiped contained within the matrix's rows. The more spread out and orthogonal the rows, the further the determinant gets from zero.

Because of equation (10), the covariance matrix for **y** is proportional to the covariance matrix of **s**, with the effects of the channel applied to it through **H**, followed by a term containing the noise. As an equation

$$\text{cov}(\mathbf{y}) = E_{\text{sT}}\mathbf{H}\,\text{cov}(\mathbf{s})\mathbf{H}^H + N_0\mathbf{I} \tag{14}$$

where E_{sT} is the average energy of the symbol per stream, and N_0 is the energy of the noise. (Note that the noise's covariance matrix is **I**, the identity matrix, because no two noise values are alike, and thus two different noise values have a covariance of 0.)

Now, we begin to see where the **HH**^{H} comes from: it is just a representation of how the channel affects the symbol covariance matrix. As mentioned before, the amount of information that is in the signal is the determinant of the covariance matrix cov(**y**). Dividing out the noise value N_0, the amount of mutual information in **s** and **y** becomes simply equal to the number of bits (base-2 logarithm) of the determinant of cov(**y**), which, using (11), gives us

$$\text{capacity} = \lg \det\left[(E_{\text{sT}}/N_0)\mathbf{H}\,\text{cov}(\mathbf{s})\mathbf{H}^H + \mathbf{I}\right] \tag{15}$$

Remember that we are looking at this from the point of view of how to make the best radio, so we can pick whatever set of symbols we want. However, symbols are usually chosen so that each symbol is as different as possible from the others, because the channel isn't known in advance and we are considering a general-purpose radio that should work well in all environments. Therefore, no two symbols have strong covariance, and so, if you think of cov(**s**) as being close to the identity matrix itself (zero off-diagonal, one on), we drop the cov(**s**) = **I** and get

$$\text{capacity} = \lg \det\left[(E_{\text{sT}}/N_0)\mathbf{HH}^H + \mathbf{I}\right] \tag{16}$$

Going a bit deeper into linear algebra, we can extract the eigenvalues of the matrix **HH**^{H} to see what that matrix is doing. With a little math, we can see that the effect of **HH**^{H} is to act, for each spatial stream, like its own independent attenuator, providing attenuation equal to the eigenvector for each spatial stream.

$$\text{capacity} = \text{sum of capacity for each spatial stream}$$
$$= \sum \lg[(E_{sT}/N_0)(\text{eigenvector for } i) + 1] \tag{17}$$

Okay, a lot of math, with a few steps skipped. Let's step back. The capacity for each of the spatial stream is based on the signal-to-noise ratio for the symbol (E_{sT}/N_0) times how much the combined effect $\mathbf{HH}^H$ of the channel matrix has on that spatial stream. This itself is the per-stream SNR, with the noise coming from the other streams. The more independent each basis for $\mathbf{HH}^H$ is, the more throughput you can get. And since $\mathbf{HH}^H$ depends only on the channel conditions and environments, we can see that capacity is directly dependant on how independent the channel is. If the channel allows for three spatial streams, say, but one of the spatial streams has such low SNR that it is useless, then the capacity is only that of two streams and you should hope that the radio doesn't try to use the third stream. And overall, this dependence on the channel was what we sought out to understand.

CHAPTER 6

Voice Mobility over Wi-Fi

6.0 Introduction

In the previous chapter, you learned what Wi-Fi is made of. From here, we can look at how the voice technologies from previous chapters are applied to produce voice mobility over Wi-Fi.

The keys to voice mobility's success over Wi-Fi are, rather simply, that voice be mobile, high-quality, secure, and that the phones have a long talk and standby time. These are the ingredients to taking Wi-Fi from a data network, good for checking email and other non-real-time applications but not designed from the start to being a voice network, to being a network that can readily handle the challenges of real-time communication.

It's important to remember that this has always been the challenge of voice. The original telegraph networks were excellent at carrying non-real-time telegrams. But to go from relay stations and typists to a system that could dedicate lines for each real-time, uninterruptible call, was a massive feat. Then, when digital telephony came on the scene, the solution was to use strict, dedicated bandwidth allocation and rigorous and expensive time slicing regimes. When voice moved to wired Ethernet, and hopped from circuit switching to packet switching, the solution was to throw bandwidth at the problem or to dedicate parallel voice networks, or both. Wi-Fi doesn't have that option. There is not enough bandwidth, and parallel networks—though still the foundation of many Wi-Fi network recommendations—are increasingly becoming too expensive.

So, let's look through what has been added to Wi-Fi over the years to make it a suitable carrier for voice mobility.

6.0.1 Quality of Service with WMM—How Voice and Data Are Kept Separate

The first challenge is to address the unique nature of voice. Unlike data, which is usually carried over protocols such as TCP that are good at making sure they take the available bandwidth and nothing more, ensuring a continuous stream of data no matter what the network conditions, voice is picky. One packet every 20 milliseconds. No more, no less. The packets cannot be late, or the call becomes unusable as the callers are forced to wait for maddening periods before they hear the other side of their conversation come through. The packets cannot arrive unpredictably, or else the buffers on the phones overrun and the call becomes choppy and impossible to hear. And, of course, every lost packet is lost time and lost sounds or words.

On Ethernet, as we have seen, the notion of 802.1p or Diffserv can be used to give prioritization for voice traffic over data. When the routers or switches are congested, the voice packets get to move through priority queues, ahead of the data traffic, thus ensuring that their resources do not get starved, while still allowing the TCP-based data traffic to continue, albeit at a possibly lesser rate.

A similar principle applies to Wi-Fi. The *Wi-Fi Multimedia* (WMM) specification lays out a method for Wi-Fi networks to also prioritize traffic according to four common classes of service, each known as an *access category* (AC):

*   AC_VO: highest-priority voice traffic

*   AC_VI: medium-priority video traffic

*   AC_BE: standard-priority data traffic, also known as "best effort"

*   AC_BK: background traffic, that may be disposed of when the network is congested

The underscore between the AC and the two-letter abbreviation is a part of the correct designation, unfortunately. You may note that the term "best effort" applies to only one of the four categories. Please keep in mind that all four access categories of Wi-Fi are really best effort, but that the higher-priority categories get a better effort than the lower ones. We'll discuss the consequences of this shortly.

The access category for each packet is specified using either 802.1p tagging, when available and supported by the access point, or by the use of *Diffserv Code Points* (DSCP), which are carried in the IP header of each packet. DSCP is the more common protocol, because the per-packet tags do not require any complexity on the wired network, and are able to survive multiple router hops with ease. In other words, DSCP tags survive crossing through every network equipment that is not aware of DSCP tags, whereas 802.1p requires 802.1p-aware links throughout the network, all carried over 802.1Q VLAN links.

There are eight DSCP tags, which map to the four access categories. The application that generates the traffic is responsible for filling in the DSCP tag. The standard mapping is given in Table 6.1.

There are a few things to note here. First is that the eight "priorities"—again, the correct term, unfortunately—map to only four truly different classes. There is no difference in quality of service between Priority 7 and Priority 6 traffic. This was done to simplify the design of Wi-Fi, in which it was felt that four classes are enough. The next thing to note is

Table 6.1: DSCP tags and AC mappings

DSCP	TOS Field Value	Priority	Traffic Type	AC
0x38 (56)	0xE0 (224)	7	Voice	AC_VO
0x30 (48)	0xC0 (192)	6	Voice	AC_VO
0x28 (40)	0xA0 (160)	5	Video	AC_VI
0x20 (32)	0x80 (128)	4	Video	AC_VI
0x18 (24)	0x60 (96)	3	Best Effort	AC_BE
0x10 (16)	0x40 (64)	2	Background	AC_BK
0x08 (8)	0x20 (32)	1	Background	AC_BK
0x00 (0)	0x00 (0)	0	Best Effort	AC_BE

that the many packet capture analyzers will still show the one-byte DSCP field in the IP header as the older TOS interpretation. Therefore, the values in the TOS column will be meaningless in the old TOS interpretation, but you can look for those specific values and map them back to the necessary ACs. Even the DSCP field itself has a lot of possibilities; nonetheless, you should count on only the previous eight values as having any meaning for Wi-Fi, unless the documentation in your equipment explicitly states otherwise. Finally, note that the default value of 0 maps to best effort data, as does the Priority 3 (DSCP 0x18) value. This strange inversion, where background traffic, with an actual lower over-the-air priority, has a higher Priority code value than the default best effort traffic, can cause some confusion when used; thankfully, most applications do not use Priority 3 and its use is not recommended here as well.

A word of warning about DSCP and WMM. The DSCP codes listed in Table 6.1 are neither Expedited Forwarding or Assured Forwarding codes, but rather use the backward-compatibility requirement in DSCP for TOS precedence. TOS precedence, as mentioned in Chapter 4, uses the top three bits of the DSCP to represent the priorities in Table 6.1, and assign other meanings to the lower bits. If a device is using the one-byte DSCP field as a TOS field, WMM devices may or may not ignore the lower bits, and so can sometimes give no quality-of-service for tagged packets. Further complicating the situation are endpoints that generate Expedited Forwarding DSCP tags (with code value of 46). Expedited Forwarding is the tag that devices use when they want to provide higher quality of service in general, and thus will usually mark all quality-of-service packets as EF, and all best effort packets with DSCP of 0. The EF code of 46 maps, however, to the Priority value of 5—a video, not voice, category. Thus, WMM devices may map all packets tagged with Expedited Forwarding as video. A wireless protocol analyzer shows exactly what the mapping is for by looking at the value of the TID/Access Category field in the WMM header. The WMM header is shown in Table 5.5.

This mapping can be configured on some devices. However, changing these values from the defaults can cause problems with the more advanced pieces of WMM, such as WMM

Power Save and WMM Admission Control, so it is not recommended to make those changes. (The specific problem that would happen is that the mobile device is required to know what priority the other side of the call will be sending to it, and if the network changes it in between, then the protocols will get confused and not put the downstream traffic into the right buckets.)

Once the Wi-Fi device—the access point or the client—has the packet and knows its tag, it will assign the packet into one of four priority queues, based on the access categories. However, these queues are not like their wired Ethernet brethren. That is because it is not enough that voice be prioritized over data within the device; voice must also be prioritized over the air.

To achieve this, WMM changes the backoff procedure mentioned in Section 5.4.8. Instead of each device waiting a random time less than some interval fixed in the standard, each device's access category gets to contend for the air individually. Furthermore, to get the over-the-air prioritization, higher quality-of-service access categories, such as voice, get more aggressive access parameters.

Each access category get four parameters that each determine how much priority the traffic in that category gets over the air, compared to the other categories. The first parameter is a unique per-packet minimum wait time called the *Arbitration Interframe Spacing* (AIFS). This parameter is the minimum amount of time that a packet in this category must wait before it can even start to back off. The longer the AIFS, the more a packet must wait, and the more it is likely that a higher-priority packet will have finished its backoff cycle and started transmitting. The key about the AIFS is that it is counted after every time the medium is busy. That means that a packet with a very high AIFS could wait a very long time, because the amount of time spent waiting for an AIFS does not count if the medium becomes busy in the meantime. The AIFS is measured in units of the number of slots, and thus is also called the AIFSn (AIFS number).

The second value is the minimum backoff CW, called the *CWmin*. This sets the minimum number of slots that the backoff counter for this particular AC must start with. As with pre-WMM Wi-Fi, the CW is not the exact number of slots that the client must wait, but the *maximum* number of slots that the packet must wait: the packet waits a random number of slots less than this value. The difference is that there is a different CWmin for each access category. The CWmin is still measured in slots, but communicated to the client from the access point as the exponent of the power of two that it must equal. This exponent is called the *ECWmin*. Thus, if the ECWmin for video is 3, then the AC must pick a random number between 0 and $2^3 - 1 = 7$ slots. The CWmin is just as powerful as the AIFS in distinguishing traffic, by making access more aggressive by capping the number of slots the AC must wait to send its traffic.

The third parameter is similar to the minimum backoff CW, and is called the CWmax, or the maximum backoff CW. If you recall, the CW is required to double every time the sender fails to get an acknowledgement for a frame. However, that doubling is capped by the CWmax. This parameter is far mess powerful for controlling how much priority one AC gets over the other. As with the CWmin, there is a different CWmax for each AC.

The last parameter is how many microseconds the AC can burst out packets, before it has to yield the channel. This is known as the *Transmit Opportunity Limit* (TXOP Limit), and is measured in units of 32 microseconds (although user interfaces may show the microsecond equivalent). This notion of TXOPs is new with WMM, and is designed to allow for this bursting. For voice, bursting is usually not necessary or useful, because voice packets come on regular, well-spaced intervals, and rarely come back-to-back in properly functioning networks.

The access point has the ability to set these four AC parameters for every device in the network, by broadcasting the parameters to all of the clients. Every client, thus, has to share the same parameters. The access point may also have a different set for itself. Some access points set these values by themselves to optimize network access; others expose them to the user, who can manually override the defaults. The method that WMM uses to set these values to the clients is through the WMM Parameter Set information element, a structure that is present in every beacon, and can be seen clearly with a wireless packet capture system. Table 6.2 has the defaults for the WMM parameters.

Table 6.2: Common default values for the WMM parameters for 802.11

AC	Client		Access Point		CWmin	TXOP limit	
	AIFS	CWmax	AIFS	CWmax		802.11b	802.11agn
Background (BK)	7	$2^{10}-1 =$ 1023	7	$2^{10}-1 =$ 1023	$2^4-1 =$ 15	0µs	0µs
Best Effort (BE)	3	$2^{10}-1 =$ 1023	3	$2^6-1 = 63$	$2^4-1 =$ 15	0µs	0µs
Video (VI)	2	$2^4-1 = 15$	1	$2^4-1 = 15$	$2^3-1 = 7$	6016µs	3008µs
Voice (VO)	2	$2^3-1 = 7$	1	$2^3-1 = 7$	$2^2-1 = 3$	3264µs	1504µs

6.0.1.1 How WMM Works

The numbers in Table 6.2 seem mysterious, and it is not easy to directly see what the consequences are by WMM creating multiple queues that act to access the air independently. But it is important to understand what makes WMM works, to understand how WMM—and thus, voice—scales in the network.

Looking at the common WMM parameters, we can see that the main way that WMM provides priority for voice is by letting voice use a faster backoff process than data. The shorter AIFS helps, by giving voice a small chance of transmitting before data even gets a chance, but the main mechanism is by allowing voice transmit, on average, with a quarter of the waiting time that best effort data has.

This mechanism works quite well when there is a small amount of voice traffic on a network with a potentially large amount of data. As long as voice traffic is scarce, any given voice packet is much more likely to get on the air as soon as it is ready, causing data to build up as a lower priority. This is one of the consequences of having different queues for traffic. As an analogy, picture the security lines at airports. Busy airports usually have two separate lines, one line for the average traveler, and another line for first-class passengers and those who fly enough to gain "elite" status on the airlines. When the line for the average traveler—the "best effort" line—is full of people, a short line for first class passengers gives those passengers a real advantage. In other words, we can think of best effort and voice as mostly independent. The problem, then, is if there are too many first-class passengers. For WMM, the problem happens when there is "too much" voice traffic. Unlike with the children of Lake Wobegone, not everyone can be above average.

Let's look at this more methodically. From Section 5.4.8, we saw that the backoff value is the primary mechanism that Wi-Fi is affected by density. As the number of clients increases, the chance of collision increases. Unfortunately, WMM provides for quality of service by reducing the number of slots of the backoff, thus making the network more sensitive to density. Again, if voice is rare, then its own density is low, and so a voice packet is not likely to collide with other voice packets, and the aggressive backoff settings for voice, compared to data, allow for voice to get on the network with higher probability. However, when the density of voice goes up, the aggressive voice backoff settings cause each voice packet to fight with the other voice packets, leading to more collisions and higher loss.

One solution for this problem is to limit the number of voice calls in a cell, thus ensuring that the density of voice never gets that high. This is called *admission control*, and is described in Section 6.1.1. Another and an independent solution is for the system to provide a more deterministic quality of service, by intelligently setting the WMM parameters *away* from the defaults. This exact purpose is envisioned by the standard, but most equipment today expects the user to hand-tune these values, something which is not easy. Some guidelines are provided in Section 6.4.1.2.

6.0.2 Battery Life and Power Saving

On top of quality of service, voice mobility devices are usually battery-operated, and finding ways of saving that battery life is paramount to a good user experience.

The main idea behind Wi-Fi power saving is that the mobile device's *receive* radio doesn't really need to always be turned on. Why the receive radio? Although the transmit side of the radio takes more power, because it has to actually send the signal, it is used only when the device has something to send. The receive side, on the other hand, would always have to be on, slowly burning power in the background, listening to every one of the thousands of packets per second going over the air. If, at the end of receiving the packet, it turns out that the packet was for the mobile device, then good. But, if the packet was for someone else, the power was wasted. Adding up the power taken for receiving other device's packets to the power needed just to check whether a signal is on the air leads to a power draw that is too high for most battery-operated devices.

When the mobile device is not in a call—when it is in standby mode—the only real functions the phone should be doing are maintaining its connection to the network and listening for an incoming call. When the device is in a call, it still should be sleeping during the times between the voice packets. Thus, Wi-Fi has two modes of power saving, as described in the following sections.

6.0.2.1 Legacy Power Save

The first mode, known as *legacy* power saving because it was the original power saving technique for Wi-Fi, is used for saving battery during standby operation.

This power save mode is not designed for quality-of-service applications, but rather for data applications. The way it works is that the mobile device tells the access point when it is going to sleep. After that time, the access point buffers up frames directed to the mobile device, and sets a bit in the beacon to advertise when one or more frames are buffered. The mobile device is expected to wake every so many beacons and look for its bit set in the beacon. When the bit is set, the client then uses one of two mechanisms to get the access point to send the buffered frames.

This sort of system can be thought of as a *paging* mechanism, as the client is told when the access point has data for it—such as notification of an incoming call. Figure 6.1 shows the basics of the protocol.

The most important part of the protocol is the paging itself. Each client is assigned an association ID (AID) when it associates. The value is given out by the access point, in a field in the Association Response that it sent out when the client connected to it. The AID is a number from 1 to 2007 (an extremely high number for an access point) that is used by the client to figure out what bit to look at in the beacon. Each beacon carries a *Traffic Indication Map* (TIM), which is an abbreviated bit field. Each client who has a frame buffered for it has its bit set in the TIM, based on the AID. For example, if a client with AID of 10 has one or more frames buffered for it, the tenth bit (counting from zero) of the TIM would be set.

Because beacons are set periodically, using specific timing that ensures that it never goes out before its time, each client can plan on the earliest it needs to wake up to hear the

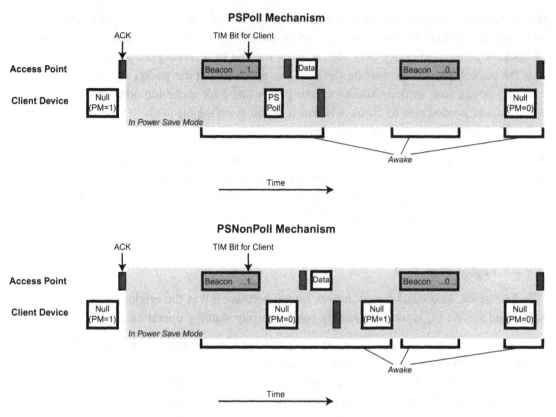

Figure 6.1: Wi-Fi Legacy Power Save

beacon. That doesn't guarantee that the client will hear the beacon at exactly that time, however. Beacons can be delayed if the air is occupied at that time. Furthermore, because beacons are sent out as broadcasts, the client might just miss the beacon or the beacon can be collided with. If the client does hear the beacon, it can then go to sleep so long as no traffic is buffered for it.

Clients may also skip beacons. They would do this to save additional battery, at the expense of increasing the amount of time the frames would be buffered. Clients usually let the access points know how many beacons they will skip by sending a *listen interval* in their Association Request messages. A listen interval of 1 means that the client will wake for every beacon; a listen interval of 10 means that the client will wake only for every tenth beacon. Be careful, however; some clients do not follow the listen interval they state, waiting either for more or less beacons than they advertise.

The client signals that it is going to sleep by using the *power management* bit in any unicast frame it sends to the access point (except for non-Action management frames). The power management bit is in the Frame Control field for the frame. When the client sends a frame with the power management bit set and when it gets an Acknowledgement in response, it knows that the access point has heard the client's change of state and can now go to sleep. From this moment on, the access point will buffer frames, until the client sends any frame to the access point with the power management bit not set. That signals that the client is now awake, and can be sent packets as usual.

While the client is in power save mode, and it wakes to find that its TIM bit is set to signify that it has frames available for it, the client has two choices on how to gather those frames. The first choice is known as the *PSPoll* mechanism, and uses the *Power Save Poll* (PS Poll) frames. After the beacon with the client's TIM bit set, the client would send a PS Poll frame to the access point. This frame, which is usually acknowledged right away, triggers the access point to deliver exactly one of the buffered frames for the client. That buffered frame is put into the transmit queue, using the appropriate access category for WMM. The frame that is sent also has its *More Data* bit in the Frame Control field set if there are subsequent frames that are buffered. Once the client has the frame, it can chose to send another PS Poll to get another frame. This one-PS-Poll/one-data-frame exchange continues until the access point's buffer is drained or the client wishes to sleep more.

The other option the client has is to use the *PSNonPoll* mechanism. This mechanism is quite simple: the client simply sends a data frame, usually a Null data frame, stating that it is no longer sleeping, by clearing the power management bit. The access point will proceed to queue all of the buffered frames, each using its own WMM access category. The client can then wait for a certain amount of time, hoping that it got all of the frames it was going to get, after which it can send another Null data frame, signifying it is going back to sleep. Any frames that may have still been in a transmit queue might get buffered again by the access point, for a later PSNonPoll exercise. The advantage of the PSNonPoll mechanism is that it is simple and doesn't require a significant back-and-forth. The disadvantage is that the client has no way of knowing if there are any remaining frames for it, without going to sleep and waiting for the next beacon.

The choice between PSPoll and PSNonPoll modes is often left up to the client's software implementation, and not exposed to you. However, some clients do give a choice up front, or have specific behavior where they will use one method or the other, depending on how aggressive you set its power save settings to (using a slider, say). It should be clear that neither mode is good for quality-of-service traffic, because the client can be forced to wait as much as a beacon interval (times its listen interval) before it finds out traffic is available. If the beacon interval is set to the typical 100 milliseconds, and the listen interval is 10, then that can be up to a second of delay.

Broadcast and multicast frames are also covered in the legacy scheme. However, no polling is necessary for those frames to be delivered. Instead, the access point sets aside a certain number of the beacons for multicast traffic. If *any client* on the access point is in legacy power save mode, the access point will buffer all multicast traffic. The special beacons known as *Delivery Traffic Indication Messages* (the poorly named DTIM) are just like regular beacons, except that they come every so many beacons—when the next one is coming is signaled as a part of the TIM in every beacon—and they signal if multicast traffic is buffered. If multicast traffic is buffered, the TIM has the zeroth bit, corresponding to AID 0, set. If clients receive a beacon with that bit set, they know that the next frames coming from the access point will be all of the multicast frames buffered. Each multicast frame, except for the last one, will have the More Data bit set. Thus, clients can stay awake to collect all multicast traffic, and then go back to sleep after the last multicast data frame, with the cleared More Data bit, comes through. (Of course, if that last frame is lost, being multicast, the clients will have to decide on their own when to return to sleep.) The consequence of the all-or-nothing multicast buffering is that multicast traffic on Wi-Fi when any device is in power save is not generally suitable for real-time traffic! Look for architectures that provide solutions for this problem if real-time multicast is a priority for your network.

Finally, I haven't gone into details on how the TIM bits are compressed. It is not easy to read the TIM bits by hand, but a good wireless protocol analyzer will be able to read them for you, and let you know which AIDs are set in any beacon.

6.0.2.2 WMM Power Save

To provide power saving while the mobile device is in a call, the Wi-Fi Alliance came up with the second power saving technique, WMM Power Save. This technique, based on the quality-of-service additions in the 802.11e amendment to the standard, acts as a parallel scheme to the legacy one, using similar concepts but in a way that avoids having to wait for beacons and can apply on a per-access-category basis.

If you notice, there is nothing in the standard that prevents clients that are using the legacy power save scheme from ignoring beacons, for the most part, and sending PS Polls whenever they want. If the client were sure that there is going to be a packet for it waiting every so often—say, 20 milliseconds—then it could just send PS Polls every 20 milliseconds, collect its data, and have real-time power save. Of course, this doesn't happen for legacy power save, because the client has no guarantee that it won't get some other frames rather than what it is looking for. However, this is the concept that WMM Power Save builds on.

WMM Power Save is optional, and support for it is signaled by the WMM information elements in the Association messages and the beacons. Unlike with legacy power save, WMM Power Save (capitalized, as it is a formal name) is aware of the WMM access categories and can apply to a subset of them. The two subsets are *delivery-enabled* access categories and *trigger-enabled* access categories.

First, let's start with the polling protocol. The client no longer checks the beacons to see if there is traffic. Instead, it is responsible for knowing that traffic is waiting for it, and how often. For phones, this is not a problem, as voice is bidirectional and consistent. Instead of sending a PS Poll frame, or using the PSNonPoll mechanism, the phone sends data frames in access categories that it has specified to be trigger-enabled. The access point looks for those data frames, and uses that as a trigger—just as it does in legacy with Power Save Poll frames—sending packets in response from the power save buffer. Those packets, however, can only come from the delivery-enabled access categories. Which categories are delivery- and trigger-enabled are usually specified in the Association Request from the client—there, a bitmask specifies which categories are legacy and which are delivery and trigger enabled together—or in TSPEC messages, which we will come to later.

Here's a common example. The phone associates, and tells the access point that it wants the voice category (AC_VO) to be delivery- and trigger-enabled. That means that the other three categories work on the legacy scheme. If packets come in for those other categories while the client is asleep, the TIM bit on the beacon will be set and the client will use legacy power save mechanisms to get the frames. But when a voice packet is sent to the access point, the access point silently holds onto the packet. The only way the client can get the voice packet is to send a voice packet of its own. When it does, that causes the access point to respond with one or more voice packets in its buffer. Unlike with legacy power save, the client can ask for more than one packet at a time. Using the concept of a *service period,* which is set at Association time by the client and specifies the number of frames the client wants to get for every trigger (either two, four, six, or all), the access point will send out the correct number of frames. The last frame, whether because the buffer is empty or the service period has been exceeded, will have a special *end of service period* (EOSP) bit set in the QoS header. Once the client gets that frame, it can go back to sleep.

As you can see, the legacy and WMM Power Save schemes operate simultaneously and independently. The only overlap is that the client goes into to power save mode for both schemes simultaneously. This means that devices that are actively using WMM Power Save should never use the PSNonPoll method during that time, because the client waking up from power save mode will cause the access point to send all frames, whether they are from the legacy or WMM Power Save access categories.

The capability to support WMM Power Save should be considered nearly mandatory for most voice equipment. Some mobile devices use proprietary mechanisms that may or may not be supported by every access point, but the trend is towards using WMM Power Save. Of course, the problem with WMM Power Save is that it works well only for voice, but that is not a concern for us in this book.

6.1 Technologies that Address Voice Mobility with Wi-Fi

The introduction of WMM into Wi-Fi allowed voice to now have a prioritized way of being carried over the air. But other basic elements of providing a toll-quality voice system needed to be put in place. Many of these newer techniques borrow from how things are done on the cellular networks, and work is only beginning now to try to standardize certain parts of them. How vendors—access point and phone—implement these features goes a long way towards determining how well the voice mobility network will work.

6.1.1 Admission Control: The Network Busy Tone

The first concept that is needed is providing a "network busy" tone. Every networking technology has its capacity limits, and given the discussion in Section 6.0.1.1, Wi-Fi can have some fairly severe ones. As the number of voice calls exceeds the network capacity, the air becomes crowded with aggressive, high-priority voice packets. This causes increased loss and can end up hurting the quality of every active call on the air in that region.

The solution is to not let in the calls that cause the capacity to be exceeded. The goal is to provide the caller with that network busy signal. (If you have never heard a network busy tone before, on standard telephones, they sound like the usual caller busy tone, but they beep at a much faster rate.)

When used for voice, admission control is often called *Call Admission Control* (CAC), pronounced "cack." There are two methods currently in use for Wi-Fi to provide this.

6.1.1.1 SIP-Based Admission Control

The first method is to rely on the call setup signaling. Because the most common mechanism today is SIP, we can refer to this as *SIP-based admission control*. The idea is fairly simple. The access point, most likely in concert with a controller if the architecture in use has one, uses a firewall-based flow-detection system to observe the SIP messages as they are sent from the phones to the SIP servers and back. Specifically, when the call is initiated, either by the phone sending a SIP Invite, or receiving one from another party, the wireless network determines whether there is available capacity to take the call. If there is available capacity, then the wireless network lets the messages flow as usual, and the call is initiated.

On the other hand, if the wireless network determines that there is no room for the call, it will intercept the SIP Invite messages, preventing them from reaching the other party, and interject its own message to the caller (as if from the called party, usually), with one of a few possible SIP busy statuses. The call never completes, and the caller will get some sort of failure message, or a busy tone.

Other, more advanced behaviors are also possible, such as performing load balancing, once the network has determined that the call is not going to complete.

The advantage of using SIP flow detection to do the admission control is that it does not require any added sophistication on the mobile devices than they would already have with SIP. Furthermore, by having that awareness from tracking the SIP state, the network can provide a list of both calls in progress and registered phones not yet in a call. The disadvantage is that this system will not work for SIP calls that are encrypted end-to-end, such as being carried over a VPN link.

6.1.1.2 WMM Admission Control

Building on even more of the specification in the 802.11e quality-of-service amendment is *WMM Admission Control*. This specification and interoperability program from the Wi-Fi Alliance, which is required to achieve Voice Enterprise certification (see Section 6.3), uses an explicit layer-2 reservation scheme. This scheme, in a similar vein as the lightly used *RSVP protocol* (RFC 2205), requires the mobile device to reach out and request resources explicitly from the access point, using a new protocol built on top of 802.11 management frames.

This protocol is heavily dependant on the concept of a *traffic specification* (TSPEC). The TSPEC is created by the mobile phone, and specifies how much of the air resources either or both directions of the call (or whatever resource is being requested) will be taken. The access point processes the request as an *admission controller* (a function often placed literally on the controller, by coincidence), which is in charge of maintaining an account of which clients have requested what resources and whether they are available.

The overall protocol is rather simple. The mobile device, usually when it determines that it has a call incoming our outgoing, will send an *Add Traffic Stream* (ADDTS) *Request* message (a special type of Action management frame) to the access point, containing the TSPEC that will be able to carry the phone call. The access point will decide whether it can carry that call, based on whatever scheme it uses (see following discussion), and send an *ADDTS Response* message stating whether the stream was admitted.

WMM Admission Control can be set to mandatory or optional for each access category. For example, WMM Admission Control can be required for voice and video, but not for best effort and background data. What this would mean is that no client is allowed to transmit voice or video packets without first requesting and being granted admission for flows in those access categories, whereas all clients would be allowed to freely transmit best effort and background data as they see fit. Which access categories require admission control is signaled as a part of the WMM information element, which goes out in beacons and some other frames.

For WMM Admission Control, it is worth looking at the details of the concepts. The main concept is one of a traffic stream itself, and how it is identified and recognized. Traffic streams are represented by *Traffic Identifiers* (TID), a number from 0–7 (the standard allows up to 15, but WMM limits this to only 7) that represents the stream. Each client gets its own set of eight TIDs to use.

Each traffic stream, represented by its TID, maps onto real traffic by naming which of the eight priority values in WMM will belong to this traffic stream (see Table 6.1). Thus, if the phone intends to send and knows it is going to receive priority 7—recall that this is the highest of the two voice AC priorities—it can establish a traffic stream that maps priority 7 traffic to it, and get both sides of the call. In order for that to work, the client can specify whether the traffic stream is upstream-only, downstream-only, or bidirectional. It is possible for the client to request both an upstream-only and downstream-only stream mapping to the same priority (different TIDs, though!), if it knows that the airtime used by the downstream side is different than the upstream side—useful for video calls—or it may request both at once in one TID, with the same airtime usage. All of this freedom leads to some complexity, but thankfully there is a rule preventing there from being more than one downstream and one upstream flow (bidirectional counts as one of each) for each access category. Thus, the AC_VO voice access category will only have one admitted bidirectional phone call in it at any given time.*

The client requests the traffic stream using the TSPEC.

Table 6.3 shows the contents of the TSPEC that is carried in an ADDTS message.

There's quite a lot of information in a TSPEC, so let's break it down slowly, using the example of a 20 millisecond G.711 (nearly uncompressed) one-way traffic flow:

- The *TS Info* field (see Table 6.4) identifies the TID for the stream, the priority of the data frames that belong to this stream, what direction the stream is going in (00 = up, 01 = down, 10 = reserved, 11 = bidirectional), and whether the AC the stream belongs to is to be WMM Power Save delivery enabled (1) or not (0). The rest of the fields are not used in WMM Admission Control, and have specific values that will never change (Access Policy = 01, the rest are 0).

- The *Nomimal MSDU Size* field mentions the expected packet size, with the highest-order bit set to signify that the packet size never changes. G.711 20ms packets are 160

---

* Of course, there had to be a catch. Some devices can carry two calls simultaneously, if they renegotiate their one admitted traffic stream to take the capacity of both. Because WMM Admission Control views flows as being only between clients and access points, the ultimate other endpoint of the call does not matter. However, this is not something you would expect to see in practice.

Table 6.3: WMM admission control TSPEC

TS Info	Nominal MSDU Size	Maximum MSDU Size	Minimum Service Interval	Maximum Service Interval	Inactivity Interval	Suspension Interval	Service Start Time	
3 bytes	2 bytes	2 bytes	4 bytes	4 bytes	4 bytes	4 bytes	4 bytes	...

	Minimum Data Rate	Mean Data Rate	Peak Data Rate	Maximum Burst Size	Delay Bound	Minimum PHY Rate	Surplus Bandwidth Allowance	Medium Time
...	4 bytes	4 bytes	4 bytes	4 bytes	4 bytes	4 bytes	2 bytes	2 bytes

Table 6.4: The TS info field

	Traffic Type	TID	Direction	Access Policy	Aggregation	WMM Power Save	Priority	TSInfo Ack Policy	Schedule	Reserved
Bit:	0	1–4	5–6	7–8	9	10	11–13	14–15	16	17–23

bytes of audio, plus 12 bytes of RTP header, 8 bytes of UDP header, 20 bytes of IP header, and 8 bytes of SNAP header, creating a data payload (excluding WPA/WPA2 overhead) of 208 = 0xD0. Because the packet size for G.711 never changes, this field would be set to 0x80D0.

- The *Maximum MSDU Size* field specifies what the largest a data packet in the stream can get. For G.711, that's the same as the nominal size. There is no special bit for fixed sizes, so the value is 208 = 0x00D0. This can also be left as 0, as it is an optional field.

- The *Inactivity Interval* specifies how long the stream can be idle—no traffic matching it—in microseconds, before the access point can go ahead and delete the flow. 0 means not to delete the flow automatically, and that's the common value.

- The *Mean Data Rate* specifies, in bits per second, what the expected throughput is for the stream. For G.711, 208 bytes every 20 milliseconds results in a throughput of 83200 bits per second.

- The *Minimum Data Rate* and *Peak Data Rate* specify the minimum and maximum throughput the traffic stream can expect. These are optional and can be set to 0. For G.711, these will be the same 83,200 bits per second.

- The *Minimum PHY Rate* field specifies what the physical layer data rate assumptions are for the stream, in bits per second. If the client is assuming that the data rate could drop as low as 6Mbps for 802.11ag, then it would encode the field at 6Mbps = 6,000,000bps = 0x005B8D80.

- The *Surplus Bandwidth Allowance* is a fudge factor that the phone can request, to account for that packets might be retransmitted. It's a multiplier, in units of 1/8192nds. A value of 1.5 times as an allowance would be encoded as 0x3000 = 001.1000000000000, in binary.

- The other fields are unused by the client, and can be set to 0.

In other words, the client simply requests the direction, priority, packet size, data rate, and surplus allowance.

The access point gets this information, and churns it using whatever algorithms it wants—this is not specified by the standard, but we'll look at what sorts of considerations tend to be used in Section 6.1.1.3. Normally, we'll assume that the access point knows what percentage of airtime is available. The access point will then decide how much airtime the requested flow will take, as a percentage, and see whether it exceeds its maximum allowance (say, 100% of airtime used). If so, the flow is denied, and a failing ADDTS Response is sent. If not, the access point updates its measure of how much airtime is being used, and then allows the flow. The succeeding ADDTS Response has a TSPEC in it that is a mirror of the one the client requested, except that now the *Medium Time* field is filled in. This field specifies exactly how much airtime, in 32-microsecond units per second, the client can take for the flow.

The definition of how much airtime a client uses is based on what packets are sent to it or that it sends as a part of a flow. Both traffic sent by the client to the access point and sent by the access point to the client are counted, as well as the times for any RTSs, CTSs, ACKs, and interframe spacings that are between those frames. Another way of thinking about it is that the time from the first bit of the first preamble to the last bit of the last frame of the TXOP counts, including gaps in between. In general, you will never need to try to count this. Just know that WMM Admission Control requires that the clients count their usage. If they exceed their usage in the access category they are using, they have to send all subsequent frames with a lower access category—and one that is not admission control enabled—or drop them.

One advantage of WMM Admission Control is that it works for all traffic types, without requiring the network to have any smarts. Rather, the client is required to know everything about the flows it will both send and receive, and how much airtime those flows will take. The network just plays the role of arbiter, allowing some flows in and rejecting others. Thus, if the client is sufficiently smart, WMM Admission Control will work whether the protocol is SIP, H.323, some proprietary protocol, or even video or streaming data. The disadvantage of that, however, is that the client is required to be smart, and all of its pieces—from wireless to phone software—have to be well integrated. That pretty much eliminates most softphones, and brings the focus squarely on purpose-built phones. Furthermore, the client needs to know what type of traffic the party on the other side of the call will send to it. Some higher-level signaling protocols can convey this, such as with SDP within SIP, but doing so may be optional and may not always be followed. For a phone talking to a media gateway, for example, the phone needs to know exactly how the media gateway will send its traffic, including knowing the codec and packet rate and sizing, *before* it can request airtime. That can lead to situations in which the call needs to be initiated and agreed to by both parties before the network can be asked for permission to admit the flow, meaning that the call might have to be terminated by the network midway through ringing, if airtime is not available. Because WMM Admission Control is so new—by the time of publication, WMM Admission Control should be launching shortly and large amounts of devices may not yet be available—it remains to be seen how well all of the pieces will fit together. It is notoriously difficult for general-purpose devices to be built that run the gamut of technologies correctly, and so these new programs might be more useful for highly specific purpose-built phones.

6.1.1.3 How the Capacity Is Determined

Through either admission control scheme, the network needs to keep track of how much capacity is available. From the previous discussions on the effects of RF variability and cellular overlap, you can appreciate that this is a difficult problem to completely solve. As devices get further away from the access points, data rates drop. Changing levels of interference, from within the network or without, can cause increasing retransmissions and easily overrun surplus bandwidth allowances.

In the end, networks today adopt one of two stands, and may even show both to the user. The more complicated stand for the network—but simpler for the user—is for the network to automatically take the variability of RF into account, and to determine its own capacities. In systems that do this, there is no notion of a static maximum number of calls. Instead, the system accepts however many calls as it can handle. If conditions change, and fewer calls can be handled in the system, the network reserves the right to proactively end a client's reservation, often in concert with load balancing.

The other stand, simpler for the network but far more complicated for the user, is for the administrator to be required to enter the maximum number of calls per access point (or some other static metric). The idea here is that the administrator or installer is assumed to have gone through a planning process to determine how many calls can be *safely* allowed per access point, while still leaving room for best effort data. That number is usually far lower than the best-case maximum capacity, and is designed to be a low water mark: barring external changes, the network will be able to achieve that many calls most of the time. This number is then manually input into the wireless network, which then counts the number of calls. If the maximum number of calls is reached on that access point, the system will not let any more in. These static metrics may be entered either as the number of calls, or a percentage of airtime. Systems that work as a percentage of airtime can sometimes take in a padding factor to allow for calls that are roaming into the network.

Setting these values can be fraught with difficulty. Pick a number that's too low, and airtime is being wasted. Pick a number that's too high, however, and sometimes call quality will suffer. Even percentage of airtime calculations are not very good, because they may not take into account airtime that is unusable because of variable channel conditions or co-channel interference that the access point cannot directly see, such as client-to-client interference (see Sections 5.4.6 and 5.4.7 for some of these problems).

All in all, you might find vendors recommending setting the values to a low, safe value that allows for voice to work even if there is plenty of variability in the network. This works well for networks that are predominantly data-oriented, but voice-only networks cannot usually afford that luxury.

6.1.2 Load Balancing

Load balancing is the ability for the network to steer or direct clients towards more lightly loaded access points and away from more heavily loaded ones. As you may recall from the discussion in Section 5.2.3.3 for scanning and will see in the discussions in Section 6.2, the client decides to which access point the client will connect. However, the network has the ability to gently influence or guide the client's decision.

First, let's recap what is meant by wireless load. The previous discussion on admission control first introduced the concept of counting airtime or calls. This is one measure of load—a real-time one. However, this counts only phones in active calls. There is likely to be far more phones not in active calls, and these should be balanced as well. The main reason for balancing inactive phones is that the network has little ability, once the phone starts a call, to transfer the phone to another access point without causing the call to fail going through. To avoid that, load balancing techniques attempt to establish a more even balance up front. The thinking goes that if you can get the phones evenly distributed when the connect to the network, then you have a better shot at having the calls they place equally distributed as well.

6.1.2.1 Mechanics of Load Balancing

Let's start with the basic mechanics of load balancing. Because the client chooses which access point to associate to, based on scanning operations, the only assured way to prevent a client from associating to an overloaded access point is for that overloaded access point to ignore the client. The access point can do this in a few ways. When the client sends probe requests, trying to discover whether the SSID it wants is still available on the access point, the access point can ignore the request, not sending out a probe response. Hopefully, the client will not enter on the basis of that alone. However, the client may have scanned before, when it could have (but chose not to) enter the access point, and may remember a prior probe response. Or, it can see the beacon, and so it knows that the access point is, in fact, providing the service in any event.

To prevent the client from associating, then, the access point has no choice but to ignore or reject Authentication and Association Request messages from the client. This will have the desired effect of preventing a burdensome load from ending up on the access point, but may not cause the client to choose the correct access point quickly. (More will be discussed in Section 6.2.)

Assuming, for the moment, that load balancing is effective in causing clients to distribute their load evenly, we need to look at what the consequences of balancing load are.

6.1.2.2 Understanding the Balance

Explicit in the concept of load balancing is that it is actually possible to balance load—that is, to transfer load from one access point to another in a predictable, meaningful fashion. To understand this, we need to look at how the load of a call behaves from one access point to another, assuming that neither the phone nor the access points have moved.

Picture the environment in Figure 6.2. There are two access points, the first one on channel 1 and the second on channel 11. A mobile phone is between the two access points, but physically closer to access point 1. The network has two choices to distribute, or balance, the load. The network can try to guide the phone into access point 1, as shown in the top of the figure, or it can try to guide the phone into access point 2, as shown in the bottom of the figure.

This is the heart of load balancing. The network might choose to have the phone associate to access point 2. We can imagine that access point 1 is more congested—that is, it has more phone calls currently on it. In the extreme case, access point 1 can be completely full, and might be unable to accept new calls. The advantage of load balancing is that the network can use whatever information it sees fit—usually, loads—to guide clients to the right access points.

There are a few wrinkles, however. It is extremely unlikely that, in a non-channel-layered environment, the phone is at the same distance from each of the two access points. It is

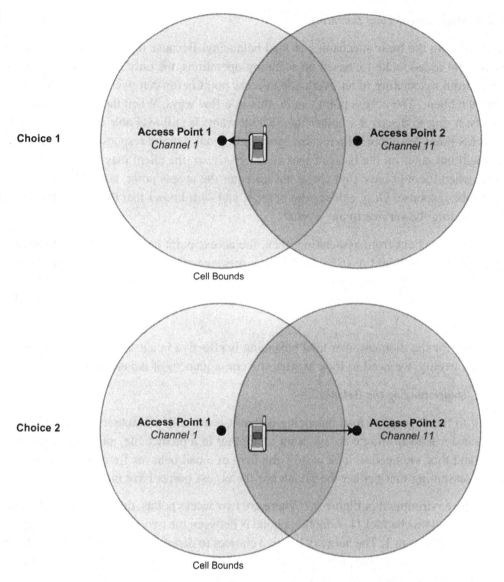

Figure 6.2: Load Balancing Across Distances

more likely that the phone is closer to one access point than another. The consequence of the phone being closer to an access point is that it can get higher data rates and SNR, which then allows it to take less airtime and less resources. It may turn out that, if the network chooses to move the phone from access point 2 to access point 1, that the increase in data rate because the phone is closer to access point 1 allows possibly two calls in for access point 2. In this case, the same call produces *unequal load* when it is applied to different access points, all else being equal.

For this reason, within-band load balancing has serious drawbacks for networks that do not use channel layering. Load balancing should be thought of as a way to distribute load across equal resources, but within-band load balancing tends to work rather differently and can lead to performance problems. If the voice side of the network is lightly used—such as having a small CAC limit—and if the impact of voice on data is not terribly important, then this sort of load balancing can work to ease the rough edges on networks that were not provisioned properly. However, for more dense voice mobility networks, we need to look further. The concept of load balancing among near equals does exist, however, with *band load balancing*. Band load balancing can be done when the phones support both the 2.4 GHz and 5 GHz bands (some newer ones do) and the access points are dual-radio, having one 2.4 GHz radio and another 5 GHz radio in the same access point. In this case, the two choices are collocated: the client can get similar SNR and data rates from either radio, and the choice is much closer to one-to-one. Figure 6.3 illustrates the point.

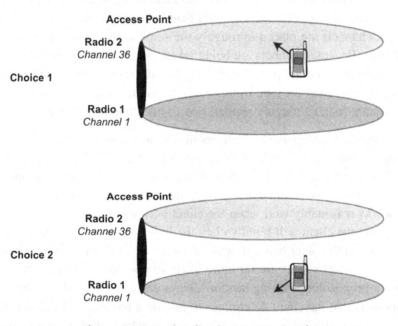

Figure 6.3: Load Balancing Across Bands

A variant of band load balancing is *band steering*. With band steering, the access point is not trying to achieve load balancing across the two bands, but rather is prioritizing access to one band over the other—usually prioritizing access to the 5GHz band for some devices. The notion is to help clear out traffic from certain devices, such as trying to dedicate one band for voice and another for data. Using differing SSIDs to accomplish the same task is also possible, and works across a broader range of infrastructures.

There are differences between the two bands, of course, most notably that the 5GHz band does not propagate quite as far as the 2.4GHz band. The 5GHz band also tends to be unevenly accessed by multiband phones: sometimes, the phones will avoid the 5GHz band unless absolutely forced to go there, leading to longer connection times. On the other hand, the 2.4GHz band is subject to more microwave interference. And, finally, this mechanism will not work for single-band phones. (The merits of each band for voice mobility are summarized later in this chapter.) Nevertheless, band load balancing is an option for providing a more even, one-to-one balance.

For environments with even higher densities, where two channels per square foot are not enough, or where the phones support only one band or where environmental factors (such as heavy microwave use) preclude using other bands, channel layering can be employed to provide three, four, or many more choices per square foot. Channel layering exists as a benefit of the channel layering wireless architecture, for obvious reasons (see Section 5.2.4.7), and builds upon the concept of band load balancing to create *collocated channel load balancing*. The key to collocated channel load balancing is that the access points that are on different channels are placed in roughly the same areas, so that they provide similar coverage patterns. Because channels are being taken from use as preventatives for co-channel interference and are instead being deployed for coverage, channel layering architectures are best suited for this. In this case, the phone now has a choice of multiple channels per square foot, of roughly similar, one-to-one coverage. Figure 6.4 illustrates this.

Bear in mind that the load balancing mechanisms are in general conflict with the client's inherent desire to gain access to whatever access point it chooses and to do so as quickly as possible (see Section 6.2.2). The network is required to choose an access point and then must ignore the client, if it should come in and attempt to learn about the nonchosen access points. This works reasonably well when the client is first powered up, as the scanning table may be empty and the client will blindly obey the hiding of access points as a part of steering the load. On the other hand, should the client already have a well-populated scanning table—as voice clients are far more likely to do—load balancing can become a time-consuming proposition, causing handoff delays and possible call loss. Specifically, what can happen is that the client determines to initiate a handoff and consults the information in its scanning table, gathered from a time when all of its entries were options, based on load. The client can then directly attempt to initiate a connection with an access point, sending an Authentication or Reassociation frame (depending on whether the client has visited the access point before) to an access point that may no longer wish to serve the client. The access point can ignore or reject the client at that point, but usually clients are far less likely to abandon an access point once they choose to associate than when they are scanning. Thus, the client can remain outside the access point, persistently knocking on the door, if you will, unwilling to take the rejection or the ignoring as an answer for possibly long periods of time. This provides an additional reason why load balancing in an

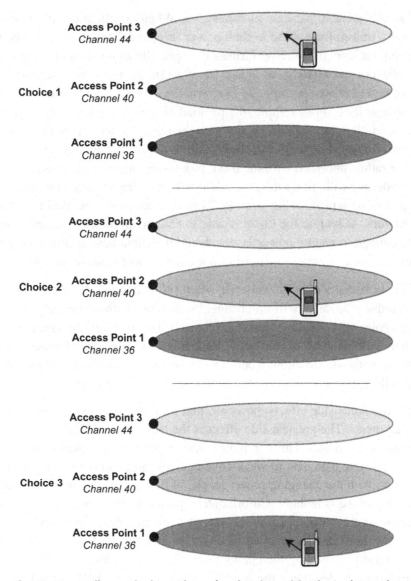

Figure 6.4: Collocated Channel Load Balancing with Channel Layering

environment where multiple handoffs are likely can have consequences for the quality of voice calls.

6.1.3 Power Control

Power control, also known as *transmit power control* (TPC), is the ability of the client or the access point to adjust its transmit power levels for the conditions. Power control comes in two flavors with two different purposes, both of which can help and hurt a voice mobility network. The first, most common flavor of power control is vested in the client. This TPC exists to allow the client to increase its battery life. When the client is within close range to the access point, transmitting at the highest power level and data rate may not be necessary to achieve a similar level of voice performance. Especially as the data rates approach 54Mbps for 802.11a and 802.11g, or higher for 802.11n, the preamble (Section 5.4.2) and per-packet backoff overhead becomes in line with the over-the-air resource usage of the voice data payload itself. For example, the payload of a voice packet at the higher data rates reduces to around 20 microseconds, on par with the preamble length for those data rates. In these scenarios, it makes sense for the client to back off on its power levels and turn off portions of the radio concerned with the more processing-intensive data rates, to extend battery life while in a call. To do this, the client will just directly reduce its transmit power levels, as a part of its power saving strategy. This mechanism can be used for good effect within the network, as long as the client is able to react to an increase in upstream data loss rates quickly enough to restore power levels should the client have turned power levels down too low for the range, or if increasing noise begins to permeate the channel.

The other TPC is vested within the network. Microcell networks, specifically, use access point TPC to reduce the amount of co-channel interference without having to relocate or disable access points. By reducing power levels, cell sizes in every direction are reduced, keeping in line with the goals of microcell. Reducing co-channel interference is necessary within microcell networks, to allow a better isolation of cells from fluctuations in their neighboring cells, especially those related to the density of mobile clients.

Network TPC has some side effects, however, that must be taken into account for voice mobility deployments. The greatest side effect is the lack of predictability of coverage patterns for the access points. This can have a strong effect on the quality of voice, because voice is more sensitive than data to weak coverage, and areas where voice performs poorly can come and go with the changing power levels, of both the access point the client is associated to and of the neighbors. Unfortunately, power levels in microcell networks usually fluctuate on the order of a few seconds or a few minutes, especially when clients are associated, as the network tries to adapt its coverage area to avoid causing the increase in packet rate and traffic caused by the clients from affecting neighboring cells. Site surveys, which are performed to determine the coverage levels of the network, are always snapshots in time and cannot take TPC into account. However, the TPC variation is necessary for proper microcell operation, and unfortunately needs to happen when phones are associated and in calls. Therefore, it can cause a strong network-induced variation in call quality. It is imperative, in microcell deployments, for the coverage and call quality to be continuously monitored, to ensure that the TPC algorithms are behaving properly. Follow the manufacturer's recommendations, as you may find in Voice over WLAN design guides, to ensure that problems can be detected and handled accordingly. (One such guide is Cisco's *Voice over Wireless LAN 4.1 Design Guide*, which has general wireless information that applies to microcell voice architectures in general and thus is a useful resource.)

6.1.4 Voice-Aware Radio Resource Management

The concept of *voice-aware* radio resource management is to build upon the measurements used for determining network capacity and topology, and integrate them into the decision-making process for dynamic microcell architectures (Section 5.2.4.6).

Basic radio resource management is more concerned with establishing minimum levels of coverage while avoiding interference from neighboring access points and surrounding devices. This is more suitable for data networks. Voice-aware RRM shifts the focus towards providing a more consistent coverage that voice needs, often adjusting the nature of the RRM process to avoid destroying active voice calls. Voice-aware RRM is a crucial leg of voice mobility deployments based on microcell technology. (Layered or virtualized deployments do not use the same type of voice-aware RRM, as they have different means of ensuring high voice quality and available resources.)

The first aspect of voice-aware radio resource management is ironically to disable radio resource management. Radio resource management systems work by the access points performing scanning functions, rather similar to those performed by clients when trying to hand off (see Section 6.2.2). The access point halts service on a channel, and then exits the channel for a short amount of time to scan the other channels to determine the power levels, identities, and capacities of neighboring access points. These neighboring access points may be part of the same network, or may belong to other interests and other networks. Unlike with client scanning, in which the client can go into power save to inform the access point to buffer frames, however, access point scanning has no good way for clients to be told to buffer frames. Moreover, whereas client scanning can go off channel between the packets of the voice call, only to return when the next packet is ready, an access point with multiple voice calls will likely not have any available time to scan in a meaningful way. In these cases, scanning needs to be disabled. In RRM schemes without voice-aware services, administrators often have to disable RRM by hand, thus nullifying the RRM benefits for the entire network. Voice-aware RRM, however, has the capability to turn off scanning on a temporary basis for each access point, when the access point is carrying voice traffic. There are unfortunately two downsides to this. The first is that RRM is necessary for voice networks to ensure that coverage holes are filled and that the network adapts to varying density. Disabling the scanning portion of RRM disables RRM, effectively, and so voice-aware RRM scanning works best when each given access point does not carry voice traffic for uninterrupted periods of time. Second, RRM scanning is usually the same process by which the access points scan for wireless security problems, such as rogue access points and various i ntrusions. Disabling scanning in the presence of voice leaves access points with voice more vulnerable, which is unacceptable for voice mobility deployments. Here, the solution is to deploy dedicated air monitors, either as additional access points from the same network vendor, but set to monitor rather than serve, or from a dedicated WLAN security monitoring vendor, as an independent overlay solution.

The second aspect of voice-aware radio resource management is in using coverage hole detection and repair parameters that are more conservative. Although doing so increases the likelihood of co-channel interference, which can have a strong downside to voice mobility networks as the network scale and density grows, it is necessary to ensure that the radio resource management algorithms for microcells do not leave coverage holes stand by idle. As mentioned in the previous section on power control, coverage holes disproportionately affect the quality of voice traffic over data traffic. Increasing the coverage hole parameters ensures that these coverage holes are reduced. Radio resource management techniques often detect the presence of a coverage hole by inferring them from the behavior of a client. RRM assumes that the client is choosing to hand off from an access point when the loss becomes too high. When this assumption is correct, the access point will infer the presence of a coverage hole by noticing when the loss rate for a client increases greatly for extended periods of time. This is used as a trigger that the client must be out of range, and informs the access point to increase its power levels. It is better for voice mobility networks for the coverage levels to be increased prior to the voice mobility deployment, and then for the coverage hole detection algorithm to be made less willing to reduce coverage levels. Unfortunately, the coverage hole detection algorithms in RRM schemes are proprietary, and there are no settings that are consistent from vendor to vendor. Consult your microcell wireless network manufacturer for details on how to make the coverage hole detection algorithm be more conservative.

The final aspect of voice-aware RRM is for when proprietary extensions are used by the voice client and are supported by the network. These extensions can provide some benefit to microcell deployments, as they allow the network to alter some of the tuning parameters that clients use to hand off (see Section 6.2). Unfortunately, the aspects of voice-aware radio resource management trade off between coverage and quality of service, and so operating these networks can become a challenge, especially as the density or proportion of network use of voice increases. Monitoring tools for voice quality are especially important in these networks such as those mentioned in Section 6.1.6.

6.1.5 Spectrum Management

Spectrum management is the technology used by virtualization architectures to manage the available wireless resources. Unlike radio resource management, which is focused on adjusting the available wireless resources on a per-access-point basis, to ensure that the clients of that access point receive reasonable service without regard to the neighbors, spectrum management takes a view of the entire unlicensed Wi-Fi spectrum within the network, and applies principles of capacity management to the network to organize and optimize the layout of channel layers. In many ways, spectrum management is radio resource management, applied to the virtualized spectrum, rather than individual radios.

Spectrum management focuses on determining which broad swaths of unlicensed spectrum are adequate for the network or for given applications within the network. One advantage of channel layering is that channels are freed from being used to avoid interference, and thus can be used to divide the spectrum up by purposes. Much as regulatory bodies, such as the FCC, divide up the entire radio spectrum by applications, setting aside one band for radio, another for television, some for wireless communications, and so on, administrators of virtualization architectures can use spectrum management to divide up the available channels into bands that maximize the mutual capacity between applications by separating out applications with the highest likely bandwidth needs onto separate channel layers.

The constraints of spectrum management are fairly simple. A deployment has only a given number of access points. The number and position of the access points limits the number of independent channel layers that can be provided over given areas of the wireless deployment area. It is not necessary for every channel layer to extend across the entire network—in fact, channel layers are often created more in places with higher traffic density, such as libraries or conference centers. The number of channel layers in a given area is called the network *thickness*. Spectrum management can detect the maximum number of channel layers that can be created given the current deployment of access points, and is then able to determine when to create multiple layers by spreading channel assignments of close access points, or when to maximize signal strength and SNR by setting close access points to the same channel. Thus, spectrum management can determine the appropriate thickness for each given square foot. For 802.11n networks, spectrum management is able to work with channel widths, as well as band and channel allocations, and is thus able to make very clear decisions about doubling capacity by arranging channels as needed.

Spectrum management also applies the neighboring-interference-avoidance aspects that RRM uses to prevent adjacent networks from being deployed in the same spectrum, if it can at all be avoided. Because there is no per-channel performance compromise in compressing the thickness of the network, spectrum management can avoid some of the troublesome aspects of radio resource management when dealing with edge effects from multiple, independent networks. Furthermore, spectrum management is not required to react to transient interference, as the channel layering mechanism is already better suited to handle transient changes through RF redundancy. This allows spectrum management to reserve network reconfigurations for periods of less network usage and potential disruption (such as night), or to make changes at a deliberate pace that ensures network convergence throughout the process.

6.1.6 Active Voice Quality Monitoring

A large part of determining whether a voice mobility network is successful is in monitoring the voice quality for devices on the network. When the network has the capability to measure this for the administrator on an ongoing basis, the administrator is able to devote attention to other, more pressing matters.

Active voice quality monitoring comes in a few flavors. SIP-based schemes are capable of determining when there is a voice call. This is often used in conjunction with SIP-based admission control. With SIP calls, RTP is generally used as the bearer protocol to carry the actual voice, SIP-based call monitoring schemes can measure the loss and delay rates for the RTP traffic that makes up the call, and report back on whether there are phones with suffering quality. In these monitoring tools, call quality is measured using the standard MOS or R-value metrics, as defined in Chapter 3 on voice quality.

SIP-based schemes can be found in a number of different manifestations. Wireline protocol analyzers are capable of listening in on a mirror port, entirely independent of the wireless network, and can report on upstream loss. Downstream loss, however, cannot be detected by these wireline mechanisms. Wireless networks themselves may offer built-in voice monitoring tools. These leverage the SIP-tracking functions already used for firewalling and admission control, and report on the quality both measured by uplink and downlink loss. Purely wireless monitoring tools that monitor voice quality can also be employed. Either located as software on a laptop, or integrated into overlay wireless monitoring systems, these detect the voice quality using over-the-air packet analysis. They infer the uplink and downlink loss rates of the clients, and use this to build out the expected voice quality. Depending on the particular vendor, these tools can be thrown off when presented with WPA- and WPA2-encrypted voice traffic, although that can sometimes be worked around.

Voice call quality may also be monitored by measurements reported by the client or other endpoint. RTCP, the RTP Control Protocol, may be transmitted by the endpoints. RTCP is able to encode statistics about the receiver, and these statistics can be used to infer the expected quality of the call. RTCP may or may not be available in a network, based on the SIP implementation used at the endpoints. Where available, RTCP encodes the percentage of packets lost, the cumulative number of packets lost, and interarrival packet jitter, all of which are useful for inferring call quality. At a lower layer, 802.11k, where it is supported, provides for the notion of traffic stream metrics (Section 6.2.6.6). These metrics also provide for loss and delay, and may also be used to determine call quality. However, 802.11k requires upgrades to the client and access point firmware, and so is not as prevalent as RTCP, and nowhere near as simple to set up as overlay or traffic-based quality measurements.

6.2 Inter-Access Point Handoffs

In a voice mobility network with Wi-Fi as a major component, we have to look at more than just the voice quality on a particular access point. The end-user of the network, the person with a phone in hand, has no idea where the access points are. He or she just walks around the building, going in and out of range of various access points in turn, oblivious to the state of the underlying wireless network. All the while, the user demands the same high degree of voice quality as if he or she had never started moving.

So now, we have to turn our focus towards the handoff aspect of Wi-Fi voice networks. Looking back on how Wi-Fi networks are made of multiple cells of overlapping coverage, we can see that the major sources for problems with voice are going to come from four sources:

1. How well the coverage extends through the building

2. How well the phone can detect when it is exiting the coverage of one access point

3. How well the phone can detect what other options (access points) it has available

4. How quickly the phone can make the transition from the old access point to the new one

Let's try to gain some more appreciation of this problem. Figure 6.5 shows the wireless environment that a mobile phone is likely to be dwelling within.

As the caller and the mobile phone move around the environment, the phone goes into range and out of range of different access points. At any given time, the number of access points that a client can see, and potentially connect to, can be on the order of a dozen or more in environments with substantial Wi-Fi coverage. The client's task: determine whether it is far enough out of range of one access point that it should start the potentially disruptive process of looking for another access point, and then make the transition to a new access point as quickly as possible. The top part of Figure 6.5 shows the phone zigzagging its way through a series of cells, each one from an access point on a different channel. Looking at the same process from the point of view of the client (who knows only time), you can see how the client sees the ever-varying hills and valleys of the differing access points' coverage areas. Many are always in range; hopefully, only one is strong at a time.

The phone is a multitasker. It must juggle the processes of searching for new access points and handing off while maintaining a good voice connection. In this section, we'll go into details on the particular processes the phone must go through, and what technologies exist to make the process simpler in the face of Wi-Fi. But first, we will need to get into some general philosophy.

6.2.1 The Difference Between Network Assistance and Network Control

If you have read the sections on cellular handoff, you'll know that there are broadly two different methods for phone handoffs to occur. The first method, *network control*, is how the network determines when the phone is to hand off and to which base station the phone is to connect. In this method, the mobile phone may participate by assisting in the handoff

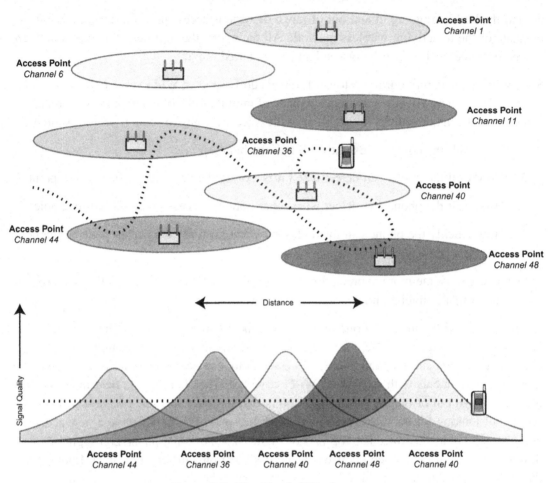

Figure 6.5: The Handoff Environment

process, usually by providing information about the radio environment. The second method, *network assistance*, is where the network has the ability to provide that assistance, but the mobile phone is fundamentally the device that decides.

For transitions across basic service sets (BSSs) in Wi-Fi, the client is in control, and the network can only assist. Why is this? An early design decision in Wi-Fi was made, and the organization broke away from the comparatively long history of cellular networking. In the early days of Wi-Fi, each cell was unmanaged. An access point, compared to a client, was thought of as the dumber of the two devices. Although the access point was charged with operating the power saving features (because it is always plugged in), the client was charged with making sure the connection to the network stayed up. If anything goes wrong and a connection drops, the client is responsible for searching out for one of any number of networks the client might be configured to connect to, and the network needed to learn only about the client at that point. It makes a fair amount of sense. Cellular networks are managed by service providers, and the force of law prevents people from introducing phones or other devices that are not sanctioned and already known about by the service provider. Therefore, a cell phone could be the slave in the master/slave relationship. On the other hand, with Wi-Fi putting the power of the connection directly into the hands of the client, the network never needs to have the client be provisioned beforehand, and any device can connect. In many ways, this fact alone is why Wi-Fi holds its appeal as a networking technology: just connect and go, for guest, employee, or owner.

This initial appeal, and tremendous simplicity which comes with it, has its downsides, and quickly is meeting its limitations. Cellular phones, being managed entities, never require the user to understand the nature of the network. There are no SSIDs, no passphrases to enter. The phone knows what it is doing, because it was built and provisioned by the service provider to do only that. It simply connects, and when it doesn't, the screen shows it and users know to drive around until they find more bars. But in Wi-Fi, as long as the handset owns the process of connecting, these other complexities will always exist.

Now, you might have noticed that SSIDs and passwords have to do only with selecting the "service provider" for Wi-Fi, and once the user has that down (which is hopefully only once, so long as the user is not moving into hotspots or other networks), the real problem is with the BSSID, or the actual, distinct identities of each cell. That way of thinking has a lot to it, but misses the one point. The Wi-Fi client has no way of knowing that two access points—even with the same SSID—belongs to the same "network." In the original Wi-Fi, there is not even a concept of a "network," as the term is never used. Access points exist, and each one is absolutely independent. No two need to know about each other. As long as some Ethernet bridge or switch sits behind a group of them, clients can simply pass from one to the other, with no network coordination. This is what I mean, then, by client control. In this view of the world, there really is no such thing as a handoff. Instead, there is just a disconnection. Perhaps, maybe, the client will decide to reconnect with some access point after it disconnects from the first. Perhaps this connection will even be quick. Or perhaps it will require the user to do something to the phone first. The original standards remain silent—as would have phones, had the process not been improved a bit.

Network assistance can be added into this wild-west mixture, however. This slight shift in paradigm by the creators of the Wi-Fi and IEEE standards is to give the client more information, providing it with ways of knowing that two access points might belong to the same network, share the same backend resources, and even be able to perform some optimizations to reduce the connection overhead. This shift doesn't fundamentally change the nature of the client owning the connection, however. Instead, the client is empowered with increasingly detailed information. Each client, then, is still left to itself to determine what to do and when to do it. It is an article of faith, if you will, that how the client determines what to do is "beyond the scope of the standard," a phrase in the art meaning that client vendors want to do things their own way. The network is just a vessel—a pipe for packets.

You'll find, as you explore voice mobility deployments with Wi-Fi as a leg, that this way of thinking is as much the problem as it is a way to make things simple. Allowing the client to make the choice is putting the steering wheel of the network—or at least, a large portion of the driving task—in the hands of hundreds of different devices, each made by its own manufacturer in its own year, with its own software, and its own applications. The complexity can become overwhelming, and the more successful voice mobility networks find the right combinations of technologies to make that complexity manageable, or perhaps to make it go away entirely.

6.2.2 The Scanning Process

Now it is time to explore the handoff process in detail. The most important, and most involved, part of a client's decision making for handoffs is not the actual handoff technique itself, but the research, so to speak, that the client must engage in *before* it attempts to make the transition.

This starts with scanning. Scanning, as mentioned in Chapter 5, is the prelude to connecting. The client starts up with no knowledge of the cells that make up the network. It must look around, browsing the channels for access points whose beacons claim to offer the service (based on that SSID) that the client is looking for. As the client looks around, it creates a repository of this information, known colloquially as a *scanning table*. (Colloquially, because there is no standard out there that dictates the scanning process itself. The act of scanning, and the results that it can produce, are mentioned in 802.11, but the text is intentionally vague and leaves the real intelligence to the implementer.)

6.2.2.1 The Scanning Table

Let's look at the scanning table in a bit more detail. This table is primarily a list of access point addresses (BSSIDs), and the parameters that the access point advertises. The 802.11 standard lists at least some parameters that may be useful to hold in the client's scanning table, as in Table 6.5.

This table contains the fields taken from the access point's beacons and probe responses. Most of the information is necessary for the client to possess before it can associate, because this information contains parameters that the client needs to adopt upon association. By looking at this table, clients can easily see which access points have the right SSID, but will not allow the client to associate. Examples are for access points that require a higher grade of security than the client is configured for, or require a more advanced radio (such as 802.11n) than the client supports. Most of the time, however, a properly configured

Table 6.5: Scanning table contents from 802.11

Field	Meaning
BSSID	The Ethernet address of the access point's service for this SSID
SSID	The SSID text string
BSS Type	Whether the access point is a real access point, or an ad hoc device
Beacon Period	Number of microseconds between beacons
DTIM Period	How many beacons must go by before broadcast/multicast frames are sent
Timestamp	The time the last beacon or probe response was scanned for this client
Local Time	The value of the access point's time counter
Physical Parameters	What type of radio the access point is using, and how it is configured
Channel	The channel of the access point
Capabilities	The capabilities the access point advertises in the Capabilities field
Basic Rate/MCS Set	The minimum rates (and MCS for 802.11n) that this client must support to gain entry
Operational Rate/MCS Set	The allowed rates (and MCS for 802.11n) that this client can use once it associates
Country	The country and regional information for the radio
Security Information	The required security algorithms
Load	How loaded the access point reports itself to be
WMM Parameters	The WMM parameters that the client must use once it associates
Other Information	Depends on the standards that the client and access point supports

network will not advertise anything that would prevent a properly configured client from entering.

In addition to all of this mostly static, configuration information that the access point reports, clients may collect other information that they may themselves find useful when deciding to which access point they should associate. This information is unique to the client, based on environmental factors. Generally, this information (not that in Table 6.5) is far more important in determining how a client chooses where to hand off or associate to. Table 6.6 contains some more frequent examples of information that different clients may choose to collect. Again, there is no standard here; clients may collect whatever information they want. Roughly, the information they collect is divided into two types: information observed about the access point, and information observed about the channel the access point is on. This split is necessary, because clients have to choose which channel to use as a part of choosing which access point to associate to. Properties like noise floor or observed over-the-air activity belong to the channel at the point in place and time that the client is in. On the other hand, some properties belong directly to the access point without regard to channel, such as the power level at which the client sees the access point's beacon frames. Furthermore, some of the per-access-point information may have been collected from

Table 6.6: Other possible scanning table contents

Field	Meaning
Signal Strength	The power level of the beacon or probe response from the access point
Channel Noise	The measured noise floor value on the channel the access point is on
Channel Activity	How often the channel the access point is on is busy
Number of Observed Clients	How many clients are on the channel the access point is on
Beacon Loss Rate	How often beacons are missed on that channel, even though they are expected
Probe Request Loss Rate	How many times probe requests had to be sent to get a probe response
Previous Data Loss Rate	If associated earlier, how much loss was present between the access point and client
Probe Request Needed	Whether the client needed to send a probe request

previous periods when the client had been associated to that access point, and measured the quality of the connection.

The scanning table is something that the client maintains over time, as a fluid, "living" menu of options. One of the challenges the client has is in determining how old, or stale, the information may be—especially the performance information—and whether it has observed that channel or access point long enough to have some confidence in what it has seen. This is a constant struggle, and different clients (even different software versions from the same client vendor) can have widely different ways of judging how much of the table to trust and whether it needs to get new information. This is one of the sources of the variability present in Wi-Fi.

6.2.2.2 The Scanning Process

As mentioned before, the scanning table's contents come from beacons and probe requests. Scanning is a process that can be requested explicitly the user—often by performing an operation that is labeled "Reconnect." "Update," or "Scan." But far more often, scanning is a process that happens in the background or when the client decides that it is needed. To understand why the client makes those choices, we will need to look at the mechanisms of scanning itself.

There are two ways that the scanning table can be updated. When a client is associated to an access point, it has the ability to gather information about other access points on that channel. Especially when the client is not in power save mode, the client will usually ask its hardware to let it receive all beacon frames from any access point. Each beacon frame is then used to update the scanning table entry for that access point.

On the other hand, the client may want to survey other channels to find out what other access point options are out there. To do this, the client clearly needs to leave the channel of its access point for at least a small amount of time. Therefore, before engaging in this process, the client will usually tell the access point that it is going into power save mode, even though it is doing no such thing. That way, the access point will buffer traffic for the client, who can then look around the network with impunity.

When the client changes channels, it has two methods it can use to find out about the access points. The quickest method is to send out the probe request mentioned earlier. This probe request contains the SSID the client desires (with the option of a *null* SSID, an empty string, if the client wants to learn about all SSIDs), and is picked up by all access points in range that support the SSID and wish to make themselves known to the client. (Remember that access points may hide from the client, as a part of load balancing. See Section 6.1.2.) Each access point that wishes to answer and that supports the SSID in question will respond with the probe response, a frame that is nearly identical to a beacon but is sent, unicast, directly to the client who asked for it. This procedure is called *active scanning*, though it can also be called *probing*, given the name of the frames that carry out the procedure. The other option is called *passive scanning*, and, as the name suggests, involves sending no frames out by the client. Instead, the client waits around for a beacon. Keep in mind that passive scanning clients do not know, ahead of time, how many access points are on a channel or when these access points may transmit the beacons. Therefore, a client may need to wait for at least one beacon period to maximize its chances of seeing beacons from every access point of possible interest.

In these two ways, the client goes from channel to channel, collecting as much information as possible about the available networks.

Clients may choose between active or passive scanning for a number of reasons. The advantage of active scanning is that the client will get definitive answers about the access points that are on that channel and in range in short order. Sometimes the client needs to send more than one probe request, just to make sure that none of those broadcast frames were lost because of transient RF effects or collisions. But the process itself concludes rather quickly. Furthermore, active scanning with probe requests is the only way to learn about which access points serve SSIDs that are hidden, where hidden SSIDs are not put in beacons and require the user to enter the SSID by hand. On the other hand, active scanning comes with two major penalties. The first one is for sheer network overhead. A probe request can trigger a storm of probe responses to the client, all of which take up valuable airtime. Especially when there is a network fluctuation (access point reboots, power outages, or RF interference), all of the probes pile onto an already fragile network, making traffic significantly worse. The second penalty is that active scanning is simply not allowed on the majority of the 5 GHz channels. Any channel that is in a DFS band cannot be used with active scanning. Instead, the client is always required to wait for a beacon (an *enabling signal*), to know that the channel is allowed for operation, does not have a radar, and thus can be used. (Note that, once a client has an enabling signal, it is allowed to proceed with a probe request to discover hidden SSIDs. However, the time hit has been taken, and the process is no faster than a normal passive scan.)

Therefore, to better understand scanning, we need to look at the timing of scanning. Active scanning, of course, is the quicker process, but it too has a delay. Active scanning is limited by a *probe delay*, required by the standard to prevent clients from tuning into a channel in the middle of an existing transmission. The potential problem is that a client abruptly tuning into a channel might not be able to detect that a transmission is under way—carrier sense mechanisms that are based on detecting the preamble will miss out, and thus produce a false reading of a clear channel. Thus, if the client were then to send a probe request, the client could very well destroy the ongoing transmission and lose out on the access points' seeing the probe request, because of a collision. As it turns out, many voice clients set that probe delay to a trivial value, in order to not have to wait. But the common value for that delay is 12ms, which is a long time in the world of voice. Passive scanning is worse. Most access points send their beacons every 102.4ms, or as close as they can get. This means that a client who tunes to a channel has a good chance of having to wait 50ms just to get a beacon, and may have to wait the entire 100ms in the worst case, for just that one access point.

The timescale that dominates, for voice mobility, is the voice packet arrival interval. Normally, that value is 20ms (though it can be 30ms in some cases). A client will usually want to get all of the scanning it can get done in those 20ms, so that it can return to its original channel and not miss the next voice packet. Certainly, the client will not want to take 100ms unless it has to, because 100ms is a long enough jitter that it can be quite noticeable (as mentioned in Chapter 3). Again, this tends to make active scanning the choice for voice clients, who are always in a hurry to learn about new access points.

If the client is going to scan between the voice packets, then the client's ability to scan will probably be limited to one channel at a time. When limited this way, the client may take up to a second, easily, to scan every possible channel. Recall from Chapter 5 that there are 11 channels in 2.4GHz, 9 non-DFS channels in 5GHz, and 11 more in the DFS bands, for a total of 31 channels to scan (or 23 channels if clients make the assumption that service is provided only on channels 1, 6, and 11 in the 2.4GHz band). Of course, scanning is also a battery-intensive process, and so a client may choose to spread out the scanning activity over time.

Furthermore, the process of changing channels is not always instantaneous. Depending on the radio chip vendor, some clients will have to wait through a multimillisecond radio settling and configuration time, reprogramming the various aspects of the radio in order to ensure proper transmission on the new channel. This adds additional padding time to the individual scanning channel transitions.

Overall, this scanning delay is a major source of handoff delays, and some methods for reducing the scanning time have been created, which we will examine shortly.

6.2.2.3 When Scanning Happens

The client's handoff is only as good as its scanning table. The more the client scans, the more accurate the information it receives, and the better decision the client can make, thus ensuring a more robust call. However, scanning can cost as much in call quality as it saves, and most certainly diminishes battery life. So how do phones determine when to scan?

The most obvious way for a client to decide to scan is for it to be forced to scan. If the phone loses connection with the access point that it is currently attached to, then it will have no choice but to reach out and look for new options. Clients mainly determine that they have lost the connection with their current access point in three different ways.

The first method is to observe the beacons for loss. As mentioned earlier, beacon frames are transmitted on specific intervals, by default every 102.4ms. Because the beacons have such a strict transmission pattern, clients—even sleeping clients—know when to wake up to catch a beacon. In fact, they need to do this regularly, as a part of the power saving mechanisms built into the standard. A client can still miss a beacon, for two reasons: either the beacon frame was collided with (and, because beacon frames are sent as broadcast, there are no retransmissions), or because the client is out of the range that the beacons' data rates allow. Therefore, clients will usually observe the beacon loss rate. If the client finds itself unable to receive enough beacons according to its internal thresholds, it can declare the access point either lost or possibly suffering from heavy congestion, and thus trigger a new scan, as well as deprioritize the access point in the scanning table. The sort of loss thresholds used in real clients often are based on a combination of two or more different types of thresholds, such as triggering a scan if a certain number of beacons are lost consecutively, as well as triggering if a certain percentage is lost over time. These thresholds are likely not to directly specifiable by the user or administrator.

The second method is to observe data transmissions for loss. This can be done for received or transmitted frames. However, it is difficult for a client to adequately or accurately determine how many receive frames have been lost, given that the only evidence of a retransmission prior to a lost frame is the setting of the Retry bit in the frame's header, something that is not even required in the newer 802.11n radios. Therefore, clients tend to monitor transmission retries. The retry process is invoked for a frame (see Section 5.4.8). Retransmissions are performed for both collisions and adapting to out-of-range conditions—because the transmitter does not know which problem caused the loss, both are handled by the transmitter simultaneously reducing the transmit data rate, in hopes of extending range, and increasing backoff, in hopes of avoiding further collisions for this one frame. Should a series of frames back-to-back be retransmitted until they time out, the client may decide that the root cause is for being out of range of the access point. Again, the thresholds required are not typically visible or exposed to the user or administrator.

Voice clients tend to be more proactive in the process of scanning. The two methods just described are for when the client has strong evidence that it is departing the range of the access point. However, because the scanning process itself can take as long as it does, clients may choose to initiate the scan *before* the client has disconnected. (This may sound like the beginnings of a *make-before-break* handoff scheme, but read on to Section 6.2.3, where we see that such a scheme does not, in fact, happen.) Clients may chose to start scanning proactively when the signal strength from the access point begins to dip below a predetermined threshold (the signal strength itself is usually measured directly for the beacons). Or, they may take into account increasing—but not yet disruptive—losses for data. Or, they may add into account observed information about channel conditions, such as an increasing noise floor or the encountering of a higher density of competing clients, to trigger the scan. In any event, the client is attempting to make some sort of preprogrammed expense/reward tradeoff. This tradeoff is often related to the problems of handoff, as mentioned shortly.

Scanning may also happen in the background, for no reason at all. This is less common in voice clients, where the desire to ensure battery life acts as a deterrent, but nevertheless is employed from time to time. The main reason to do this sort of background scanning is to ensure that the client's scanning table is generally not as stale, or to serve as a failsafe in case the triggered scanning behavior does not go off as expected. One of the chief problems with determining when to scan is that the client has no way of knowing whether it is moving or how fast it may be moving. A phone held in the hands of a forklift driver can rapidly go from having been standing still for many minutes to racing by at 15 miles per hour in a warehouse. This sort of scanning, not being triggered, is the least likely to lead to a change in access point selection, but may still serve its appropriate place in a network. For data clients, as a comparison, this form of background scanning, triggered for no reason, is often driven by the operating system. Windows-based systems often scan, for example, every 65 seconds, just to ensure that the operating system has a good sense of the networks that are available, in case the user should want to hop from one network to another. This sort of scanning causes a noticeable hit in performance for a short period of time on a periodic basis.

6.2.2.4 The Decision

Whether the client has an updated scanning table, or whether it has been triggered to scan because of a disconnection or performance-limiting event, the decision to leave the access point and connect to a new one is entirely the client's.

This decision is driven by the same factors that trigger the scan in the first place. But it may also happen for other reasons, often a direct result of the updates made to the scanning table by a scan that might not have been triggered for the purposes of selecting a new access point. For voice clients, load can be an issue, and a scan result that shows a significantly lighter load on a different channel or access point can trigger a decision to hand off. So can varying signal strengths, even when the connection quality is more than adequate.

It bears repeating ad nauseam that clients determine which access point they wish to connect to entirely on their own, for reasons neither specified by the standard nor available to the user of the device. The only influence the network has is to invoke harsh load balancing techniques (Section 6.1.2) to simply deprive the client of an otherwise legitimate choice in the name of better balance, or to use channel layered techniques to create a virtualized access point (Section 5.2.4.7), where the client is not aware of any transitions.

Why do vendors insist on leveraging proprietary methods to determine when to scan, how to rank access points, and when to make the final transition? This is surely a perplexing question, because this very fact of hidden client control is what leads to one of the greater complexities for voice mobility.

The best way to answer this is to look at it from the point of view of a vendor. Client manufacturers, especially those for voice mobility devices and phones, stake their reputation and brand value on being known for the quality of the voice calls made from their device. The behavior of a voice call over Wi-Fi, when the handset is not in motion, is already well defined, with WMM completely specifying how the voice traffic can gain priority, SIP and other signaling protocols establishing how the call is made and ended, and RTP describing how the voice traffic is encoded into UDP. Therefore, for handset hardware and software vendors to be able to differentiate themselves from the manufacturers of other devices in their market, they have an incentive to produce value by creating proprietary methods for improving handoff. Even with standards (such as 802.11k, Section 6.2.6) that we will see help with the information exchange, there is pressure on each phone manufacturer to focus on creating unique methods for trying to get the handoff decision-making process "just right," and to hold the parameters that go into those methods close to the vest. Unfortunately, this greatly complicates the job of anyone who must manage a voice mobility network.

Generally, the documentation with any voice mobility device will be noticeably vague as to the procedures and controls that the administrator or user may be able to employ to influence the handoff behavior of the client. However, some vendors do recognize that a "one-size-fits-all" approach is not likely to work in all cases, and therefore offer some general-purpose settings for handoff behavior. Going by terms similar to "roaming aggressiveness" or "handoff aggressiveness," these settings often are scaled from low to high or in stages and change the behavior of the client in ways that are not made public, but are intended to allow the administrator to favor, in the low case, having the client avoid roaming unless necessary, and in the high case, having the client hand off whenever it may perceive any benefit from doing so.

Because a handoff is a juggling act, and because even one network will vary immensely from point to point within its environment, it is impossible for the client to strike an optimum balance between roaming aggressiveness levels that will work within a large-scale voice mobility deployment. There are mitigation strategies to help determine settings that may provide better results than others, and we will explore those in Section 6.4).

To understand the process better, you first need to understand where things can go wrong. In some situations, clients choose to initiate the handoff process too late. This is referred to as the *sticky client* problem (see Figure 6.6). There are two fundamental origins of this sticky client problem. The first is when the client is not able to adequately judge when voice quality is suffering. This is a problem more common to multi-purpose devices, such as smartphones or laptops, where voice quality on the Wi-Fi link is not the primary design concern. This can be especially true on devices that are running proprietary voice client extensions, as can be expected with enterprise FMC offerings. In these cases, the lack of complete integration of the voice over Wi-Fi application and the underlying Wi-Fi handoff decision engine can cause the voice quality to suffer. Imagine a smartphone with a Wi-Fi engine that is primarily designed for data uses, running a proprietary voice over IP application. To the user, the device appears to be a cohesive, well-functioning whole. Voice over Wi-Fi applications provide dialing keypads and address book functions very similar to that which the phone provides natively for cellular dialing. Furthermore, these applications will attempt to use the same microphone and speaker that is used by the cellular phone application, thus allowing the Wi-Fi calling experience to be as similar as possible to the cellular one. However, because the Wi-Fi engine probably does not take any input from the voice application, and is thus unaware that a true voice application is running (even though voice packets are being sent and received), there is no way for the Wi-Fi engine to become aware of changing or suffering voice quality. Ideally, a phone would take voice quality as one of the major inputs in the ranking of the entries in the scanning table, as well as using them as triggers to cause scanning to occur. Voice-aware client scanning—the process of scanning between the voice packets mentioned before—will also not be possible, because the phone is unable to know when the next packet will come in, and thus may easily miss returning to the channel in time. As a consequence of all of this, the phone's handoff process is entirely dictated by the data-driven behavior of the Wi-Fi engine. Voice quality can begin to suffer rapidly when a channel becomes oversubscribed, or when the phone moves far enough out of range of an access point that multiple retransmissions of the voice packets are required to complete communication. Data-oriented Wi-Fi engines are less likely to choose lower transmit rates to avoid retransmission, which reduces latency, and are more likely to choose higher transmit rates, which increases throughput. Thus, voice quality

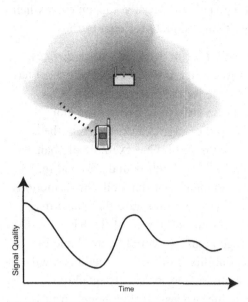

a) The irregular cell boundary, produced by normal RF effects and shown by the ligher shading, signifying less coverage, can cause a phone to experience abrutply-varying voice quality.

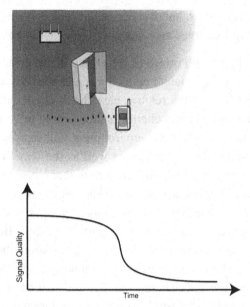

b) Stepping behind a filing cabinet can cause sudden shadowing effects, where the signal strength is suddenly diminished.

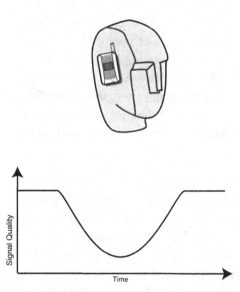

c) The caller's head can block 5dB of signal. Combined with the effects of the asymmetry of the internal antennas in the phone, the phone can lose up to 10dB of signal just by the caller turning her head.

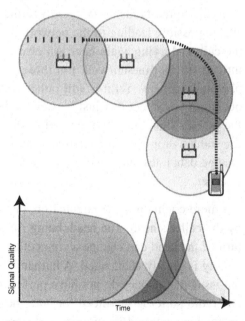

d) Running from a standing start can dramatically change the dynamics of the apparant RF environment.

Figure 6.6: The Origins of Sticky Client Behavior

may suffer, and the phone will not commence scanning. This may easily happen even when the loss rates and access point signal strength is more than adequate for data traffic.

The second origin of the sticky client problem can easily plague dedicated voice devices with well-integrated technologies that ensure that voice quality is measured. The reason has to do with the relative ease that a client can underestimate its radio environment. What happens is that clients cannot simply tell when they are in an area or being used in a manner where the environment will rapidly change. Rapid changes can come from the physical environment as well as the way the phone is being used. One example of changes to the physical environment is that caused by irregularly shaped cell boundaries and the direct influence of the phone and the user shaping the boundary for that cell. Small motions in the handset can cause large changes in the link quality, simply because the phone may move across an irregular part of the cell. Because phones are not aware of the RF environment—they have no information about coverage areas, as would be available in a detailed site survey—they cannot plan for areas of variability. In this way, the phone will not see the changes coming, and will be suddenly caught having to react to quickly reestablish a solid connection, by kicking off the scanning process. If the phone's user is lucky, and the phone had happened to have performed a random background scan recently, this might be a shorter process. But it is far more likely not to be the case, and the very act of trying to reestablish service will further disrupt the call whose quality is now suddenly shaky, possibly leading to the network dropping the call or either caller giving up, ending the call, and trying again later. An obvious example of this effect is when someone steps into an elevator or stairwell. But those may be thought of as areas of poor coverage planning, perhaps. Yet it is still rather easy to imagine a caller stepping behind a metal bookcase or cabinet that just happens to be in the line of sight between the access point and the phone. There may be plenty of coverage, as measured by signal strength for the call, but the sudden drop in signal strength may still cause poor wireless behavior, such as slow-to-change data rates and rapidly increasing retransmissions, and a handoff may nonetheless be in order.

Far more common is the attenuation, or loss of signal strength and quality, caused simply by the caller's head. The head, being full of water, does cause quality to suffer. A quick turn of the head can change a connection that was quite fine for voice into one where voice quality is greatly challenged. A human head provides about 5 dB of attenuation. Add to this the issue that phones do not have perfect radiation patterns. A phone is not a large device, and radio purity is sacrificed for both the size and the aesthetics of the phone. In terms of antenna coverage of the phone itself, the antennas are usually tiny, highly folded bits of metal that are wedged somewhere inside the phone. These are not high-performance antennas, by design. Moreover, they suffer from significant asymmetry: they just do not work as well in every direction. Some of the antennas that are used in mobile phones can vary their antenna gain by 5 dB, depending on the angle. Together, the effects of the asymmetric antenna, the caller's head, and even the caller's hand can cause signal loss that goes over 10dB, which can easily take off many yards of range. These effects are likely to be transitory, of course, but action is still often required to prevent the caller from getting a sense that the network or the phone is flaky.

Another example of changes to the physical environment comes from that the client does not and cannot know how fast it is moving. Past performance is no indication of future behavior, and even the most adaptive and intelligent phone device can have its algorithms lulled into a false sense that the phone is not experiencing dramatic, nontransitory changes. In this case, a sudden movement that results in a permanent exiting of the coverage area of the access point can leave the phone unprepared, in terms of scanning, and force it to engage in the same sudden and disruptive repair processes as before. Hospitals and warehouses are two obvious types of deployments that can have voice mobility devices that can remain stationary for long periods of time, yet are followed by sudden and extreme movements, such as a forklift speeding down the aisle or a nurse running down a hall.

In either case, sticky clients are not helped by the fact that phones tend to transmit at lower power levels as the access points serving them. Because of this, the phone may have the false sense that it is in range, getting sufficient signal strength to adequately support the *downstream* part of the voice call. However, the *upstream* part—the return link—may not arrive with a high-enough fidelity at the access point. This can result in what is known as *one-way audio*, in which the phone user can hear the other party but cannot be heard in return. For handoffs, this makes stickiness worse, as the phone sees a strong access point but is weak in return.

The opposite problem is equally vexing: in some situations, clients choose to initiate the handoff process too early and often. This is referred to as the *frisky client* problem. This problem comes about more often because the phone designers may have been aware of the sticky client problem, and decided to increase the aggressiveness of the phone's roaming behavior or the sensitivity of the phone to audio or radio variations. Frisky client behavior results in a higher than necessary number of handoffs. But more than that, frisky behavior can result in phones handing off to access points that are *less* capable of serving the phone and providing high-quality audio. This results when the phone correctly detects the variation—perhaps an increase in the noise floor on the channel or encountering higher densities of devices using the network—but incorrectly decides to act on it. The phone may choose to trigger an aggressive scan, thinking that the call quality is going to suffer or will shortly. If the caller is lucky, this aggressive scan will cause the phone quality to suffer for a shorter amount of time, while the scan progresses, but the phone may choose to remain on the same access point. But more likely, phones are tuned to make the transition, cutting bait on the original access point. This can be a very poor decision for one very simple reason. The phone was reasonably likely to have been associated to a close access point, with higher signal strength. Any change of access point can result in the phone being associated with a far more distant access point—this is especially true with microcell deployments and not generally so on layered deployments. (See Figure 6.2 in Section 6.1.2.) The consequences of the phone transitioning to a more distant access point are significant. The lower signal strength of the further access point increases the chance for RF interference, by reducing the link budget and SNR, as well as increasing the chance that the phone will go out of range of the newer, more distant access point more quickly than if it had stayed put. Furthermore, the data rate that the client and access point can use will be lower, which causes the voice packets to take more time and causes more interference with the clients in the cell the distant access point is generating. Finally, because the client is further away from the access point, its perceived RF environment is going to be more different from the access point and those of the other clients on that access point than if the phone were closer to the access point. The access point's reporting of information using features such as 802.11k, its own load-balancing and decision-making properties, and the headroom that the access point is reserving for voice will all be incorrect for a distant client. Even worse, dynamic microcell architectures may be forced to increase the power level of the distant access point to cover the frisky client, increasing intercellular overlap and causing co-channel interference and 802.11 noise to rise.

In short, the network and RF variability that leads to poor local audio performance with sticky clients can lead to bad decisions by frisky clients. The variations of the environment for a frisky client can lead those overly sensitive clients to make changes needlessly and to the detriment of the caller, as mentioned earlier. But overly aggressive frisky clients can make poor handoff decisions even when the existing connection has not changed appreciably. The same RF variations can cause neighboring access points' signals to arrive at the phone, which may become increasingly tempting. Or access points that were ruled out earlier can become more tempting based on load variations. This is more true in denser deployments, and can result in negative effects caused by *herd mentality*. With herd mentality, the behavior of other phones affects the behavior of the phone in question. Unlike with real animal herds, it is highly unlikely that a frisky client will make a decision on the basis of directly observing the decisions of other phones and copying them. Rather, the phone is likely to make *simultaneous*, and sometimes *identical*, decisions, based on the indirect effects the other clients' decisions have on the channel and on the access point's reporting. Let's look at some examples.

One source of herd behavior comes from the load reporting that access points perform. As will be mentioned in Section 6.2.6, access points are allowed to report the load of the network. They may report the capacity available, determined by the load balancing and admission control operations (Section 6.1.1). They may also measure the average access delay—how long it takes for its packets to get access to the air. The longer the delay, the busier the channel. They also can report the raw number of clients supported. These numbers are based not on anything intrinsic to the access point, but on the numbers and behavior of the clients within the cell. More intelligent voice clients will often base their roaming decisions on the basis of these numbers. Frisky clients will be especially sensitive to them. When the network is in steady state, and very little is changing, a client can quite effectively use this information to make a good decision. However, the herd mentality problem comes into play when the dynamism of the network increases, and these numbers start varying.

Picture two access points within range of the client. Figure 6.7 illustrates this. There is some distribution of clients associated to each access point. Assuming that nothing changes, this situation is likely to be stable. However, let's now imagine that a small handful of the clients move from one access point to another. A good reason for this could be that the holders of these phones have left a meeting, and have moved closer to the new access point. Immediately, the clients who are associated to the old access point may notice a difference. The load reporting information from the access point will announce the arrival of these new clients. Even if there is enough capacity on the new access point, and the admission controller allowed the clients in, phones' having to compete for resources is not necessarily an ideal situation. On the other hand, the old access point suddenly has room to spare. This seems like an ideal situation, and the frisky clients may determine that they want to pack up and move for greener pastures. Of course, there is no way to know exactly how many will make this decision. The more aggressive ones will flood over to the new access point, tipping the balance back. If only a few move, then there is not likely to be an enduring issue. However, because many voice mobility deployments for the enterprise use the same devices with the same configurations, a large number of devices may make the transition. The grass is not always greener, and instead, the devices may start interchanging, producing a significant amount of handoff activity that eats up resources that are more needed for the sending of data. These sorts of positive feedback loops cause *ping-ponging* between the access points and are quite easy to encounter whenever feedback is offered without dampening and without memory. However, the design of 802.11 is such that this feedback is provided in precisely this manner. Moreover, even if none of the feedback is offered, clients can and do still use their own ability to measure over-the-air load to make these decisions.

Overall, you will notice that both sticky and frisky clients are just opposite sides of the same coin. If they're too passive, the call may suffer as detrimental effects that demand action go unnoticed or ignored. If they're too sensitive or too aggressive, the call quality of both the current call and of those around the phone may suffer as too much changing · occurs, causing cascading handoffs or increased network waste.

There are strong parallels here between the handoff behavior of clients and the stability problems that can occur in dynamic routing protocols—or to any reactive, feedback-based system. Too much dampening, and needed changes don't occur. Too little, and network

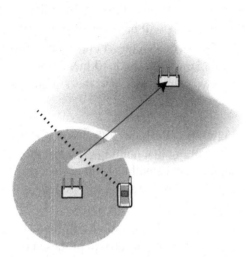

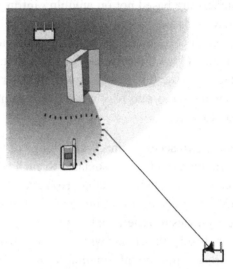

a) Wireless variations, such as irregular cell boundaries, can cause a frisky client to be stolen away, for the possible detriment to the call quality. In this case, instead of the irregular coverage causing the connection to the connected access point to look bad, it causes neighboring access points to look good—especially if the more distant access point is lightly loaded.

b) The wireless environmental variations that, when strong, can lead to sudden onsets of poor performance for sticky clients can, when minor, fool frisky clients into making needless transitions.

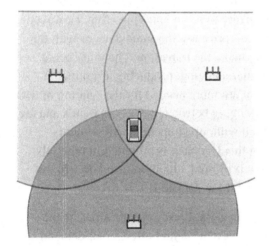

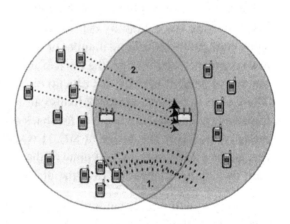

c) Overly-aggressive frisky clients can become trapped when they have multiple choices that are all very similar, especially when the choices are not good, such as when caught in the weak spot between three access points.

d) Herd mentality behavior cna be triggered by fluctuations in the steady-state distribution of clients, especially when density is higher. For example, the sudden motion of a few clients can trigger a cascading reaction, one that might not settle down.

Figure 6.7: The Origins of Frisky Client Behavior

flapping and oscillations can set it. Overall, the issue is that every client things for itself. Each client's handoff behavior is a fundamentally *local* process. Only what the client can see—whether it is observed directly or indirectly, with aid of the access point—determines the client's behavior. On the other hand, there is a stable, *global* optimum. This global optimum would take into account all of the effects—transient, permanent, important, irrelevant, direct, and indirect. The network has a better chance to have this view than the client, but Wi-Fi is as it is, and the client owns the decision.

Because of the vast complexity of interactions available (not only the first-order factors based on which access point a client should hand off to from its own observations, but the increasing-order factors based on client-to-client interactions and indirect observations), the production of workable—never mind optimal—handoff schemes that apply to dense deployments is an area of active research and development for client manufactures.

If you have a voice mobility deployment in which the above-mentioned handoff issues are prevalent or are a concern, you are encouraged to ask your phone vendor specifically as to the degree and meaning of the handoff controls they do offer, and to press them for recommended settings for your environment.

6.2.3 The Wi-Fi Break-Before-Make Handoff

Basic Wi-Fi handoffs are always either *break-before-make* or *just-in-time*. In other words, there is no ability for a wireless phone to decide on a handoff and establish a relationship with a new access point without disconnecting from the previous one. The rules of 802.11 are rather simple here: no client is allowed to associate (send an Association message to one while maintaining data connectivity to another) to two access points at the same time. The reason for this is to remove any ambiguity as to which access point should forward wireline traffic destined to the client; otherwise, both access points would have the requirement of receiving the client's traffic, and therefore would not work in a switched wireline environment.

However, almost all of the important protocols for Wi-Fi happen only after a data connection has been established. This prevents clients from gaining much of a head start on establishing a connection when the old one is at risk.

Let's look at the contents of the Wi-Fi handoff protocol itself step by step. It will be helpful to consult Section 5.2.3.3 for further information.

1. Once a client has decided to hand off, it need not break the connection to the original access point, but it must not use it any longer.

2. The client has the option of sending a Disassociation message to the old access point, a good practice that lets the old access point free up network resources.

3. At this point, if the new access point is on a different channel, the client will change the channel of its receiver.

4. If the new channel is a DFS channel, the client is required to wait until it receives a beacon frame from the access point, unless it has recently heard one as a part of a passive scanning procedure.

5. The client will send an Authentication message to the new access point, establishing the beginnings of a relationship with this new access point, but not yet enabling data services.

6. The access point will respond with its own Authentication message, accepting the client. A rejection can occur if load balancing is enabled, and the access point decides that it is oversubscribed, or if key state tables in the access point are full.

7. The client will send a Reassociation Request message to the access point, requesting data services.

8. The access point will send a Reassociation Response message to the access point. If the message has a status code for success, the client is now associated with and connected to this access point, and only this access point. Controller-based wireless architectures will usually ensure this by immediately destroying any connection that may have been left over if step 2 has not been performed. The access point may reject the association if it is oversubscribed, or if the additional services the client requests (mostly security or quality-of-service) in the Reassociation Request will not be supported.

At this point, the client is associated and data services are available. Usually, the access point or controller behind it will send a broadcast frame, spoofed to appear as if it were sent by the client, to the connected Ethernet switch, informing it of the client's presence on that particular link and not on any one that may have been used previously.

If no security is employed, skip ahead to the admission control mechanisms, towards the end of the list. If PSK security is employed, skip ahead to the four-way handshake. Otherwise, if 802.1X and RADIUS authentication is employed (WPA/WPA2 Enterprise), we'll continue immediately next. For any security mechanisms, you may wish to flip to Section 5.6 for more details on the mechanisms.

9. The access point and client can only exchange EAP messages at this point. The client may solicit the EAP exchange with an optional EAP Start message.

10. The access point will request the client to log in with an EAP Request Identity message.

11. Depending on the EAP method required by the RADIUS server on the network, the client and access point will continue to exchange a number of data frames, all EAPOL.

12. The access point relays the RADIUS server's EAP Success or EAP Failure message. If this is a failure, the access point will also likely send a Deauthentication or Disassociation message to the client, to kick it off of the access point.

At this point, the client and access point have agreed on the pairwise master key (PMK), based on the key material generated during the RADIUS exchange and sent to the access point when the authentication process concluded. But, as Section 5.6.1.2 showed, the access point and client still need to generate a per-connection, pairwise transient key (PTK), which will be used to do the actual encryption. Pre-shared key (PSK) networks skipped the listed EAP exchanges, and use the PSK as the master key.

13. The access point send the first message in the RSN (802.11i) four-way handshake. This is an EAPOL Key frame.

14. The client sends the second message in the four-way handshake.

15. The access point sends the third message in the four-way handshake.

16. The client sends the fourth message in the four-way handshake.

At this point, all data services are enabled, and the client and access point can exchange data frames. However, if a call is in progress, and WMM Admission Control is enabled, the client is required to request the voice resources before it can send or receive a single voice packet with priority. Until this point, both sides may either buffer the packets or send the voice packets as best-effort. Section 6.1.1.2 has the details on WMM Admission Control.

17. The client sends the access point an ADDTS Request Action frame, with a TSPEC that specifies the over-the-air resources that both the upstream and downstream part of the voice call will occupy.

18. The access point weighs whether it has enough resources to accept or deny the request. It sends an ADDTS Response Action frame with the results.

19. If the request was successful, the client and access point will be sending voice traffic and the call successfully handed off. On the other hand, if the request fails, the client will disconnect from the access point with a Disassociation message, because, although it is allowed to remain on the access point, it can't send or receive any voice traffic.

Hopefully, everything went well and the handoff completed. On the other hand, if any of the processes failed, the connection is broken. The old connection was abandoned early on—in step 8 for sure and step 2 for more charitable clients. In order to not drop the phone call, the phone will need to restart the process from the beginning with another access point—perhaps the original access point it just left, if none is available.

You will notice that the client has a lot of work to do to make the handoff successful, and there are many places where the procedure can go wrong. Even if every request were to be accepted, any loss of some of the messages can cause long timeouts, often up to a second, as each side waits to make sure that no messages are passing each other by.

If nothing at all is done to optimize this transition, the handoff mechanics can take an additional second or two, on top of the second or so taken by the scanning process before the handoff decision was made. In the worst case, the 802.1X communication can take a number of seconds.

Part of the issue is that the mechanisms are nearly the same for a handoff as they are for when the client initially connects. This lack of memory within the network within basic Wi-Fi prevents any optimizations and requires a fresh start each time.

6.2.4 Reducing Security Handoff Overhead with Opportunistic Key Caching

The good news is that the 802.1X mechanisms can be taken out of the picture for handoffs, for wireless architectures with a controller (or large number of radios in one access point). This mechanism, available today for many vendors, is known as *opportunistic key caching* (OKC). The name comes from the main concept underlying the technology. Once a client performs the authentication with the RADIUS server, and has a PMK, there is no reason for it to have to negotiate a new one just to handoff and create a new PTK just for that access point. The term "opportunistic" is used because the mechanism was designed to be a simple extension of 802.11i, and the client is not made aware that OKC is enabled. If it works, it works. If not, no problems arise except the increased time required for doing the handshake.

The main protocol for OKC is identical to the ordinary key caching mentioned in Section 5.6.2.3. The only difference is that whereas ordinary key caching requires that the client is returning to an access point where it had already performed 802.1X, opportunistic key caching requires only that the new access point somehow have access to the PMK, even though it was created on a different access point.

How can this work? The PMK, if you recall, does not have any information unique to the wireless network within it. It is a function purely of the EAP protocol in use between the wireline RADIUS server and the wireless client. There is no intrinsic reason that the same PMK cannot be used for different access points, as long as the following two restrictions are held to: the PMK must never be transmitted as plaintext or using weak encryption, and the PMK must not have expired.

In practice, opportunistic key caching implementations never move around the PMK. Instead, these implementations take advantage of the architecture of the WPA2 protocol and how it interacts with 802.1X. 802.1X doesn't know about clients and access points. Instead, it uses a different language, in which the role of the user is held by a supplicant, and the role of the network is held by an authenticator. These terms and roles were described in Section 5.6.2, when 802.1X was introduced. The mapping of the supplicant to real devices is clear: the supplicant is a part of the client. The authenticator, on the other hand, has flexibility built in. For standalone access point architectures, the authenticator is a part of the access point. For controller-based architectures, however, the authenticator is almost always in the controller.

Now we get a sense for the scale of opportunistic key caching. The PMK was originally created in the authenticator, and most opportunistic key caching architectures leave the PMK inside the authenticator, never to come out. For controller-based architectures, the controller generates the PTK within the authenticator, and then distributes it to the encryption engine, which may be located locally in the controller or in the access points. With opportunistic key caching, then, the only change is to allow a client with a PMK to associate to a new access point, and to use the PMK for the new connection as if it had been negotiated on that access point.

There is no addition of protocols or state changes in opportunistic key caching, which explains why it is so prevalent within network implementations. The only changes are to clients, who have to create a new PMKID, based on the original PMK, when they associate to a new access point, and to the authenticator, which needs to look past that a PMKID was not created for the PMK, create the new one, and then continue as if nothing unusual had happened.

You should look for wireless clients and network infrastructure that supports opportunistic key caching when rolling out a voice mobility network. OKC has been generally embraced by the industry, though there are a few notable exceptions, and is generally used as the solution to the 802.1X overhead.

6.2.5 An Alternative Handoff Optimization: 802.11r

As good as opportunistic key caching sounds, it does have its flaws. The major flaw is that the client has no way of knowing whether a new access point it is associating to has the PMK already. This is not a major flaw, in that the opportunistic nature of the key caching doesn't require the client to make the correct guess. However, in areas where access points from multiple authenticators overlap, such as at the border area between two controllers' domains, it is possible for the client to be unable to take advantage of the optimizations.

For this reason, as well as some detailed interest in having a key caching mechanism that is specified in the 802.11 standard—opportunistic key caching is a well-known mechanism, but neither IEEE nor the Wi-Fi Alliance officially recognize it nor test for interoperability or performance of the mechanisms—the IEEE created an effort to produce a standard version. This standard version is known by the amendment 802.11r.

802.11r, entitled "Fast BSS Transition," is fundamentally a more elaborated-upon version of opportunistic key caching. The general goal and concept is the same: eliminate the overhead of 802.1X by allowing multiple access points to use the same PMK to have their own PTKs created. The difference is that where opportunistic key caching specifies nothing about how this happens, 802.11r specifies some structure so that there can be a better sense of what is happening.

However, 802.11r is more ambitious. In addition to standardizing key caching, 802.11r made two optimizations. The first optimization is to allow the original Authentication and Reassociation messages be used to piggyback the four-way handshake, eliminating the need for those extra frames. WMM Admission Control is piggybacked as well, allowing the elimination of the ADDTS protocol for handoffs. The second optimization is to allow the client to start the first part of the handoff before the handoff actually occurs, providing a semblance of make-before-break behavior into Wi-Fi.

6.2.5.1 802.11r Key Caching

Let's first look at the key caching changes. WPA2 has a very simple *key hierarchy*, or a nesting of keys. The 802.1X exchange creates the PMK, and the PMK is used to create a unique PTK for each association of that client. 802.11r expands that key hierarchy to allow for intermediate steps and provide ways for the client to know which PMK is shared between two access points.

The key hierarchy for 802.11r is more complicated. There is the original master key from the RADIUS authentication, followed by a *PMK-R0*, a *PMK-R1*, and finally by the PTK. Let's start with the PMK-R0. This is the first level of the hierarchy, and is stored in a central authenticator that is to be shared across the entire *mobility domain*, which is the advertised set of access points that can share key state. The PMK-R0 is derived using the same master key that was generated with 802.1X, but also includes the SSID and the client's Ethernet address, as well as some other identifiers. This PMK-R0 is forever confined within the mobility domain, and remains in the central location where the RADIUS server is accessed from. (For controller-based architectures, this is always within the controller.) This holder is named the *R0 key holder* (R0KH).

When an access point is introduced into the mobility domain, a corresponding PMK-R1 is derived by the R0 key holder. The PMK-R1 adds to the PMK-R0 the Ethernet address of the access point or the controller—wherever the group keys are generated from. This entity is named the *R1 key holder* (R1KH).

When a client associates with a specific BSS, the R1 key holder creates the PTK. This PTK serves the identical function as the PTK from WPA2, and, in fact, plugs directly into the WPA2 encryption in the same way as the PTK from the original 802.11i four-way handshake.

6.2.5.2 802.11r Transitions

Stepping back, we can see what this means for clients. Every access point that belongs to the same controller (R0 key holder) can belong to the same mobility domain. The mobility domain is advertised in the beacons, within the *Mobility Domain IE*, shown in Table 6.7.

Table 6.7: Mobility domain information element

Element ID	Length	Mobility Domain ID	FT Capabilities and Policy
1 byte	1 byte	2 bytes	1 byte

The Mobility Domain ID is a two-byte field, chosen by the network administrator to uniquely identify the mobility domain, and to distinguish it from other mobility domains that may be in the area. The FT Capabilities and Policy field specifies two bits. The first bit, when set, enables *over-the-DS transitions*, which will be described shortly. The second bit, when set, enables the use of pre-authentication resource reservations.

First, let's look at the changes to the transition, for a completely over-the-air scenario. 802.11r eliminates the frames that carried the four-way handshake, WMM AC resource requests, and the 802.1X exchange for a handoff. The process for handoff thus changes from the steps mentioned in Section 6.2.3 to the following, starting from that section's step 5.

1. The client will send an Authentication message to the new access point, establishing the beginnings of a relationship with this new access point, but not yet enabling data services. This Authentication message is called an Authentication Request, and has the structure shown in Table 6.8.

Table 6.8: Authentication request/response frame body

Algorithm Number	Authentication Sequence	RSN IE	Mobility Domain IE	Fast BSS Transition IE
1 byte	1 byte	variable	variable	variable

The Algorithm Number specifies equals 2 to use 802.11r. The RSN IE contains a similar RSN IE to the Association Request for the original WPA2, except that one PMKID will be given (just as with opportunistic key caching) but with the ID for the PMK-R0, and the key management suite will be given as 00:0F:AC, 3, meaning to use 802.11r for the key derivation, rather than WPA2. The Mobility Domain IE matches the one in the beacon. The Fast BSS Transition IE contains the client's nonce, and thus establishes a similar purpose as the first message in the 802.11i four-way handshake.

2. The access point, if it will accept the client, sends an Authentication Response message. This Authentication message has the same structure as the Authentication

Request, except that the contents of the Fast BSS Transition IE are different. That IE now contains the authenticator's nonce, as well as repeats the client's nonce. This corresponds to the second message in the Four-Way handshake. At this point, the 802.11 "Authentication" process is done, but 802.11r is only halfway through.

3. The client sends the access point a Reassociation Request message. This message has all of the same fields as a normal Reassociation Request message, but also includes the fields given in Table 6.9.

Table 6.9: Additional 802.11r fields for the Reassociation Request

RSN IE	Mobility Domain IE	Fast BSS Transition IE	RIC Descriptor IE	TSPEC	...
variable	variable	variable	variable	variable	

Here, the RSN IE's PMKID has now changed to that for PMK-R1, signifying that this phase of the communication is now for the R1 key holder, and does not need to involve R0 at this point. Furthermore, the Fast BSS Transition IE holds a MIC, or secure hash, of the information elements added to the frame for 802.11r, thus protecting the admission control requests from being forged or modified. The RIC Descriptor IE states that one or more TSPECs are following, and that these TSPEC information elements, which would ordinarily be a part of the ADDTS Request for WMM Admission Control, are actually related to this transaction. Note that the access point will choose one and only one of the multiple possible TSPECs given, as each one is considered an alternative. In this way, the protocol of requests and responses for TSPECs have been embedded in the Reassociation Request. This message serves as message three of the four-way handshake.

4. The access point, if it accepts the client and the client has not failed the security handshake, will send a Reassociation Response. The Reassociation Response carries the usual fields, but also the same fields as in the previous Reassociation Request. The RIC Descriptor IE from the request is returned, but with the status code within properly set to inform the client of whether the request was accepted. On an accepted resource request, the TSPEC that corresponds to the accepted request is returned. On a failed request—which is still a part of a valid, successful Reassociation Request—a TSPEC that might still work can be returned by the access point, to give the client an option to submit an ADDTS with something that might succeed.

At this point, the client is associated, data is flowing, and the admission control request has been either accepted or rejected. This only took four frames, and so the over-the-air overhead has been dramatically reduced.

We can notice, however, that the first two frames of the exchange do not change where the client is associated or how its traffic flows. They affect only the establishment of the security keys. Because of this, there is one more change that can be made, and this one can ease the transition burden even more. Instead of the first two Authentication frames going over the air, on the channel of the new access point, the client may have the option of sending the contents of those frames to the *current* access point. This is allowed only if the current access point states that it supports this feature in the FT Capabilities field. This type of transition is called an *over-the-DS* transition. Over-the-DS transitions, named because of the term *distribution service* (DS) referring to the nebulous wireline interconnection between access points that, for standalone access point systems, is normally only a switched Ethernet network, requires that the current access point be able to forward the messages intended to the new access point. The added contents of the Authentication frames specified in the steps, starting from the RSN IE, are, instead, placed into an *FT Action* frame. The FT Action frame has the format given in Table 6.10.

Table 6.10: FT action request/response frame body

Category	Action	Client Address	Target BSSID	RSN IE	Mobility Domain IE	Fast BSS Transition IE
1 byte	1 byte	6 bytes	6 bytes	*variable*	*variable*	*variable*

The main difference between the Authentication frames and these Action frames, besides the obvious one that the Action frames go to and from the current access point, is that the Action frames contain the BSSID of the access point that the client wishes to hand off to. In a mobility domain, the current access point will invoke some sort of unspecified mechanism to get the message to the target access point, with exactly the same effects as if the message had been sent directly to the target access point over the air. The previous list of steps reads the same, if one just substitutes FT for Authentication and remembers that the frames are sent to the current access point for those FT messages.

This sort of handoff is not a part of a make-before-break scheme, because there is no commitment on the part of the client to make the transition when performed over the air or the DS, and the data path is unambiguously owned by only one of the two access points.

6.2.5.3 Preauthentication Resource Allocation

802.11r allows for one more mode. Preauthentication resource allocation allows a client to request resources on the new access point before leaving the old one. These resources are requested by inserting two additional steps into the handoff, right after the Authentication/ FT Response.

These two steps, known as Authentication/FT Confirm and Authentication/FT Acknowledgement, serve a somewhat similar purpose for resource reservation as the

Reassociation steps do. Essentially, the resource reservation part of the protocol is moved forward, from the Reassociation messages to the new Confirm and Acknowledgement messages. In this manner, the resources can be requested *before* the client makes the transition, thus allowing the client to have more confidence that the resources will be available before the transition.

An important part of the protocol is that the target access point is not required to specifically allocate the resources that were a part of the accepted resource requests, and can specifically deny the resource request when the client finishes the protocol with the Reassociation Request, by rejecting the Reassociation. This flexibility exists to allow the network to avoid certain greedy client behavior, where a client who wishes to hand off may feel compelled to allocate resources for a second and third backup, thus hogging valuable airtime without using it. The network, however, is allowed to take these essentially provisional resource allocations, known as *accepted* rather than *active* resources, into account when admitting or denying other clients.

Even with this mechanism, the handoff scheme is still one of break-before-make, as the client is forbidden from using over-the-air data resources on two access points simultaneously.

6.2.5.4 802.11r in Wireless Architectures

With a better understanding of the mechanisms employed in 802.11r, we can now look back and answer the question of how this fits in across the variety of different architectures. Because 802.11r has such a strong focus on a central authority, one might ask the question of whether 802.11r requires a controller or can still be used for standalone access points.

The mapping to controllers is rather obvious. The controller needs to be the R0 Key Holder, and communicates directly with the RADIUS server, just as with WPA2. The R1 Key Holder doesn't need to be located elsewhere, so it can be present on the controller as well, usually in the same module. This way, the two levels of PMK holders collapse, and work just as with the one PMK holder with opportunistic key caching. When the client hands off to a second access point, that access point requests the PTK from the controller, which generates a new PMK-R1 along the way, and stores it internally if needed for later use. Figure 6.8(a) shows one such setup. Minor alterations can be encountered, such as having the PTK holder in the controller for architectures who encrypt their packets centrally.

So, how can this possibly map to a standalone access point model? Figure 6.8(b) shows one method. The major change is for the removal of any one centralized PMK holder. Instead, access points are both R0 and R1 key holders, but with the R1 key holders being different access points from the one R0 key holder. When a client performs 802.1X for the first time, that access point becomes the R0 key holder. Along the way, it generates a PMK-R1 and a PTK. But when the client hands off, the new access point will also become an R1 key holder. The challenge in the architecture is for the system to have some sort of protocol where the new access point can find out which other access point is the R0 key holder. Once that is found, the new access point requests a PMK-R1 from the old access point, generates a PTK, and starts working. The identity of the R0 key holder does not change, until the key expires or 802.1X is performed again. In this way, there are many R1 key holders to each R0 key holder, and every access point can be an R0 key holder.

So, the good news is that standalone architectures can also participate in the 802.11r protocol.

One final note about key exchanges. There is no standard in 802.11i, 802.11r, or WPA2 that specifies just how these PMKs and PTKs get moved around. Every vendor is allowed to use its own proprietary protocol, and they do. The only requirement is that whatever protocol is used to move keys around provide for privacy and integrity. Controller-based architectures already have these protocols, and so there should be no concern. For standalone access points, a protocol will have to be created. You may have heard of 802.11F, the *Inter-Access Point Protocol* (IAPP), in previous years. This withdrawn recommendation—it was not even a standard—had tried to describe some earlier methods for communication. But it was not adequate for more modern technologies, and the controller-based architectures had lead to its being abandoned. Therefore, whatever protocol is created will likely be proprietary or lightly used.

An issue also arises about 802.11r mobility domains across vendors. Mobility domains do not extend across access points from two different vendors, even if both vendors support 802.11r. Because the key exchange protocols are vendor-specific, there is no way to connect the access points of multiple vendors together into one mobility domain. This is not likely to change, as there is quite a bit more to managing the security state of a client than swapping keys, and so there would not be a one-size-fits-all protocol.

6.2.6 Network Assistance with 802.11k

In Section 6.2.1, we considered the difference between network assistance and network control. Here, we will go through the details of the technologies that allow network assistance to happen.

802.11k, labeled "Radio Resource Measurement" (similarly named as radio resource management [RRM], but actually distinct), is a collection of protocols designed to improve the flow and exchange of information between the client and the access point. The basic mechanism of 802.11k is designed around the concept of reports. These reports are requested by one side of the conversation and furnished by the other. Many of the reports require time to produce, mostly because they require the reporter to engage in some sort of scanning behavior.

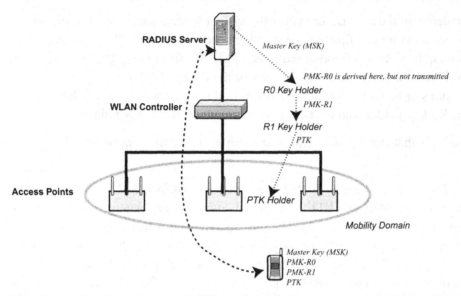

a) With a wireless controller architecture, locating where the 802.11r key holders can be rather simple. The mapping is, in fact, almost the same as for WPA2, with the PTK (connection key) holder being the access point, and the PMK holders all being in the controller. On a handoff to a new access point, that access point only needs to request a new PTK from the controller, which will produce a new PMK-R1 along the way, using the original PMK-R0. The main thing to remember is that there is only one PMK-R0 for the entire mobility domain.

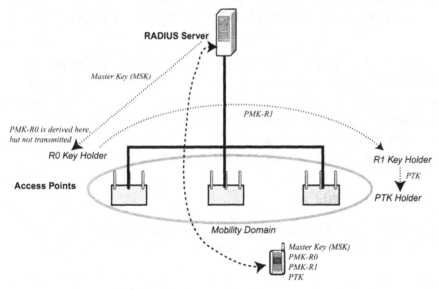

a) With a standalone access point architecture, the mapping is quite different. There is no centralized PMK-R0 holder. Instead, the first access point that the client performs 802.1X becomes the R0 Key Holder. When the client moves to another access point, the new access point needs to discover which original access point is the R0 Key Holder, and then request a PMK-R1 from it. The new access point will then generate the PTKs as needed, from its local store. In this way, each access point still has a PMK and a PTK, just as with WPA2, but there may be a higher-order PMK holder elsewhere in the network if the cliient has handed off.

Figure 6.8: Ways of Embedding 802.11r in Wi-Fi

802.11k has many potential uses beyond voice, but because voice network depend on the efficient functioning of the underlying technology, it is worth looking into 802.11k here. The reader is warned, however, that the material here can become a little dense, and you may wish to skip around to get a better sense of how all of these new capabilities fit together.

6.2.6.1 The Capabilities of 802.11k

802.11k contains a long list of features, as it is more a framework and a collection of independent uses of that framework than it is a solution to one specific problem. Many of these features come with variants, based on the willingness of vendors to implement the features. The way that the access point and client can determine which subset of 802.11k the devices may support is through the exchange of an *RRM Extended Capability* information element.

Table 6.11 gives the complete list of capabilities that can be expressed. This extensive list is overwhelming, belying the fact that 802.11k is made of multiple methods of performing roughly the same thing. That does not meant that there is no usefulness within the standard, but rather that interoperability among vendors is more important than ever. The following sections will expand on the features more likely to apply to a voice mobility network with Wi-Fi as one of the legs.

6.2.6.2 Requests and Reports

Not every capability within 802.11k is a request and report, but most of it is. The idea of the request and report structure came originally from 802.11h, which is the amendment concerning performing radar detection for DFS bands, first in Europe and now applicable to the United States and Canada as well. In the context of DFS, the goal is to request or require stations to scan in order to determine the presence of radar. 802.11k has broadened out the purpose of the mechanism, allowing for a broad number of properties to be measured and reported.

A request for measurement is generally requested for in a Radio Measurement Action frame, generically described in Table 6.12. The dialog token specifies which of the possibly multiple outstanding requests a response matches up to, and is specified by the requester and copied by the responder. The number of repetitions specifies how many times the report is to be produced from new information and sent to the responder, with a value of 65,535 (all bits set) specifying to continue forever, until explicitly cancelled.

The contents of each Measurement Report element are given in Table 6.13. The measurement token is a unique number of this specific measurement, to identify it if multiple measurements are processing in the background. (They serve a different purpose from the dialog tokens, and the two numbers are not related.) The measurement request

Table 6.11: The 802.11k RRM extended capability information element

Link Measurement	Neighbor Report	Parallel Measurements	Repeated Measurements	Beacon Passive	Beacon Active	Beacon Table	Beacon Reporting Conditions
Bit: 0	1	2	3	4	5	6	7

Frame Measurement	Channel Load	Noise Histogram	Statistics Measurement	LCI Measurement	LCI Azimuth	Transmit Stream	Triggered Transmit Stream
Bit: 8	9	10	11	12	13	14	15

AP Channel	RRM MIB	Operating Channel Max Measurement Duration	Non-operating Channel Max Measurement Duration	Measurement Pilot	Measurement Pilot Information	Neighbor Report TSF Offset	RCPI
Bit: 16	17	18–20	21–23	24–26	27	28	29

RSNI	BSS Average Access Delay	BSS Available Admission Capacity	Antenna Information	Reserved
Bit: 30	31	32	33	34–39

Table 6.12: 802.11k Radio measurement request action frame

Category	Action	Dialog Token	Number of Repetitions	Measurement Request Element
1 byte	1 byte	1 byte	2 bytes	*n* different copies

Table 6.13: Measurement request element format

Element ID	Length	Measurement Token	Measurement Request Mode	Measurement Type	Measurement Request
1 byte	1 byte	1 byte	1 byte	1 byte	*variable*

mode specifies how the measurement is to be made. The Parallel bit (bit 0) states that this measurement is to run in parallel with the next measurement request within the same frame. The Enable bit (bit 1), Request bit (bit 2), and Report bit (bit 3) are used in concert to determine whether the sender is requesting a new background (or *autonomous*) report series to begin or end with this request, or whether the request is to run right away. The Duration Mandatory bit (bit 4) specifies that the measurement is to run for exactly a certain amount of time, specified later in the Measurement Request itself.

The notion of autonomous or background reporting exists to allow the requester to have the other device send reports as needed. A further refinement allows the requester to specify specific triggers, which the receiver uses to determine just when to send the autonomous reports. In this way, the request protocol can be used to set up some sort of background reporting agreement. How each report works is specified with the reports.

The response to a request comes in the format shown in Table 6.14, with the contained Measurement Report shown in Table 6.15. The measurement report mode specifies the type of report, and can be value 2 to indicate that the reporter is *incapable* of generating this sort of report, or the value 4 to indicate that the reporter *refuses* to generate this sort of report, probably given an unspecified decision that the reporting work is too burdensome for the client at the time.

One can see that there is a large number of moving parts to this apparatus. The use for this complexity will hopefully become clearer as we progress.

Table 6.14: 802.11k Radio measurement report action frame

Category	Action	Dialog Token	Measurement Report Element
1 byte	1 byte	1 byte	*n* different copies

Table 6.15: Measurement report element format

Element ID	Length	Measurement Token	Measurement Report Mode	Measurement Type	Measurement Report
1 byte	1 byte	1 byte	1 byte	1 byte	*variable*

6.2.6.3 Beacon Reports

The first useful mechanism for 802.11k, especially as it applies to voice, is the *beacon report*. The beacon report is a report from the client, to the access point, about the entries within its scanning table. From the discussion on scanning in Section 6.2.2, one can imagine that learning about the contents of the scanning table can have some positive specific uses.

The beacon report request comes in three variants: *active reporting*, *passive reporting*, and *beacon table reporting*. These map directly to the scanning types: an active report is one in which the client was demanded to perform an active scan first; a passive report is one in which the client was demanded to perform a passive scan first; and a beacon table report is where the client is only required to give what it currently has in its scanning table, no matter how stale.

Table 6.16: Beacon request format

Regulatory Class	Channel	Randomization Interval	Measurement Duration	Measurement Mode	BSSID	SSID	Reporting Detail	AP Channel Report
1 byte	1 byte	2 bytes	2 bytes	1 byte	6 bytes	*(optional)*	*(optional)*	*(optional)*

Table 6.16 specifies the contents of the beacon request. The regulatory class and channel together specify which channel the client is to scan or report on. (Regulatory class is needed with the channel, because some channel numbers are reused in different bands.) The value of zero for the channel means to report on all channels in the band; the value of 255 means to report on all channels listed in the AP Channel Report. The randomization interval and measurement duration are used to specify whether the report should start now or after a delay, and how long the scan should take. The measurement mode specifies whether the request is for a passive scan (0), active scan (1), or beacon table (2). The BSSID is used to limit the request to produce the entry for just the given BSSID; this restriction is ignored if the BSSID given is all zeros. Furthermore, the optional SSID limits the scanning for SSIDs that match only, thus allowing the requesting access point to ignore other SSIDs. Finally, the requester can specify just what it gets back about each entry. The entry is not returned as arbitrary information; rather, a subset of the fields in each access point's beacon or probe response frame is returned, based on the contents of the reporting detail. The reporting

detail can be for nothing extra (0), or for requested parts of the beacon, using an added IE that appears at the end of the request and would be inappropriate to go into that level of detail here. (A good wireless protocol analyzer will be able to satisfy any curiosity you may have, if a device is using this protocol.)

The client that receives a request may choose to perform the request, or reject it for being too busy. If the client accepts the request, it must perform the actions and report on them. Now, there is no requirement that the client modify its real scanning table, the one it uses to make handoff decisions on, or that it report back the contents of that table to the access point. It is allowed to keep a separate table just for the purposes of 802.11k. However, the hope behind the amendment is that the scanning and beacon table processes are related.

Table 6.17: Beacon response format

Regulatory Class	Channel	Measurement Start Time	Measurement Duration	Reported Frame Information	RCPI	RSNI	
1 byte	1 byte	8 bytes	2 bytes	1 byte	1 byte	1 byte	...

	BSSID	Antenna ID	Parent TSF	Frame Body
...	6 bytes	1 byte	4 byes	*variable*

Table 6.17 shows the structure of the report. Each report provides information on exactly one BSSID. A Beacon Report fr-ame, a type of Management Report frame, will have multiple Management Report information elements, with one element containing one beacon report, to allow for reports to be sent back on multiple access points. The channel specifies what channel the access point was on. The measurement start time specifies when this measurement began, and the duration reports how long the measurement took. The reported frame information is set to zero. The *RCPI* provides a measure of the signal strength of the access point; the *RSNI* provides the signal-to-noise ratio for the same frame. The BSSID field contains the BSSID of the access point. The antenna ID field specifies which antenna the receiver heard the access point on, and is not terribly useful. The parent TSF field specifies the lowest four bytes of the timestamp clock on the client when it precisely heard the beginning of the beacon. Finally, the frame body contains all of the requested and fixed field bits of the access point.

Considering the tremendous size of the requests and responses, especially when there is a high amount of density in the environment, one might wonder what the purpose of such a beacon report really is. The goal is to allow the access point to assign idle clients with the task of informing it what its neighboring access points are, for use in the neighbor report.

6.2.6.4 Neighbor Reports

The *neighbor report* is requested for by a client from the access point, and specifies a list of neighboring access points. A client would request a neighbor report in order to supplement its scanning list, most likely to avoid having to stay off channel for a potentially long period of time. The information in the neighbor report comes from no particular source, however, and clients are not expected to replace their scanning table with the limited information returned in the neighbor report, or even to augment it in any significant way.

Together with the beacon report mechanism, however, one can picture an architecture where clients do request neighbor reports to learn about off-channel access points, and where access points do assign idle clients with the task of keeping them up to date.

Table 6.18 specifies the *Neighbor Report Request* Action frame. This frame is substantially simpler than the one used for beacon requests, because the client is not allowed to change the access point's behavior or to require it to do work. Instead, the client can simply ask for a neighbor report list, and can specify an optional SSID to limit the results down to the given SSID, rather than see results for every SSID.

Table 6.18: 802.11k Neighbor report request action frame

Category	Action	Dialog Token	SSID
1 byte	1 byte	1 byte	*(optional)*

The access point will respond right away with a *Neighbor Report Response* Action frame, as shown in Table 6.19. This response includes a number of *Neighbor Report* elements, each element corresponding to one neighbor.

Table 6.19: 802.11k Neighbor report response action frame

Category	Action	Dialog Token	Neighbor Report Elements
1 byte	1 byte	1 byte	*multiple*

The format of the neighbor report element is shown in Table 6.20. The BSSID is that of the access point. The *BSS Info* field provides information as to whether the neighbor supports the same security as the current access point, can participate in opportunistic key caching (Section 6.2.4) with this access point, or supports a subset of 802.11 capabilities. The channel and radio type of the neighbor is reported. Additionally, if the access point knows the timestamp clock of the neighbor, the country of the neighbor, the RRM capabilities of the neighbor, or how many other BSSIDs the neighbor supports, these can be communicated in the last four optional information subelements.

Table 6.20: Neighbor report element

Element ID	Length	BSSID	BSS Info	Regulatory Class	Channel	PHY Type	TSF	Country	RRM Capabi-lities	Multiple BSSID
1 byte	1 byte	6 bytes	4 bytes	1 byte	1 byte	1 byte	*(opt.)*	*(opt.)*	*(opt.)*	*(optional)*

Again, this is a somewhat involved protocol, and the meaning of most of what you see here will not necessarily become required knowledge, as a good protocol analyzer will shed light on what the protocol is communicating. But the intent is clear: the neighbor report is designed to provide a way for the client to find a neighbor that it could very well hand off to, without itself having to scan. In a perfect world, this can reduce scanning times, and it still remains to be seen whether 802.11k will be successful in reducing times. Unfortunately, some part of the process cannot be eliminated—DFS bands, for example, always require a passive scan of the channel, whether the client is told about the access point's presence on that channel or not. However, more information can be better than less, and the neighbor report mechanism aims to provide it.

Now, we can see that the neighbor report can be filled in from information gotten by a beacon report.

6.2.6.5 Link Measurement and Power Reporting

It is not possible for a client or access point to guess what power the other is using. Especially for systems that engage in dynamic transmit power control (see Section 6.1.3), if the client or access point wants to make a decision on the basis of the power level at which the other device is transmitting, it needs to find this information out explicitly.

To accomplish this, 802.11k provides for a method for requesting a link measurement. A *link measurement* is an exchange where the requester tells the power level it is sending at, and the reporter returns with what power level and SNR it saw that frame received at, thus letting the requester know its link margin to the other device. Table 6.21 shows the contents of the *Link Measurement Request* frame.

Table 6.21: Link measurement request frame

Category	Action	Dialog Token	Transmit Power Used	Maximum Power
1 byte	1 byte	1 byte	1 byte	1 byte

The responder puts together an immediate response formatted as specified in Table 6.22. The *TCP Report* element, given in Table 6.23, specifies the power that the responder is sending at. Its transmit power field is that of the link measurement response, and the link margin field is the expected link margin as measured in a possibly proprietary way by the responder. The receive and transmit antenna IDs are to help determine which antenna is being used in complex, sectorized solutions. The RCPI reports the signal strength of the original Link Measurement request at the responder, and the RSNI reports the signal-to-noise ratio.

Table 6.22: Link measurement response frame

Category	Action	Dialog Token	TPC Report Element	Receive Antenna ID	Transmit Antenna ID	RCPI	RSNI
1 byte	1 byte	1 byte	1 byte	1 byte	1 byte	1 byte	1 byte

Table 6.23: TPC report element

Element ID	Length	Transmit Power	Link Margin
1 byte	1 byte	1 byte	1 byte

With this much information available, no device lack knowledge of the transmit power or link margin of the other.

6.2.6.6 Traffic Stream Metrics

802.11k also allows for one device to ask the other about the quality of the admission control flows that it has. This is useful for inferring the voice quality of the link, and thus allowing either side to take action to improve performance if something is going wrong.

Table 6.24 shows the information that is placed in the measurement request for these metrics. The peer address is the address of the device whose performance wishes to be measured. The traffic ID is the TID of the admitted stream that the requester would like to find out about. The *Bin 0 Range* field is used to specify the properties of the histogram that is returned. Finally, if a triggered reporting element is given, the requester can specify that

Table 6.24: Traffic stream request

Randomization Interval	Measurement Duration	Peer Address	Traffic ID	Bin 0 Range	Triggered Reporting
2 bytes	2 bytes	6 bytes	1 byte	1 byte	(optional)

the reporter send a report whenever conditions cross certain thresholds. The sorts of thresholds which can be used include reporting because of excessive average errors (lost packets), and consecutive errors.

	Discarded Frame Count	Failed Frame Count	Multiple Retry Count	CF Polls Lost Count	Average Queue Delay	Average Transmit Delay	
...	4 bytes	4 bytes	4 bytes	4 bytes	4 bytes	4 bytes	...

	Bin 0 Range	Bin 0	Bin 1	Bin 2	Bin 3	Bin 4	Bin 5
...	1 byte	4 bytes	4 bytes	4 bytes	4 bytes	4 bytes	4 bytes

The report format is given in Table 6.25. This report includes both exact counts and a histogram of the nature of the errors. The transmitted frame count specifies the number of frames sent successfully for the flow. The discarded frame count specifies the number of frames that could not be delivered and were thrown away without any retransmissions having been knowingly delivered. The failed frame count specifies how many frames were lost, just like the discarded count, but excluding frames that timed out because they were in a buffer for too long. The multiple retry count specifies the number of frames that took more than one try to be successfully delivered. The CF Polls lost field can be ignored, as CF Polling belongs to a part of the standard not used in WMM. The average queue delay is the amount of time packets tend to be delayed until they start transmitting on the air. The average transmit delay is the amount of time packets tend to be delayed until they are successfully completed, including retransmissions. The histogram measures delay, and is divided up into five bins, where each bin measures a range of values double that of the previous bin, starting at Bin 1. The Bin 0 Range specifies the amount of time the delay value for Bin 0 covers. For example, if Bin 0 Range is equal to 10, then Bin 0 covers delays from 0 to 10.24ms. That would make Bin 1 cover the range from 10.24ms to 20.48ms, and Bin 2 would cover the range from 20.48ms to 40.96ms, and so on.

With this information, each side can measure the loss rates of the other, as well as the delay profile.

6.2.6.7 Other Features of 802.11k

802.11k includes a plethora of other features, many of which are not likely to be as useful for voice mobility deployments.

802.11k includes a provision for reporting on the noise levels that a client sees. It also includes a provision for reporting on the activity of each device in the air, based on frame counts. A peculiar mechanism known as *measurement pilot frames* exist, which serve as frequent, miniature versions of beacons. This could have been useful for aiding in the

Table 6.25: Transmit stream report

Measurement Start Time	Measurement Duration	Peer Address	Traffic ID	Reporting Reason	Transmitted Frame Count	
8 bytes	2 bytes	6 bytes	1 byte	1 byte	6 bytes	...

	Discarded Frame Count	Failed Frame Count	Multiple Retry Count	CF Polls Lost Count	Average Queue Delay	Average Transmit Delay	
...	4 bytes	4 bytes	4 bytes	4 bytes	4 bytes	4 bytes	...

	Bin 0 Range	Bin 0	Bin 1	Bin 2	Bin 3	Bin 4	Bin 5
...	1 byte	4 bytes	4 bytes	4 bytes	4 bytes	4 bytes	4 bytes

scanning across channels, by reducing mandatory scan times for DFS, but also has an enormous consequence for load on the network.

Most of these features may find usage as time progresses, or may wait for future standards to refine them or produce a compelling application—there is some compelling location-reporting capabilities in 802.11k that are itching for an application with the usage of emergency call reporting. However, in the mean time, the complexity of the features means that these technologies are less likely to be encountered in products implemented within the next two years as of the time of writing.

6.2.6.8 What 802.11k Is Not

With all of the tools that 802.11k provides, there is a feeling among some people that it must have enough to solve the major problems in wireless. Unfortunately, this view falls far short of the actual state of affairs.

The main benefit of 802.11k for voice is that it can provide assistance for clients in their scanning procedure. However, although there has been some speculation that the neighbor reporting feature has the ability to direct clients to the most optimal access point, 802.11k cannot actually do more than provide additional information to clients. The decision-making ability is still firmly held by the client. The problem has to do with how neighbor reports would be handled by clients. Neighbor reports, because of their size, are unlikely to contain more than a couple of options for the client. However, there is nothing in the standard that states how the access point should, or whether it ought to, cut down on the number of neighbor report entries from the likely far higher number of neighbors expressed in the beacon reports. Notice how beacon reports can be gathered from stations anywhere in the cell, and those stations on the edge of the cell can hear access points that are out of range of the access point those clients are associated with. The net effect is that beacon reporting can produce neighbor reports that cover a cell size over twice that of the access point. This is no mistake, however. The optimal choice set of access points for clients, if optimal is restricted to distance, is based entirely on where the *client* is, and not where the access point is. This means that the access point is not necessarily in a good position to judge which neighbors a client can see or would want to use. The best way for the access point to determine what the client sees is to ask it with a beacon request. However, any information the client has would already be used by the client in its handoff decision making, and the access point cannot add anything.

Looking at the same problem another way, a client just entering a cell and asking for neighbors has an excellent chance of being told about neighbors that are out of range of it, because they lie on the far side of the cell from the client. The answering access point can try to pick an optimum for the client, but that would require the network tightly tracking the location of the client in real time. Doing so is not a bad idea, but it may require a different architecture than is typical for microcell environments.

More to the point, even if the neighbor reports were optimal, the client has no way of knowing what type of network it is connected to, or whether the network is providing optimal results, useful results, or just anything it feels like. So the designers of clients have a strong incentive to not treat the neighbor report as definitive, and to just add the information provided into the mix of information the client already has. In fact, if the client vendor thinks that it has done a good job in producing the scanning table, then following the lines of the discussion in Section 6.2.2.4 on the handoff decision-making process, then it would be wise to not depend on neighbor reports in any way.

This tension makes it difficult to know whether the 802.11k mechanisms will finally eliminate most voice handoff issues, or whether they are adding a degree of complexity without the same degree of value. It this sense, it is unfortunate that clients are left in control of the process, with no specification as to why they should hand off. Cellular technologies have been successful in producing this sort of assisted handoff (though reversed, with the network making the decisions and the clients providing the candidates it might like), mostly because the end-to-end picture is adequately known. Wi-Fi will need to overcome its challenges for a similar scheme to be as effective.

Nevertheless, the presence of network assistance greatly improves the operation of networks compared to those with neither assistance nor control, and is necessary for high-quality voice operation for microcell deployments. Section 6.4 will explore better ways to tune the network for voice deployments.

6.2.7 Network Control with Channel Layering and Virtualization

As mentioned earlier, there are two broad options for improving upon the original Wi-Fi mechanism for independent, client-driven handoffs. Network assistance seeks to improve the accuracy and adequacy of the client-driven decision process, by offering the client more information than it would have on its own, in hopes that pathological decisions can be excluded, and better decisions can be made. The protocols can be quite sophisticated, and the client is required to become significantly more intelligent in order to take advantage of them. The other option is to remove the client's ability to make poor decisions, by limiting the client's choices and transitioning the significant portion of the handoff control function into the network.

Channel layering is capable of performing the latter. The concept is straightforward: handoffs go wrong when clients make the wrong choices. To eliminate the client's ability to make the wrong choice, channel layering reduces the number of choices to exactly one per channel. Let's look at this in a bit more detail. When a client is roaming throughout a microcell Wi-Fi voice deployment, it is capable of seeing a number of different physical radios. Each physical radio has a unique BSS, with a unique Ethernet address—the BSSID. As the client leaves the range of one radio, it uses its scanning table of the other unique BSSIDs to determine which access point it should transition to. After the decision is made, the client exercises the Wi-Fi association protocol to establish a new connection on the new access point. Overall, the process is dominated by the property that a BSS can be served by only one radio, constant for the life of the BSS.

This property is not a requirement of Wi-Fi itself, but rather a convenience chosen by access point vendors to simply the design and manufacture of the access point. The main addition channel layering provides is to sever the static connection of the BSS to the access point, thus virtualizing the access point end of the Wi-Fi link to encompass potentially the entire network by allowing for BSSs to migrate from radio to radio. The result is that the client is no longer required to change BSSs when it changes radios. Instead, the network will migrate the client's connection from one access point to another when it is appropriate. When a handoff occurs, the access point the client is leaving ceases to communicate with the client. The network end of the connection is relocated by the controller to the second access point, which resumes the connection from where it was originally left off.

This is clearly a network-focused solution to the problem, rather than a client-focused solution. The difference is that the network, rather than the client, adopts the intelligence needed to and the responsibility for making the correct decision on which physical access point a client should be connecting with. This has a few distinct advantages. The first is that this introduces a measure of client independence into the handoff behavior (and other behaviors) of the network. When clients are required to make the decision, each client will act as its own independent agent, each different client behaving differently under this architecture. But when the network makes the decision, it has the ability, being one agent in common for every client in the network, to act consistently for each client. Clients can no longer be sticky or frisky, and a greater number of clients are able to participate in more uniform, seamless handoffs. The second advantage is that the one centralized handoff engine can be monitored and managed more simply and readily, being one agent network-wide, rather than there being the multitude of distinct engines. In many ways, this is a furthering of the notion behind wireless controller architectures, with a measure of client behavior able to be centrally managed and monitored along with access point behavior. The third advantage is that clients are not required to carry the sophistication necessary to make effective handoff decisions, and thus there is no penalty for clients that are less sophisticated. In general, network control can greatly simplify the dynamics of the mobile population.

One can understand the dynamics of network control by looking at how CDMA-based cellular systems provide it. In a CDMA system, unlike time-division cellular systems, each client maintains an association with a unique network identity, known as a *pseudonoise code* (PN code). This code refers to the code division property of the CDMA network, and its individual function is not appropriate to describe here, except to state that each client has a unique PN code, and that code directly represents the connection. When the network wishes to hand off the client, rather than having to create a disconnection and a reconnection as in time-division systems, no matter how fast the reconnection is, the network can simply transfer or migrate the PN code from the old base station to the new one. This gives rise to the concept of *soft handoffs*, in which the handoff can be performed in a make-before-break manner. In make-before-break handoffs, the entirety of the connection state can be duplicated from the old base station to the new one. Both base stations are capable of participating in the connection, and the degree with which they do is determined by the network. The same concepts apply to a virtualized Wi-Fi network, where the unique per-connection PN becomes the unique per-connection BSSID. The radio for Wi-Fi still operates based on discrete time packets, rather than on continuous code streams, and so the downlink aspect of code division cannot be practiced. However, the uplink reception processing can be performed simultaneously by both access points, if the network desires, and certain transmit functions can be performed by both access points when it makes sense to do so. For layered architectures, the BSSID is shared among all connections, but the same properties of soft handoff remain.

Channel layering effects this network control on each channel. The term "channel layering," however, evokes the second important property of the approach. Microcell architectures work to reduce the number of access points that are in close range to a client to one in each band. The reason is that minimizing cross-channel overlap—the overlap in square feet of the cells from access points on two different channels—reduces the co-channel overlap—the overlap of cells from access points on the same channel. Channel layering architectures decouple co-channel and cross-channel coverage characteristics, however. The result is that each channel can be thought of as being entirely independent of the others, and thus more than one channel can be covered in the same band. In fact, channel layering architectures tend to recommend, though not require, that multiple channels, when desired, be covered by access points on each channel in a similar manner. The goal is to make sure that the coverage of the multiple channel layers appears similar to the clients, with the major difference in the layers being only the channel.

The client still has an important role to play in the channel layering scheme—one that it is better suited for. By channel layering's reduction of the client's per-channel search space to just one BSS, it falls out from the behavior of channel layering that the client's scanning process becomes one of choosing the appropriate channel. Because within each channel, the client's choice is constrained to one and only one BSS, the client's scanning table will be filled with information that really applies to the channel. In this case, clients are able to measure reasonable information about the coverage and RF properties of the channel as a whole. Assuming that the network is making the optimal decision of access-point-to-client on each channel, the client is able to use the access point properties to deduce the best available performance it will be able to achieve on that channel, with a greater likelihood than it had when access points bore distinct BSSIDs.

For example, let's look at the signal strength of the beacons. As mentioned in Section 6.2.2, the signal strength of beacons can be used by the client to determine how far in or out of range it is from the access point. When a client, in a microcell environment, begins to move to the transition region between two clients, it will start to perceive a drop in signal strength of the access point's beacons, and will begin to invoke the scanning and handoff process, at some arbitrary and likely unpredictable time, to try to choose another access point. However, this situation looks identical to the situation where the client is exiting the coverage area of the wireless network in general, and yet the proper resolutions to these two different scenarios can be quite different. With channel layering, however, the client will only perceive a severe drop in signal strength when it is truly exiting the coverage area of the network.

Another area of information the client can act upon is channel noise. Because microcell networks minimize high-performing cross-channel alternatives, sudden variations in the amount of non-Wi-Fi interference on a channel requires that the network detect and adapt to the noise by shuffling the channel settings on the access points in the area of the noise to attempt to avoid the noise source. Clients also detect the noise, and initiate the handoff process, but because the network is reconfiguring, the scanning tables are incorrect, even if they were gathered just before the reconfiguration event. Thus, clients can miss the access point's reconfiguration, and the network can fragment, taking possibly substantial lengths of time to converge. Channel layering is more proactive than reactive, and noise that is introduced into and affects one channel layer may avoid the other channels, thus allowing clients to detect the noise and initiate a cross-channel handoff as needed.

Of course, channel layering architectures may also alter the channel assignments, which they may do to avoid neighboring interference or at an administrator's request. However, channel layering architectures do not need to reconfigure the network as a primary line of defense against network fluctuations, especially transient ones, and thus any reconfiguration works at far longer timescales and provides more consistency and invariance to the network. Thus, because channel layering provides a more stable coverage of channels, it allows the client's scanning table to be more useful.

In terms of over-the-air behavior of a given channel layer, there are broadly two methods for performing the virtualization of the BSS across the layer of access points. The first method involves replicating the BSS across the multiple radios simultaneously. This method allows every client to associate to the same BSS. The second method involves assigning each client to a unique BSS dedicated to it only. When the client approaches a transition, the BSS itself, along with the connection state, is migrated from access point to access point. Both methods have similar effects in terms of the client's lack of perception of a handoff. However, the second method, which is unique to the virtualized over-the-air architecture (Section 5.2.4.8) rather than the channel layering architecture (Section 5.2.4.7), provides an increased element of network control by extending the control from handoffs to over-the-air resource usage itself. Most Wi-Fi devices present do not and are not able to respect or create admission control requests (Section 6.1.1.2) before accessing the air. Instead, they perform their own categorization of whether traffic should be given the priority for voice, video, data, or background, and then use WMM mechanisms to directly compete with their neighboring clients to access the air. The access point is extremely limited in what it can do, short of disconnecting the client, in controlling its over-the-air resource utilization. WMM does provide an excellent way of altering the behavior of every client on an access point, providing methods of prioritizing one cell over its neighbor. Virtualization for Wi-Fi extends that control by segmenting the client population into unique BSSs, one per client. These BSSs each have their own WMM parameters. Thus, WMM can be leveraged directly to adjust resource usages of clients relative to each other, even when associated to the same SSID. This next-order level of network control has its advantages for ensuring that voice mobility traffic is unaffected by other devices, no matter what the load or in what direction the load is offered.

6.2.7.1 The Mechanics of Channel Layering Handoffs

Because the channel layering architectures do not require client action, we can describe the handoff procedure within a channel from the point of view of the network. Compare this procedure to that of Section 6.2.3, which describes an inter-BSS handoff without 802.11r, and Section 6.2.5.2, which describes an inter-BSS handoff with 802.11r.

1. The client approaches an area of the physical wireless network where it would be better served by a different access point than it is already being served by.

2. The network reevaluates the decision for the client to be connected to the first access point, and decides that the client should be connected with the second.

3. The connection state of the client is copied to the second access point.

4. The first access point ceases servicing the client. At the same time, the second access point initiates service for the client, continuing where the first left off.

The method that is used to determine whether a client should be handed off may still be proprietary, as it is with client-directed handoffs. The difference, however, is that there is only one consistent and managed agent that is performing the decision, so the network behavior will be similar across a widely differing array of clients.

Note that client movement is not the only reason that the network may choose to migrate a client's connection. The network may migrate the connection based on load factors, such as that the client might experience better service being on the new access point, rather than being on the old one. Or, the old access point may be going down for administrative reasons, and the network is ensuring seamless operation during the downtime. In any event, the advantage the network has in making these decisions is that it can do so based on a global optimal for the client, ensuring that the client is not forced to chose between close second and third alternatives, and poor or pathological behavior such as herd mentality is eliminated, as decisions are not made for each client in isolation. By reversing the control and consolidating it into one entity, the dynamics of the system become more predictable.

6.2.7.2 The Role of 802.11k and 802.11r

Network assistance is still useful in the context of channel layering, but in a better-defined, well-constrained method that actually improves the behavior of the assistance protocols. Because "horizontal" handoffs, or handoffs between access points due to the spatial motion of the client, is already addressed by the channel layering network, the only handoff left is "vertical," between channels due to load. This means that load balancing, as mentioned in Section 6.1.2, becomes the main focus of the client handoff engine.

Under channel layering, the 802.11k neighbor report, mentioned in Section 6.2.6.4, now serves the purpose of identifying the channel layers available to the client at its given position. The inherent location-determining behavior of channel layering architectures allows the neighbor report to be more appropriate for client at its given position, eliminating the problem in microcell deployments of providing more neighbor entries that are out of range than are in range.

802.11r (as well as opportunistic key caching) can also be leveraged, allowing the network to make explicit load-rearrangement operations while minimizing the service disruption to the clients. Clearly, there will be some service disruption whenever an 802.11r transition occurs, as compared to the seamless handoff of channel layering. However, the ability to use the multiple channel layers, combined with fast inter-BSS handoff techniques, allows the network to shuffle load far more quickly than with either technique alone. Furthermore, the 802.11k reports allows the network to gather more information about the RF environment than it can otherwise gain. Unlike clients, which have limited processing resources and limited ability to exchange necessary information for an optimal handoff without affecting overall network performance, the network has comparatively overwhelming resources to analyze the 802.11k reporting data and use that not to offer better assistance, but to make better controlled decisions upfront. Note that the primary mechanism for mobility-induced handoffs is the soft handoff, and the 802.11r handoff is reserved purely for load balancing.

In general, network assistance works well with network control in producing a more accurate and efficient operation, yet is not necessary to produce a high-quality voice mobility environment.

6.3 Wi-Fi Alliance Certifications for Voice Mobility

As voice has taken off, the Wi-Fi Alliance has created a number of certifications that are of benefit for determining whether an access point or wireless phone is more likely to be able to support high-quality voice.

Figure 6.9 shows an example Wi-Fi Alliance certificate. Certificates for all products which are certified by the Wi-Fi Alliance are available at the Wi-Fi Alliance's website at http://www.wi-fi.org.

The certificate is organized into a few sections. The Wi-Fi logo is color-coded and shows the amendment letters corresponding to the radio types that the device supports. The letter "a" corresponds to 802.11a, "b" to 802.11b, "g" to 802.11g, and "n" with the word "DRAFT" follwing it to 802.11n Draft 2.0. The certification date, category of the device (Enterprise Access Point or Phone for our purposes), manufacturer, and model number are also available on the top.

The columns list the certifications that the device has achieved. The first column lists the radio standards that the device has passed certification on, repeating the information in the color-coded logo. Additionally, the amendments 802.11d and 802.11h are shown for devices which have been submitted for the optional country code certification. The second column shows the security specifications that the device has passed. WPA and WPA2 are shown, each with Enterprise and Personal variations, based on what the device has passed. If the device has passed WPA or WPA2 Enterprise, there will also be a list of EAP types that were used. For clients, seeing an EAP type means that the client should be capable of using this EAP type in live deployments. Currently, this list includes EAP-TLS, EAP-TTLS with MSCHAPv2 password authentication inside the tunnel, PEAPv0 with EAP-MSCHAPv2

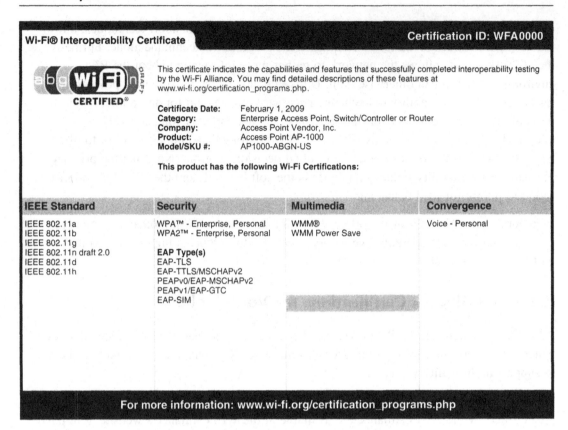

Figure 6.9: Example Wi-Fi Alliance Certificate

inside the tunnel, PEAPv1 with EAP-GTC inside the tunnel, and EAP-SIM. Under the third column comes, at the top, quality-of-service specifications. WMM should always be listed for voice devices. Expect to find WMM Power Save as well, and WMM Admission Control for devices which support it. The bottom half of the column is for special features, and is not present in this example certificate, as those features are not typically used for enterprises. The final column specifies voice and mobility certifications, and may contain Voice Personal or Voice Enterprise.

6.3.1.1 WMM Certifications

The WMM protocol makes up the very foundation of voice over Wi-Fi. The Wi-Fi Alliance tests WMM devices to ensure that they are able to provide that differentiation for all four priority levels, with a battery of tests which ensure that performance is preserved based on the presence of background traffic. All Voice- and 802.11n-certified devices support WMM.

The WMM Power Save certification continues by ensuring that the WMM Power Save protocol is followed, allowing for power savings to be applied for voice mobility devices. All Voice devices are WMM Power Save–certified.

The WMM Admission Control certification tests to see that the admission control protocol is followed by clients and access points, ensuring that clients do not seek to access the air with priority without an admission for a resource request, or that, if they do access the air without permission or after having exceeded their resource bounds, that they access the air in a nonprioritized manner. WMM Admission Control is required for Voice Enterprise–certified devices.

6.3.1.2 Voice Certifications

There are two certifications for voice within the Wi-Fi Alliance. These two programs are both mixtures of interoperability and performance tests to ensure that voice quality is likely to be maintained by the devices. These are the first certifications within the alliance to be focused on a nondata application, and thus are set up in specific ways to maximize the amount of voice testing coverage without increasing the complexity.

Both programs establish a set of observable over-the-air criteria that must be met for the access point and the client to pass the test. Specifically, the tests require a one-way jitter less than 50ms, from client to a wireline device connected on a low-latency network to the access point or vice versa; a maximum jitter also less than 50ms; a packet loss rate of less than 1%; and no more than three consecutive packet losses. These numbers are applied to simulated voice streams, generated by the test tools to produce packets with the approximate sizes and the exact timings of typical G.711 and G.729 encoded bidirectional voice flows. Both programs also test for a certain number of voice calls while generating a high-bitrate video stream, as well as an unbounded best-effort TCP data stream, to ensure that voice quality operates well in the presence of converged applications. Devices are placed into WMM Power Save and non–power save modes and are exercised with different security settings to ensure a more uniform test.

The Voice Personal test includes having four voice clients simultaneously, and all four clients must have voice flows that pass the above criteria for the test to pass, even if only one of the four clients is a voice client being certified. (The rest are already-certified devices being used to test with.) Furthermore, the Voice Personal certification requires that devices already be certified for WPA2 Personal, WMM, and WMM Power Save. The test is primarily focused on consumer-grade devices, but a small handful of enterprise-grade vendors have also passed the Voice Personal test, allowing a wider range of certified phones to potentially be paired with the network, if certification is desired for both sides.

The Voice Enterprise test is more appropriate for voice mobility networks. Based on the Voice Personal test, the Voice Enterprise test increases the density of voice clients from four to ten. More interestingly, however, is that it includes portions of 802.11k (Section 6.2.6) and 802.11r (Section 6.2.5), to increase the chances of handoff success. The 802.11k and other measurement features publicly mentioned as important foundations for the certification, as of the time of this writing are:

- Neighbor Reports

- Beacon Reports

- Transmit Stream Metrics

- Quiet Time

- Power Constraint (for setting the transmit power of clients)

Finally, the Voice Enterprise test carries over the voice quality test used for the ten immobile clients, and uses the same metrics to measure the quality of one call that is forcibly handed off between two access points of the same vendor. The handoff need not have the same voice quality during the handoff as before, but it must not exceed thresholds of 100ms loss or delay, as of the time of writing. The test will be performed using automated, sophisticated handoff and measurement systems designed to allow for repeatability in measurement. Note that the test is not measuring the handoff performance under density or variability, and so is intended to ensure the correctness of the implementation of the underlying protocols, and not to measure whether any client handoff decision-making scheme is better than another.

The Voice Personal program is new, as of this writing, but a handful of products are available for purchase that have been certified. The Voice Enterprise certification is still in progress—802.11k and 802.11r were not ratified amendments until fall 2008—but may be ready by 2010. Voice Enterprise certification is expected to be available as a software upgrade for vendors that seek the certification.

6.4 Real Concepts from High-Density Networks

Chapter 5 introduced the basics of wireless coverage and Wi-Fi operation, and this chapter has already covered some of the technologies that have been bolted onto Wi-Fi to improve its usefulness for voice mobility. This next section will explore steps that voice mobility planners can take to improve the quality of the network.

6.4.1 RF Modifications for Voice Mobility

One of the keys for voice mobility, especially when using a microcell-based architecture, is to ensure that the RF plan and various parameters to the automatic planning tools are both set to improve, not disrupt, the voice quality of the network.

The mobility of voice traffic requires two competing properties from the network: ensure that voice traffic is interfered with or disrupted with as rarely as possible, and yet ensure that coverage is high enough to increase the chances of successful handoffs. These two are set against each other, because the techniques for mitigating the first increase the risks for the second, and vice versa. Therefore, the correct approach depends entirely on the density of voice traffic, the ratio of voice traffic to data, and the likelihood of data congestion.

6.4.1.1 Less Voice than Data

Let's look at what to do when data is the primary traffic on the network. When voice is not expected to be used at a high density—for example, networks that do not include call centers, auditoriums, or dense cubicle or desk farms—the biggest impact upon voice quality is inconsistent and variable coverage. One reason this is true is that voice traffic is shorter in bytes than data traffic, therefore exposing it to a lower likelihood of interference-related packet error rates than longer packets, for the same bit error rate. Another is that the background transmissions—even the co-channel transmissions that cannot be interpreted but can be carrier-sensed—are overwhelmingly data transmissions, and not voice transmissions. In this context, the voice traffic can be thought of, as a limiting approximation, as the only traffic on the network, with just a higher noise floor from the background, irrelevent data traffic. The topology and raw coverage of the network begin to matter more than the capacity of the network.

When this holds, the two biggest concerns are to avoid network variation and ensure the highest possible coverage. Voice traffic, not being buffered and not being sent reliably at the transport layer as is data traffic, is remarkably good at exposing areas of weak coverage that site surveys and data usage patterns have missed. Depending on the deployment and the degree of weakness, large areas of the network can be found that perform tolerably on data, but produce MOS values (see Chapter 3) that are not acceptable, let alone of toll quality. In these areas, it is worth considering whether deploying an extra access point in that area will raise the SNR enough to alleviate the problem. A good approach to test this out is to run a long Ethernet cable to the location from a data jack and place the access point high enough to peer over the tops of furniture or short walls. If the voice quality improves, then it is worth considering a permanent installation.

Before installing that access point, check on the power levels of the neighboring access points is recommended. Depending on whether the network is using RRM, or whether installers or administrators have decided to turn down the power levels manually, the problem may be fixed by simply reverting those changes and either setting a minimum (for RRM) or static power level that is higher than the one currently in use. It may take some trial and error to determine how many and which of the access points which cover the area need to be powered up to improve the quality.

Network variance—especially power instability—is a major concern for voice mobility deployments. The power instability arises from RRM or autotuning, which attempts to back off the power levels from the maximum to avoid co-channel interference. The systems that do this produce a power reduction on the basis of the signal strength detected from neighboring access points, with those access points also adjusting their power levels dynamically. When one access point detects enough neighbors with higher power levels, it will retract the size of its cell by a predetermined threshold. This variation can happen as often as once a minute on adaptive microcell systems. Because of this, adaptive power control can become a problem for voice. One longstanding recommendation from at least one of the microcell vendors has been to perform a site survey after an adaptive power control run has been performed, to verify that the voice coverage is still sufficient. This is sound advice. In general, it is best, on an ongoing basis, to either constrain or disable adaptive power control in these voice mobility environments, erring on the side of higher power to prevent the power fluctuations from causing momentary or long-lasting areas of weak coverage.

In terms of access point placement and channel usage, it is important to plan for the best handoffs possible. Given the complexities of the scanning process mentioned in Section 6.2, voice mobility installers would do well to plan specifically for voice, to help avoid some of the problems that occur when coverage is tightened up. The two areas that the installer or administrator can make the biggest impact is with increasing coverage overlap, and carefully choosing the deployed channel set.

Handoffs tend to cause quality problems or lost calls when the phone is able to be rapidly moved from an area of high quality to an area of low quality within timescales shorter than it is tuned for. Furthermore, monopolistic coverage patterns—in which access points are spaced, channels are chosen, and power levels are set—lead to higher risk when a phone decides to attempt a transition, as fallback options are reduced or eliminated. Avoiding monopolistic coverage can be performed by adjusting the ratio of the spacing of access points to their power levels, by ensuring a higher-than-necessary minimum SNR when performing a site survey or RF plan. Stated as one rule, the goal is to increase the signal strength of the second strongest access point in any region where the strongest access point's signal is waning. Figure 6.10 shows the before and after.

The top part of the figure shows the goal of typical RF planning and RRM exercises. This plan provides the most *efficient* coverage for the given SNR that is used to define the boundary of the cells. Not a square foot is wasted, and the area of overlap between cells is minimized. This efficiency is good for lean networking, but enforces the sort of monopolistic coverage that is bad for voice. The graph on the right shows how the power levels of one access point falls to the minimum acceptable SNR just as the other begins to rise. This transition is as bare as can be, and at almost any location in the valley, the phone has at least one of the choices being a poor one, resulting in bad signal quality and a broken call. (In fact, with monopolistic coverage, the phone will have many choices, but *all of them* but one are poor choices, for any given band.)

The bottom of the figure shows the results when the minimum SNR requirements for RF planning are raised significantly above the actual minimum SNR requirements for voice.

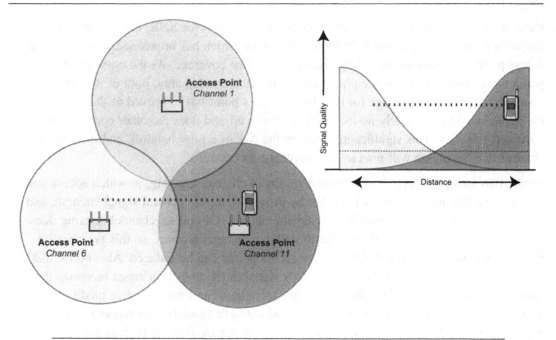

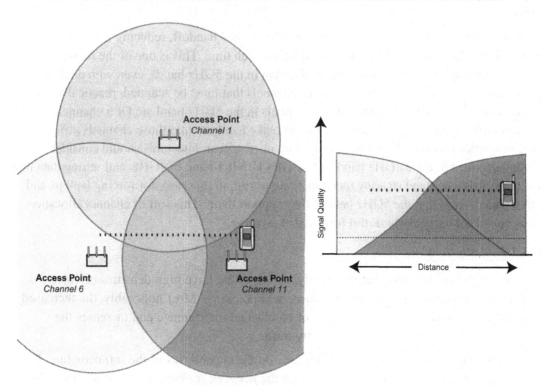

Figure 6.10: Reducing Monopolistic Coverage

Here, with the boundaries of the cell being the minimum voice SNR, the coverage overlap has been increased substantially. In the overlap area, which has broadened, more than one access point is capable of providing high-quality voice coverage. As the corresponding graph on the right shows, as the phone moves into the overlap area, both of its choices are good. Only when the coverage for the closest access point has improved to the point that the phone would have nearly no incentive to search around does the other access point's coverage give way. This significantly reduces the risk of a poor handoff, and increases the chances that the phone will scan to find coverage successfully.

Notice that taking this process to its limit results in channel layering, in which access points on the same channel cover for each other by providing higher general signal strength, and access points on different channels provide alternatives. Of course, channel layering does not make use of client handoff procedures within a channel anyway, so this prescription becomes less necessary and the number of access points can be reduced. Also notice that increasing the second strongest access point's signal in the transition zones increases the interference, as well, and thus performing this technique can improve voice quality but impact data. For this reason, some installers and network vendors may recommend to deploy side-by-side access points, one for voice and one for data, or to dedicate one band to voice and the other to data.

As a part of planning the network to improve the client's handoff, reducing the channel set in use provides the benefit of reducing the client's search time. This is one of the reasons why voice in the 2.4GHz band is more successful than in the 5GHz bands, even with dual-band phones, as the 2.4GHz band has only three channels that must be scanned, practically. Especially given that the majority of the channels in the 5GHz band are DFS channels, which have complex scanning requirements that can make handoffs into those channels difficult to perform, voice mobility deployments based on microcell architectures should consider deploying voice in the 2.4GHz band or non-DFS U-NII 1 band in 5GHz, and setting data into the DFS channels. Band steering techniques are useful, in this case, for forcing laptops and other data devices into the 5GHz band if they support them. This sort of channel allocation, however, is not particularly useful for channel layering.

6.4.1.2 Mostly Voice

In voice mobility networks that are mostly voice, or that have high densities of voice clients, the previously mentioned techniques will not work. Most noticeably, the increased density for voice both increases the risk of co-channel interference and increases the problems of collisions for the high-priority traffic.

What to do in a dense voice situation depends on the capabilities of the infrastructure. For microcell architectures, it is critical to design the network for capacity over coverage. This means increasing the aggressiveness of the RRM engine, shrinking power levels in an attempt to minimize the cellular overlap. Clearly, this will sacrifice handoffs and cause edge effects, but the tradeoff must be made, and in dense voice networks, the concern is to produce a network that can support voice first. Handoffs and frequent mobility may have to be traded for the ability to support infrequent mobility and flash crowds.

Increasing the power control aggressiveness has the effect of allowing the client's transmit power to become dominant. It is crucial to ensure that the microcell infrastructure that is being tuned this way can set the power constraint for the clients. If this is possible, the power constraint should be adjusted downward to match that of the network. Doing so will prevent cell size mismatches and link imbalance, which increases the interference each phone causes to its neighbors.

WMM parameters may have to be adjusted. There are two WMM parameters of importance: the minimum contention window, and the maximum contention window. Increasing the maximum contention window insures that highly dense networks can recover, when collisions occur. Making this increase can push up latency and jitter a bit, but it does so by reducing loss rates, which is the bigger problem to be solved. Especially when power save is in use, it is critical that uplink packets be given every chance to arrive at their destination, as these are the triggers for the downlink traffic. Adjusting the minimum contention window may also need to be done. This is a trickier problem, as the minimum contention window for voice must be altered in lockstep with the minimum window for data, to prevent data from gaining a higher prioritization. The reason that the window must be increased at all has to do with the probabilistic game of backoff that Wi-Fi uses to avoid contention. Recall that in Chapter 5, we saw how each client picks a random nonnegative integer less than the contention window. If the number of clients is high and the contention window low, the clients have fewer numbers to choose from, and thus have a higher chance of picking the same number and colliding. Increasing the minimum contention window increases the range of possibilities and choices for each client, restoring the balance. Keep in mind that the contention window is measured in powers of two, and thus an increase by just 1 can make a difference. Unfortunately, it is difficult to apply any hard and fast rules to this process. Therefore, worst-case planning is in order. For this reason, most vendors will not recommend or necessarily support networks where the minimum contention windows have been adjusted.

Additionally, when the density is high, the admission control parameters may turn out to be wrong. Admission control matters the most when the network is crowded, and the more crowded the access point's neighbors are, the more resources may need to be set aside as headroom on the access point to accommodate. WMM Admission Control access points will often provide the ability for the administrator to cap the available voice capacity as a percentage of airtime. Call admission control access points will provide the ability for the administrator to cap the maximum number of calls. In either case, it may, ironically, be necessary to reduce the maximum established call capacity for a given cell, as the number of calls using the network rises.

Virtualized wireless architectures avoid the need for setting these WMM and admission control parameters, as the network determines the correct values as a part of ordinary operation and can target individual values to individual phones simultaneously.

6.4.2 Site Survey

Site survey has a useful role in determining whether the network is adequate. When planning for voice, or doing regular inspections, it is critical to ensure that the site survey is done with voice in mind.

Most site surveys are performed by walking around with a laptop running site survey software. This software is the inverse of RF planning. Rather than requiring the operator to input the location of the walls and access points, and the tool spitting out the coverage expected, the site survey tool requires the operator to stand at every point in the building and input the location that she is standing in. The tool will then return the actual coverage that is being produced. This is a laborious task, and may not be necessary for all areas of a building, but will certainly be valuable for any areas where coverage has been historically weak and has not yet been corrected for.

When performing a voice mobility site survey, the site survey tool should be set to subtract a few dB from the signal that it provides. The reason for this is that phones tend to have lower transmit powers than the network, because of both the smaller antenna design and the need to reduce power for battery savings. Because of this, a site survey can give the false impression that coverage is good because the laptop, with a good antenna, can hear the downstream transmissions of the access point. A phone in that position may not get the same coverage levels. Furthermore, the site survey tool does not measure the uplink. As the phone may have a weaker uplink than the network, the result of this lack of measurement is that the site survey may entirely miss areas where the access point cannot hear the phone.

For this reason, it is recommended to check with the vendor of the phones for any voice mobility deployment on how the phone itself can be used for a site survey. Wi-Fi phones have advanced modes that allow them to report back on the signal strength in each area. Although not as fancy as using a site survey tool that has built-in mapping features, using the phone itself allows the administrator to get a more accurate sense of the network, and to immediately see where signal may be waning. When performing this sort of site survey, the voice mobility administrator for a microcell infrastructure should check with the vendor of the access points to see what they recommend that RRM should be set to during the site survey. Because RRM changes the coverage, and thus invalidates the site survey, it may become necessary to disable RRM completely and use a historical average for the network settings. Again, refer to the microcell vendor to see if they support reverting to historical RF settings. If not, it may be desirable to record the power levels that RRM sets manually (by exporting the current configuration, for example) and to roll back by hand.

6.4.3 Continuous Monitoring and Proactive Diagnostics

Running populated and well-used voice mobility networks can shift the focus to monitoring for good voice quality. This is especially true for networks that have a history of experiencing fluctuating voice quality, rather than having held steady for the network. In these cases, diagnostics are in order.

There are two types of diagnostics which may be available, again depending on the infrastructure technology chosen. *Reactive diagnostics* are concerned with measuring the state of the network and reporting on the conditions as they change, or offering the administrator a view into the state of the network using visualization, statistics, and reporting tools. *Proactive diagnostics*, on the other hand, are concerned with active measurement techniques to detect problems before they start.

The types of reactive diagnostics available to voice mobility networks are quite broad. Many of these tools are basic to Wi-Fi networks in general, but some of the tools are focused specifically on voice.

Many Wi-Fi networks that support voice allow the administrator to monitor the usage of voice on the network. Loss and error rates, air time utilization, and lists of the currently active voice calls and registered phones all provide invaluable information about the state of the network. It may take some time and experience to learn how these numbers and statistics correlate with good voice quality. However, once learned, they can provide a window into the operation of the network that helps establish what went on at the time in question. As usual, it is not as important, for many of these values, what the absolute values are at the time in question, but rather how they have changed when problems occurred.

Infrastructure-based client tracing and logging activity can be used to watch devices that are currently experiencing trouble. This information can then be compared with the behavior of a client that is not experiencing trouble to provide insight on what the problem might be. Further analysis can be done with wireless network management reporting tools. These tools can provide a summary of what each of the clients has been doing while on the network, or can filter through information to provide useful aggregates. It is recommended that any voice mobility administrator become intimately familiar with the network monitoring tools and platforms that each vendor offers for its products.

When non-802.11 wireless noise becomes an issue, such as in areas with radio laboratories or industrial microwave equipment, portable spectrum analyzers can come in handy. These tools may not be as useful in Wi-Fi-only networks, but when non-802.11 noise is a concern, these tools can be used to help classify the type of noise and allow the administrator to track down the source or test better shielding methods for the equipment in general.

Passive protocol analyzers with voice capabilities, as mentioned in Chapter 3, can come in handy for identifying problems in areas when these problems are occurring. These tools, whether they are integrated into the infrastructure or placed separately in portable laptop software, can deduce the quality of ongoing calls as they occur. Their greatest use is in tracking down situations where some unknown factor is intermittently causing call quality problems in an area. By placing these tools in that area and recording, the protocol analyzer may be able to capture the problem as it is occurring, and using the voice quality metrics may point the way for narrowing down the time windows that need to be searched for without requiring the network administrator to stand at the spot with a phone in her hand, waiting for the problem to occur.

Visualization techniques can be powerful complements to diagnosis. Two-dimensional visualization can quickly reveal basic problems with wireless, such as excessive loss or density. Three-dimensional "virtual reality" visualization can add the extra effects of inter-floor issues and help lead to where coverage may be improperly applied. Site survey tools can be used as a part of the visualization process when they offer remote modes that allow a laptop to be placed in a location for remote monitoring.

Proactive diagnostics go a step further. Network-based proactive monitoring allows the network itself to run trial phone calls and access services, recording success rates and measuring quality automatically. The reports can then be analyzed for signs of upcoming or newly introduced problems. Voice mobility administrators with highly critical network locations may be able to use the PBXs proactive call quality measurement tools to test phones, placed in strategic locations. These active call quality tools, associated with PBX monitoring tools, place test calls and then report on the quality the endpoint perceived. Although involved, this process can determine the behavior of the network and help head off any problems.

6.4.4 When All Else Fails

When all else fails, and the voice mobility network is generally suffering, there are a few options. The first option is probably the most painful, but can at least lead to stability. This option is to turn on call admission control, if it is not already, and to set the capacity limits to very low. This will result in busy tones for most calls that go through, but will help lead to stability of the network itself, and thus provide a potential way forward.

Once the network is stabilized, the problem becomes one of adding capacity. If the network is a mixed voice and data network, and the data traffic or the network settings necessary to accommodate the network is causing the problem, then a parallel network may be in order. If the network has room to grow (bands not filled, channel layers not deployed) and the capacity is reaching its limit, adding additional access points to expand the raw capacity onto those unused wireless resources can help. If all channels are used, and there is no more room to grow, then segregating traffic can make sense. The use of traffic shaping for data can provide enough headroom for voice to operate more reliably, especially for networks that do not properly account for voice resources in their operations or radio tunings.

CHAPTER 7

Voice over Cellular and Licensed Spectrum

7.0 Introduction

Voice mobility is centered around the concept that the call should follow the user. What happens when the Wi-Fi network runs to the end of its coverage, and yet people still want to place a call? When the callers are out in the city, or driving to their next meeting, they need to retain their access to the voice services. The only way to get the phone to work is if it hooks up to a licensed mobile operator who provides coverage around the wider areas beyond the building or the campus.

Cellular networking remains the predominant method that people get voice mobility today. Adding in-building, privately managed wireless extensions such as Wi-Fi for voice to operate on is still a new concept. In this chapter, we will see what makes cellular networks work and then explore how to combine that technology with Wi-Fi to create a comprehensive voice mobility network.

7.1 Anatomy of a Cellular Phone Call

Cellular networks the calls they provide address three basic and somewhat independent concerns for voice mobility:

1. How the landline phone can lose its wires and be portable

2. How the phone number can remain the same, wherever the phone is actually located

3. How the call can remain connected, without disruption, as the phone moves from area to area

Each of the three areas figure into the architecture of mobile telephone networks. Figure 7.1 shows the anatomy of a cellular phone call.

Most cellular networks share this architecture, though with different names and fancier pieces hanging off the sides. From the bottom up, we can see the cellphone, which is an advanced phone that uses digital technology to sample the audio, compress it, and send it on its way. The radio features of the mobile phone ensure that the phone seeks out and connects to the network.

doi:10.1016/B978-1-85617-508-1.00001-3.

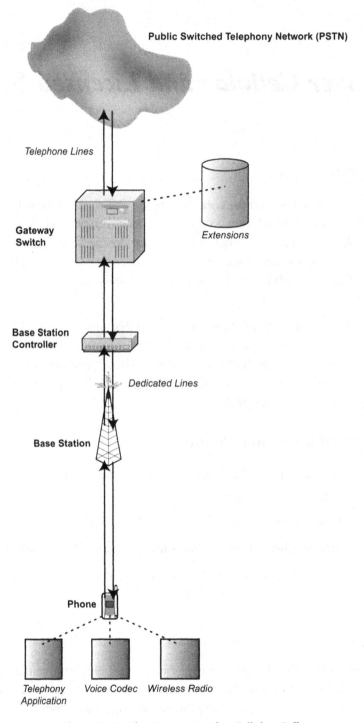

Public Switched Telephony Network (PSTN)

Telephone Lines

Gateway Switch

Extensions

Base Station Controller

Dedicated Lines

Base Station

Phone

Telephony Application   *Voice Codec*   *Wireless Radio*

Figure 7.1: The Anatomy of a Cellular Call

The *base station* is the antenna tower and network radio itself. These are located, with overlapping coverage, everywhere the network needs to provide service. Usually, these towers are on poles next to highways, on billboards, or on the sides or tops of buildings. The mobile operator leases the sites that the towers are placed, and runs cables from the tower to the shed, where the base station is located. The base station takes care of generating the signals that run the cell, and making sure that each of the phones are properly connected to the network, just as an access point does with Wi-Fi.

Located in a more central area is the *base station controller*. The base station controller provides the intelligence behind the cell. A base station controller can have multiple base stations attached to it, with dedicated digital links. The controller manages the base stations, and controls the mobile phones associated to each base station. The radio behavior of each of the base stations is determined and enforced by the controller, including channel assignments coding operation. The compressed audio from the telephones also terminates at the base station controller, which decompresses the audio, converting it into streams of PCM.

The base station controller aggregates these audio lines, only some of which will have phone calls active on them, onto trunks and sends the calls to the centralized gateway switch. The gateway switch is the first true telephone switch, and straddles the border between the *public switched telephone network* (PSTN) and the private network of the mobile operator in the area. Gateway switches usually serve a geographic region, and serve the purpose of bridging the call from the wireless portion of the network directly into the standard telephone system. The gateway switch is the mobile operator's "PBX," not surprisingly, and owns and manages the devices registered to it. The gateway switch connects, through a trunk large enough to carry the voice load coming in from the network, as the one point of contact to the PSTN.

Cellular uses the three-level architecture of gateway switch, base station controller, and base station to separate out the functions related to telephones from those related to operating a wireless service. The three-level architecture is, not by coincidence, very similar to the one we have already seen for a Wi-Fi voice mobility network in the enterprise, with an access point, wireless controller, and PBX mapping pretty closely. But the mobile operator's network is tightly integrated for the purpose of carrying voice calls. But in a general sense, the gateway switch signals the calls "directly" to the mobile phone, with the base station controller and base station proxying as needed to steer calls in the right direction.

You may notice that the description in these sections is intentionally kept to a high level, to apply to the different underlying cellular technologies and not draw too much from each architecture's specific terms and roles. In this way, the descriptions apply to all of the major second-, third-, and fourth-generation cellular networks.

7.1.1 Mobility in Cellular

Probably the most interesting feature of the cellular architecture is the notion of portability beyond the geographic region. It is fairly easy to see how mobility works within the geographic region served by the gateway switch. A phone could simply show up on a different base station another day, and the public telephone network would still interface directly with the gateway controller and not be aware of the change. There is tremendous value to this architecture, of course, in providing mobility within the geographic region, as that is where most phones will spend their time. However, the question still remains of what happens when the user takes his phone out of the geographic area and travels to another one, with another gateway.

This intergateway scenario is known as a roaming scenario, because the phone is no longer in the network operated by its gateway switch. Just as with PBX connections and enterprise private telephone networks, the mobile network anchored by the gateway switch is assigned the phone numbers for the phones that belong to that geographic region. When a phone call is placed, from the public network to the mobile phone, the public network routes the call to the gateway switch in the same way as when a phone call is placed to an enterprise extension. Once the call arrives at the gateway switch, the public phone network no longer needs to know where the call goes.

When the mobile phone is not in the network, however, the call must go somewhere. The gateway switch usually provides, through another piece of equipment attached to it, voicemail service. That works for when the mobile phone is off or its location is not known. But the feature we are looking for here is roaming. To make roaming work, when the mobile phone shows up in a geographic area not belonging to its mobile gateway, it needs to have its location updated.

The roaming mobile phone first connects to the base station, and through to the base station controller and gateway switch. The gateway switch knows the phone numbers that belong to it, and therefore knows that this mobile phone does not belong locally. However, the phone also reports information that lets the visited gateway switch determine where the home gateway switch must be on the network. Then, the visited gateway sends a message over the signaling system of the public network, SS7, to the home gateway, letting the home know that the phone has relocated. The home gateway maintains a database, commonly called a home location register (HLR), of all of the extensions that are owned by the home gateway. This database includes all phones that are operated by that carrier in that geographic region, wherever the phones actually are. In addition, the visiting gateway maintains a separate database, called a *visitor location register* (VLR), which maintains the list of phones currently associated to the gateway, regardless of whether they are managed by that operator in that geographic region. The roaming phone registers with the gateway, which updates the VLR with the phone's information and the PSTN address of the home

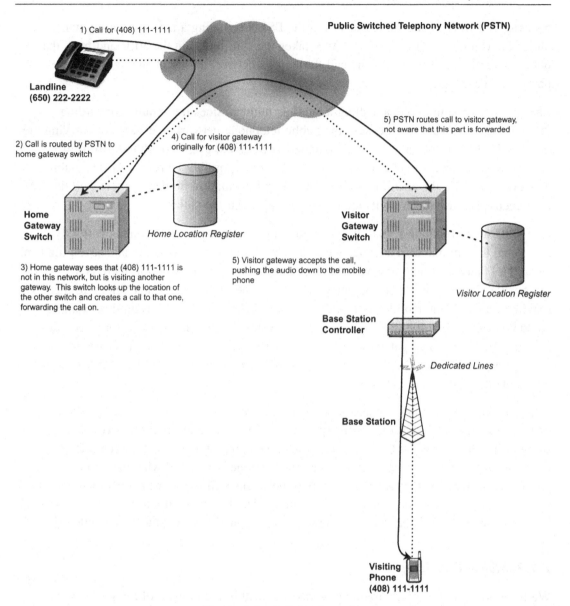

1) Call for (408) 111-1111

Public Switched Telephony Network (PSTN)

**Landline
(650) 222-2222**

5) PSTN routes call to visitor gateway,
not aware that this part is forwarded

2) Call is routed by PSTN to
home gateway switch

4) Call for visitor gateway
originally for (408) 111-1111

**Home
Gateway
Switch**

Home Location Register

**Visitor
Gateway
Switch**

3) Home gateway sees that (408) 111-1111 is
not in this network, but is visiting another
gateway. This switch looks up the location of
the other switch and creates a call to that one,
forwarding the call on.

5) Visitor gateway accepts the call,
pushing the audio down to the mobile
phone

Visitor Location Register

**Base Station
Controller**

Dedicated Lines

Base Station

**Visiting
Phone
(408) 111-1111**

Figure 7.2: Cellular Roaming

gateway. The gateway then sends the message to the home gateway, which marks the HLR to indicate that the phone has roamed and records the PSTN address of the visited gateway.

Figure 7.2 shows the process for roaming. Now, when a call comes in for the mobile phone at the home gateway, the gateway looks up its HLR and finds that the phone is located at the visited gateway instead. The home gateway needs to forward the call over to the visited gateway. To route the call, it places another call directly to the phone number of the visited gateway, but when doing so, it attaches in the PSTN signaling a message saying which phone the call is for. The visited gateway takes the call, looks up the VLR and finds that the phone is local, and then directs the call through its network to the phone. It is as if the phone had never left.

The reverse, a roaming phone calling a landline number, does not require forwarding through the home gateway, because the public telephone network is capable of carrying the outbound call with the caller's phone number represented, even though the call itself is coming from an area where that phone number cannot possibly be routed to. The public network is trusting the operators to not put wrong information in the caller ID, and thus lets equipment place whatever digits are desired to represent the caller.

One thing about roaming is that the dialing plan of the gateway the phone is currently connected to may matter. This comes into play with international dialing and seven-digit (no area code) dialing. If both gateways belong to the same mobile operator, there is a reasonable chance that the visited gateway will use the dialing rules that belong to the roaming phone. If the mobile phone in the figure dialed "333-3333" using this example, it could connect with "(408) 333-3333". But it is just as likely, and more so for international roaming, that the visited gateway will enforce the local dialing plan. In this case, the phone dialing "333-3333" may be forced to connect to "(XXX) 333-3333", where "XXX" is the area code that the visited gateway is in.

When phones are roaming between systems of different mobile operators, it is up to each operator to decide whether the visiting phone should be accepted into the network, and whether it will be able to place or receive calls. Roaming agreements between different operators are written to keep track of when these phones roam and what they do on the roamed networks, and the technology of both networks will monitor the operation and bill the home network, and thus the user, accordingly. The fact that cellular roaming works over international legs, where the visited gateway is an ocean away, is quite remarkable.

7.1.2 Mobile Call Setup

We have now looked at how roaming works. But within the domain of the gateway, the phone must still register to the network, identify itself, and be able to both place and receive calls.

At an abstract level, the concepts are not different from what we have seen earlier. The base stations send out beacons, advertising their presence and what network they belong to. The phone is built in with a preference for the network of the mobile operator who sold the phone, and the phone will scan the various frequencies that may be employed for the technology it uses, until it locates a beacon. At that point, the phone associates itself to the base station and registers itself with the network directly to the gateway it is visiting.

The gateway authenticates the phone to make sure it is not a spoof and then updates its VLR and the appropriate HLR, either its own or another.

When the phone wishes to place a call, it must gain access to the signaling channel. In cellular technologies, as we will see for each of them, signaling and bearer are generally kept separate, unlike with packet-based technologies. The signaling channel is packet-based, however, and allows the phone to send out the call setup request. This signaling goes to the switches, gets translated into a format the PSTN understands, and is routed to the final destination. This continues for the remainder of the call setup. Once the call is established, the signaling protocols ensure that there is a bearer channel—the specific form of which we will leave for the specific discussion of architectures—and the call flows.

For incoming calls, things are slightly different. The phone does not have its radio constantly powered on, waiting for an incoming call. Rather, the phone turns the radio off to save power. When a call comes in, the gateway routes it through to the base station anyway, which introduces a message into the *paging channel*. The paging channel is a broadcast channel where all messages go for phones that may or may not be awake at the moment. The paging channel repeatedly sends information informing the phone that it has an incoming call. Once the phone wakes up for the moment and sees that there is a call incoming, it comes out of power save and fully reconnects to the network. At this point, it is now able to communicate back and forth, participating in the signaling activities. If the call is accepted, the phone is granted a bearer channel, which it uses to encode the voice upstream and decode the voice downstream.

7.1.2.1 Handoff

Of course, mobile phones move. The caller typically has no idea where the coverage cell of one base station's radio ends and another one's begins, and so calls need to be able to be carried, seamlessly, across cell boundaries. The handoff mechanisms in cellular networks are designed to do just that.

The three-level hierarchy (Figure 7.3) of the mobile network allows for a small variety of handoff scenarios. When a phone is moving around the area of a base station, it may come in and out of coverage of the multiple radios that the base station has. This is one type of handoff that is very local. As the phone changes locations by significant amounts, however, it might go out of strong coverage of one cell, and become parked between multiple other cells, where it must then hand off across base stations. If the new base station is under the same base station controller, than this can be a second type of handoff. If the phone crosses to a new base station controller, however, the handoff occurs higher up the hierarchy.

As mentioned before, handoffs within one gateway switch can be handled without involving the public network. The three types of handoffs beneath the gateway can usually take place in the highest-level device in common on both paths. This is because the bearer channel

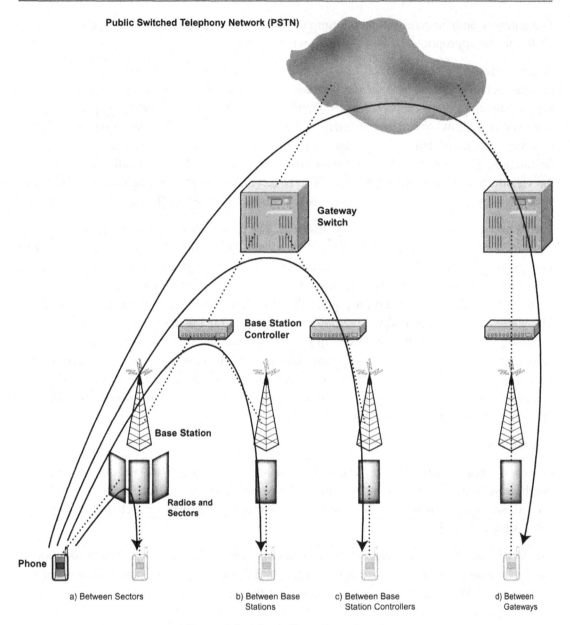

Public Switched Telephony Network (PSTN)

Gateway
Switch

Base Station
Controller

Base Station

Radios and
Sectors

Phone

a) Between Sectors   b) Between Base   c) Between Base   d) Between
                              Stations   Station Controllers   Gateways

Figure 7.3: Handoffs at Every Level

need only swivel, or pivot, from one line to another, and as long as the pivot device can get the new leg allocated without breaking the old one—a make-before-break scenario—the handoff can occur cleanly and without major changes.

When the handoff is between gateways, however, there is no central pivot point. One might notice that the calling party always stays put, and so it could be conceivable that the calling party, or some public network switch, could just request a new line to the new gateway. But the public network, by design, is not built for mobility and does not participate in the handoff process at all. Therefore, the home gateway is required to follow the same steps as in the previous section. It places a new call out to the new gateway, which then dedicates lines down to the base station the phone is handing off to. Once everything is set up, the phone drops its connection to the old base station, which frees up the lines up to the gateway. The home gateway then immediately transfers the call from the old lines in its domain to the call going across the PSTN, thus completing the circuit.

The major unknown factor in this process, for the mobile network's point of view, is whether the line connecting the two gateways will be available. Sometimes, it is not. When that happens, the call will be dropped on the handoff, and the user will be forced to redial. This transfer procedure is where the most can go wrong, but it is also supposedly a less common event, considering that gateways are planned out to encompass areas of population.

When a call that has already been anchored in the above way moves to yet another gateway, the first visited gateway does not forward the call to the second. That would lead to a rather pointless daisy-chaining of calls. Instead, the home gateway sets up the all directly over the network to the new visited gateway and drops the call from the old visited one.

7.2 Cellular Technologies

Having covered the basic concepts of cellular networking, we can now look at the specific forms that the technologies take on.

Cellular technology ranked in terms of generations. There are three well-defined generations, and a nebulous, not well agreed-upon designation for the fourth generation.

The first generation of cellular technology was that of analog cellular phones. These phones started people to be interested in the market for mobility, and provided the convenience of wireless coverage, although with the possibility of getting static and noise in the call as the caller left the range of the network. These have been surpassed by the next generation.

7.2.1 2G Technologies

The second generation, or *2G*, of cellular technologies embarked upon use of digital phones and signals to ensure that good quality was possible along a wider range of coverage. Furthermore, the transition to digital allowed the introduction of security into the network, with encryption to prevent eavesdropping.

There are two types of 2G technologies, and these two make up the basis for the cellular networks worldwide today.

7.2.1.1 GSM

One digital 2G technology is known as the *Global System for Mobile Communications* (GSM). GSM is the basis for most of the cellular systems worldwide, and introduced a new concept into the world of cellular. Each GSM phone requires a *SIM* (*Subscriber Identity Module*) card, which is provided by the mobile operator and includes the cryptographic identity tied to the telephone number and user account. The SIM card also provides for limited storage, intended originally to hold the address book that the caller can use for dialing others. The advantage the SIM card holds for the user is clear: the user can pull the card out of one phone, insert it into another GSM phone, and immediately, the second phone picks up the phone number for the user and can be used to send and receive calls. (The address book function is far less useful today, as address books are more sophisticated now and the phone cannot convert all of the information to the SIM card without loss.)

For the network, the advantage of the SIM card is that it stores the user's identity. Each SIM card is given a unique *International Mobile Subscriber Identity* (IMSI, pronounced "im-zee"). This number represents the subscriber uniquely, in every country and in every location. The IMSI is a 15-digit number. The first six digits name the mobile operator who supplies the SIM and provides service. The first three of these digits specifies the *Mobile Country Code*, and the second three specifies the *Mobile Network Code*. With these six digits, a gateway that has a roaming phone connecting to it can find the home network. For example, the first six digits of 310 170 would refer a phone to the former Pacific Bell Wireless system in California, now the AT&T network.

GSM provides names for the three levels of architecture we looked at earlier. The gateway switch is called the *Mobile Switching Center* (MSC). The base station controller keeps its name, and is referred to as the BSC. The base station itself gets a new word, and becomes the *Base Transceiver Station* (BTS).

The phones themselves are given a unique number as well, called an *International Mobile Equipment Identity* (IMEI). This number identifies the phone to the network.

Between each of the devices in the network, GSM defines and labels each link. We will not need to go into the names of these links here, but the concept is that each interface is well-defined, so that mobile operators can (in theory) substitute one vendor's equipment for another. The most important interface for which this is true is the air interface (called *Um*, for the detail-oriented). This is where multivendor interoperability comes into play.

MSCs communicate over the public signaling network using a protocol called *Mobile Application Part* (MAP). This protocol takes advantage of the extensibility of the public signaling network's protocols to add, to the messages that travel from MSC to MSC, the information necessary to run the mobile network.

7.2.1.2 GSM Radio

The GSM radio itself is based on the concept of *time division multiple access* (TDMA). The idea is that the cell's airtime is divided up into strict slots, or periods of dedicated access. Devices using a bearer channel are assigned a particular, repeating sequence of slots, one for each direction of the voice call. While the call is in progress, the phone is required to send its traffic in one slot and receive in the other. The times in between allow the phone to power down its radios in a predictable way to save power. The overall repeating pattern of slots is called a *frame*. Each frame repeats nearly every five milliseconds, which means that voice samples have to be packaged up into these short units.

The strict time sharing that GSM requires adds some complexity to the radios. The system is not forgiving if a device gets the time slot wrong, because of the precise, periodic nature of the system. Much of GSM's complexity comes into play ensuring that every device is synchronized so that they do not step on each other, even though the devices may at different, long distances from the base station, and so the speed of light and propagation delay impact the timing.

Each GSM channel provides only about 270kbps of total capacity, in which each frame asks for the traffic from eight full-rate voice streams (and thus, phones in calls). Each full-rate direction of a voice call encodes to 13kbps. This is the GSM codec that was referred to in previous chapters, and is used by some IP-based telephone equipment as well.

GSM is defined in a few licensed bands. There are four bands available worldwide. The United States and Canada use the 850MHz and the 1900MHz bands. The 850MHz band has 25MHz (barely wider than one Wi-Fi channel) for each direction: uplink is sent starting a 824MHz, and downlink is sent starting at 869MHz. These channels are spaced every 200kHz, and so there are 124 channel numbers, representing a pair of uplink and downlink narrowband slices. Voice over cellular networks is incredibly efficient in terms of spectrum, compared to voice over packet-based networks, although it must be pointed out that a link to a phone provides enough bandwidth for only one voice call, and sending data packets on that system is more complicated. The rest of the modern world uses the 900MHz and 1900MHz bands, which provide for more channels in total. Phones that are built to work worldwide on GSM must be *quad-band*; that is, they must support both the North American and international pair of bands.

7.2.1.3 GSM Data

GSM data services are provided by allowing the phone to allocate air resources that would be used for other bearer channels, and instead is used to drive a different radio and protocol for sending packet data.

The first method is called *General Packet Radio Service* (GPRS). GPRS uses an uplink and a downlink GSM channel, but drives a packet radio over those bands. This provides throughputs of around 80kbps, by occupying portions of those channels. The downlink channel does not require any contention, as it is point-to-multipoint, but the uplink channel has to allow for initial uncoordinated access. Thus, a contention-based scheme is used. GPRS provides for a radio with four different data rates, from 8 to 20kbps per slot. GPRS is sometimes called *2.5G*, because it is an addition to the 2G GSM.

To improve on the available throughput for data users, GSM has the *Enhanced Data Rates for GSM Evolution* (EDGE) technology. This is a higher-throughput way of sending traffic over the air for packet downloads. EDGE upgrades the radio to include more sophisticated coding techniques. (For a more detailed introduction to coding techniques, please refer back to Chapter 5's description of Wi-Fi radios.) EDGE defines nine data rates from 8.8 to 59.2kbps. Furthermore, EDGE offers better error correction coding methods, to recover from noise or out-of-range conditions. It is impressive to see how little bandwidth the licensed carriers have to work with, given that there is not a wide range of available frequencies to begin width, and the signals travel so far that spatial reuse becomes a challenge at every level. EDGE is sometimes referred to as a 2.75G technology, because it improves upon GPRS but isn't as modern as the newer 3G technologies.

7.2.1.4 CDMA

CDMA—the name of the mobile technology, not the radio encoding—is defined in IS-95. IS-95 does use the *code division multiple access* scheme to provide separation between devices on the channel.

CDMA, as a radio technology, works by using the notion that streams of bit patterns can be orthogonal to one another mathematically. For IS-95, each phone is given a unique pseudorandom sequence, known as a *PN sequence*. The point of the sequence is that no two sequences should correlate to each other, or agree in a statistical way, over any reasonable period of time. The insight into CDMA is that the user can use these sequences by applying an exclusive-or operation to the bit sequence with a bit sequence for the data it wishes to send, a sequence created at a much slower data rate than the PN sequence. The PN sequences' randomness is preserved by the exclusive-or operation (compare to the discussion on WEP security in Chapter 5). Once this is done, all of the phones transmit at once, at the same time, into the same 1.25MHz channel.

This is quite a surprising use of the air. Instead of having each device coordinate or be controlled, taking turns and not transmitting at the same time, to avoid collision, CDMA lets them transmit at the same time, and uses the PN sequences to sort it out. How do the PN sequences accomplish their goals? Because two different PN sequences are (pseudo)random to each other, the two streams are not expected to correlate. In other words, the inner product of the two streams should tend towards zero. Therefore, the receiver, knowing the

PN sequence of the transmitter that it wishes to listen in on, applies the PN sequence back to the signal. The sequence of the sender falls away, leaving the original data stream, and the sequences and bitstreams of the other devices becomes lower-intensity noise that must be filtered out.

This mechanism, rather than slotting, is how phones are given access to one cell for IS-95. One of the nice advantages of using CDMA is that IS-95 can perform another type of handoff. Although the backend handoff sequence is similar to that of GSM, the over-the-air part of the handoff can be made to disappear entirely. What happens is that a phone can come in range of two base stations at the same time. When this occurs, the second base station can be told of the PN sequence of the phone, and now both base stations together can transmit the same PN sequence and data to the phone. In reverse, both base stations can extract the phone's traffic. This allows these two (or more) base stations to hold the signal together, as the phone transitions from one physical area to another. This type of handoff is called a *soft handoff*, because it does not involve breaking the connection in any way, but rather in just transitioning state and resuming where the previous one left off. (Note that CDMA is not the same concept as MIMO, which also works by having multiple devices transmit simultaneously. MIMO uses the linear properties of systems with multiple antennas, rather than the properties of statistically independent sequences carried on one antenna.)

CDMA uses much of the same architectural thinking as GSM does. The major difference on the backend side is that the protocols are not the same for the links within the private mobile operator network. CDMA is popular among two carriers (Verizon and Sprint) in the United States, and is not anywhere near as common as GSM is worldwide.

7.2.2 3G Technologies

With the demand for better throughputs for the data side of the picture, the cellular industry is moving towards the next generation of technologies. These third-generation (3G) technologies, are designed to offer throughputs that start to become reasonable when compared to the consumer Internet service provider market, and therefore don't seem as antiquated to mobile users.

7.2.2.1 3GSM: UMTS and HSPA

GSM's next phase was to undergo a radical change in radio type, to allow a different type of network to be deployed. This generation of technology is named the *Universal Mobile Telecommunications System* (UMTS).

UMTS uses a new type of radio, called *Wideband-CDMA* (W-CDMA). This is a CDMA technology, and thus marks a shift away from the time-division scheme of second-generation GSM technology. W-CDMA uses 5MHz channels—considered to be wideband in the cellular world—rather than the far narrow channels in GSM. These channels are wide enough for over 100 voice calls, or 2Mbps of throughput.

The networking technology behind the W-CDMA air interface is based on the GSM way of thinking, however. SIM cards still exist, and have been expanded upon, to produce a concept called *Universal Subscriber Identity Module* (USIM). The phones use the same signaling protocols, and much of the backend infrastructure and concepts remain the same.

Because UMTS requires new spectrum, the United States allocated the 1700MHz and 2100MHz bands (getting close to the 2.4GHz of Wi-Fi, meaning that the signal starts having difficulty penetrating into buildings).

UMTS is the networking technology of choice for converged, data-enabled handsets and smartphones that are based on the GSM line of technology today. That does not mean that UMTS networks are available in most places. Some urban centers have access to strong UMTS coverage, but many other areas still remain with EDGE or even GPRS coverage only.

UMTS includes a way forward, for even higher throughput. This way forward is known as *High-Speed Packet Access* (HSPA). The first and most promising step is to optimize the downstream direction, as that's where most of the Internet traffic flows to. This is called, not surprisingly, *High-Speed Downlink Packet Access* (HSDPA). HDSPA allows downlink speeds of over 1Mbps, and as high as 14.4Mbps. To get this, HDSPA uses a radio and protocol that keeps tight tabs on the signal quality between every phone and the base station. Each phone constantly, every two milliseconds, reports on how the channel is doing. The base station then chooses who to send to, based on what will maximize the throughput of the cell. HDSPA also uses some of the higher-end encoding technologies, up to 16-QAM, to get better bits packed onto the air.

7.2.2.2 CDMA2000: EV-DO

The IS-95 architecture was not to left out of the improvements. CDMA2000 is the next step up from CDMA, and provides higher throughputs.

The first step was for a 2.5G scheme, and is called by the unfortunately complicated name of *1xRTT*. 1xRTT, for "One times Radio Transmission Technology," uses the same channels as the base IS-95, but adds an additional one times ("1x") the number of orthogonal codes, to allow more throughput to be packed onto the channel. (There is a limit to how often this can be done, but CDMA had room to grow.)

That was a stopgap, however, so CDMA had to grow more. This is where the *Evolution-Data Optimized* (EV-DO) technology steps in. EV-DO uses the same bands and channels as CDMA, but completely changes the radio underneath. EV-DO provides downlink of 2.4Mbps to 3.1Mbps, depending on the version. EV-DO changes the radio by going back the other way, to a time division scheme. EV-DO uses slots of fixed times, and for that slot, the phone is the only device accessing the air. These slots are filled with up to 12 data rates, using up to 16-QAM encodings to make better use of the air. Furthermore, EV-DO uses a similar technique to HSDPA, in that the clients with the best signal quality will get the most data, but in a way that preserves fairness across the phones.

EV-DO is being pursued by the U.S. CDMA vendors, the same way that UMTS is being pursued by the U.S. GSM vendors. As the two paths borrow ideas from each other, they seem to maintain their separateness, to allow the carriers to avoid having to completely retool their networks or run parallel networks to support a common technology.

7.2.3 4G: WiMAX

WiMAX, one of two competing 4G technologies, is a different sort of thing entirely. It is a mobile broadband technology. In some senses, WiMAX is quite a bit like Wi-Fi. WiMAX belongs to the same family of wireless Ethernet, and is defined in IEEE 802.16e (as opposed to 802.11). It is a packet-only network, designed to hook together mobile devices into a base station and deal with the handoffs between them.

The major difference, and why I have categorized WiMAX as a cellular technology, is that it runs in the licensed spectrum. WiMAX requires mobile operators.

7.2.3.1 Basics of WiMAX

WiMAX is derived around the concept of a connection. Its greatest strength is that it is a tightly controlled point-to-multipoint protocol in which the base station runs the network and the clients fit into it. Compare this to Wi-Fi, where the base station and access point are nearly equal in role, when it comes to channel access. Furthermore, WiMAX allows for the downstream and upstream to use different channels at the same time, whereas Wi-Fi is based entirely on one channel of usage, the same for every direction.

WiMAX accesses the channel using a mixture of mostly contention-free slotting, but with a small pool of contention slots for new stations to enter and reserve bandwidth. The slots are not fixed duration, as they are in GSM TDMA. Instead, the slots' lengths and timings are determined by the base station, based on the resource requirements of each of the devices. Clients must register with the base station, and then must request resources from there on out.

The WiMAX frame format is defined in Table 7.1, with the flags defined in Table 7.2. The HT field is the type of the header, and is set to 0 for all but bandwidth requests. The EC is the encryption bit, and is set to 1 if the payload is encrypted. Furthermore, the EKS will be set to the encryption key sequence. The HCS is the header check sequence, and is a checksum across the header. The LEN field is the length, in bytes, of the entire frame, including the header and the CRC. The ESF bit signifies if there are extra headers in the frame.

Table 7.1: WiMAX frame format

Flags	CID	HCS	Payload	CRC
3 bytes	1 byte	2 bytes	*optional*	4 bytes

Table 7.2: The flags format

	HT	EC	Type	ESF	CI	EKS	Reserved	LEN
Bit:	0	1	2–7	8	9	10–11	12	13–23

WiMAX can send a number of different types of frames, as determined by the Type field. The Type field designates whether there is a mesh subheader, a retransmission request, packing headers for submitting real data, or other operational controls. Some of the frames are management, and some are data. The subsequent headers are used to define what type of management or data frame is being used.

There are no addresses in the WiMAX header. WiMAX is connection-oriented. A client has to request to come into the network. Once it associates to the base station, it is assigned a connection ID (CID). This CID will be used to address the client, as needed.

WiMAX base stations transmit the WiMAX version of beacons. These frames, called *DL-MAPs*, for downlink maps, specify the base station's assumed ID, made of 24-bit operator ID and a 24-bit base station ID. Once the station finds such a DL-MAP, it knows that it has found service, and can look at the DL-MAPs and related messages to configure its radio. After this point, it needs to associate to the network. It does so by waiting to learn, on the downlink channel, the properties of the uplink channel that it will need to use to connect. If it does not like the properties, it continues on to another base station. But if it does like the properties, it will wait for the next uplink contention period, which is advertised in the downlink channel, and then start transmitting on the uplink. These exchanges are designed to let the station get tight radio timing with the base station, ensuring that the contention-free part of the channel access works without interference—a similar concern as with GSM.

Finally, the client can register with the network. The client's registration message starts of IP connectivity with the network.

The entire architecture of WiMAX is centered around the possibility that WiMAX stations are managed entities, and the network needs to run like a cellular one, with strict management controls onto the network and the devices in it.

Quality of service is strictly enforced by WiMAX. All quality-of-service streams are mapped onto the connections, which then have resources granted or removed as necessary to ensure proper operation.

7.2.3.2 Uses of WiMAX

WiMAX is not the sort of technology that most organizations will get to set up and use as a part of a private voice mobility network. In fact, as of the time of writing, WiMAX is just beginning to get rolled out to laptops, as an inbuilt option next to Wi-Fi for access.

The WiMAX model is to compete with 2G and 3G services for wireless broadband data access. WiMAX can carry much higher throughputs, as high as 70Mbps divided up more efficiently among the users, and so gives 3G a good run. Because of this, if WiMAX manages to get rolled out in any significant scale, it could become an interesting component in any enterprise mobility network, voice or data.

WiMAX's biggest challenge is in market acceptance and coverage areas. One could argue that WiMAX is just beginning to get started as a wide-area technology. And this statement has a lot of truth. Unfortunately, if rollouts continue at the rate at which they have been, and if the sour economic climate that exists at the time of writing continues into the coming years, it is not clear whether organizations should expect to rely on WiMAX making enough of a presence to be a factor.

For voice mobility specifically, WiMAX has another problem. It is a data network. Voice mobility is not just about voice; it is about access to the entire services of the enterprise when inside the building or out. But voice mobility makes sense mostly on converged telephone devices, smartphones and business pocket devices that let people do work and be reachable from anywhere. WiMAX is currently being looked at as a laptop option, to provide better coverage within metropolitan areas and to possibly avoid having users feel the need to scramble to find a Wi-Fi hotspot. Its greatest potential use is to fill the market that emptied when metropolitan Wi-Fi failed to take off. Because of this goal, if WiMAX achieves the ubiquity that cellular networks have and the quality and pricing models that metropolitan Wi-Fi was to have, it could be an incredible tool for voice mobility deployments in the upcoming years. But for now, we must sit on the sidelines and watch the developments, and rely on 3G technology to provide the bridge for voice and data in one convenient device.

7.2.4 4G: LTE

The other competing 4G network is known as *Long Term Evolution* (LTE). LTE is built on top of the UMTS style of network but is also a mobile broadband network designed for packet-based transmissions. LTE offers more than 100Mbps of throughput as a possibility, which is a major breakthrough for a cellular technology because it equals the throughput of wireline Fast Ethernet.

LTE is based on many of the same ideas as Wi-Fi but applied to a cellular network. This means that using MIMO to achieve over 300 Mbps for four antennas, while applying the necessary flexibility in how much spectrum a channel occupies, to be able to adapt to the particular license that a mobile operator carries.

LTE further offers the option of using MIMO but distributing the antennas across multiple users, producing something known as *Space Division Multiple Access* (SDMA). SDMA lets the simultaneous nature of MIMO on the base station divide the same channel into multiple transmissions to and from clients. Base stations using SDMA have a large number of antennas but increase the efficiency of the channels used. Because carriers have very little spectrum to work with, this also is a major benefit of LTE.

It remains to be seen whether LTE and WiMAX will start to compete for 4G networks in a serious way or whether one will win out. As it stands now, WiMAX is heavily favored by nontraditional carriers such as cable companies, with an investment mostly by one U.S. carrier. GSM-based carriers are unlikely to adopt WiMAX and appear to be favoring LTE at this point.

7.3 Fixed-Mobile Convergence

To this point, we have covered the two opposite ends of voice mobility: provide a private network, using private branch exchanges, private wireline networks, and private wireless coverage within a building; or use an existing mobile operator network, letting them take care of the entire management of the technology itself, and therefore providing more time to focus on managing the users themselves.

But each option, alone, has a number of downsides. Mobile-operator networks are expensive, and if the mobile population spends time within the boundaries of the enterprise campus, transitioning to mobility adds a cost that, whether per-minute or with bundled packages, may not justify the transition away from stationary—but free to use—desk phones. This is especially true if most of the calls are from user to user, and not to the outside. Furthermore, the mobile operator's network may not provide sufficient coverage within the campus to allow this sort of mobility to work in the first place. Larger buildings made of concrete and steel shield the cell tower's signal from penetrating deep within the building, and the cellular coverage may just not be sufficient for mission-critical applications to run there.

On the other hand, Wi-Fi networks are local to the campus. It is difficult enough to extend them to outside areas between buildings, but it is impossible to count on coverage once the user leaves the domain of the administrator and walks out to the street. If the population is mobile enough to need access inside or outside of the campus—whether they are road warriors or occasional but critical resources who must be reachable wherever they go—then Wi-Fi alone will not suffice.

For this reason, it makes sense to explore the *convergence* (an overused term) for sure, of the two networks. Driving this is that many enterprise-grade (and high-end consumer grade) mobile phones now have both cellular and Wi-Fi radios. Users are able to, and are coming to expect that, the phone will provide the same features, at the same level of utility, wherever the user is. For these users, the line between in-building and out-of-building networks has become artificial, a detail that should be no more intrusive or problematic than that of the user being able to access email from a desktop personal computer or a laptop.

Thus, fixed-mobile convergence (FMC) can be approached from three different angles. Those voice mobility planners who have traditionally provided their users with cellphones can look towards the economic advantages of having a potentially large percentage of their in-building phone calls be transferred off of the for-a-fee cellular network, into a free-to-operate Wi-Fi network. Those planners who have thought of their workforce as only needing in-building coverage, and had traditionally looked to phone technologies such as Digital Enhanced Cordless Telecommunications (DECT) before Wi-Fi, may find that the productivity increases brought on by allowing the phone to operate outside of the buildings can justify pursuing FMC. Finally, planners who have strong in-building Wi-Fi and an equally strong cellular solution may find that combining the two helps remove the frustrations of users who cannot understand why there is a difference in their phone's capabilities when they are in the office than outside of it.

Fundamentally, FMC means settling on the use of a dual-mode phone, one that supports both Wi-Fi and cellular, and providing both a strong Wi-Fi network and a well-thought-out mobile workforce package, with remote email and mission-critical service access. Therefore, both the carrier and the in-building network will be involved, as major players.

There is quite a bit in common between the two approaches, especially at the level of requirements that are placed on voice mobility administrators. They differ in the planning and provisioning that is required, as well as in the choices available.

7.3.1 Enterprise-Centric FMC

However, FMC itself has two major approaches, different in how involved the carrier is in the in-building network. The first approach, which we will call *enterprise-centric FMC*, works by excluding the carrier in the understanding of the mobility. This model is very similar to the way that mobile operators exclude the public telephone network from having to know about the mobility in the first place. In an enterprise-centric FMC solution, the very fact that the phone is mobile is only marginally included into the equation. The typical enterprise-centric FMC architecture is shown in Figure 7.4.

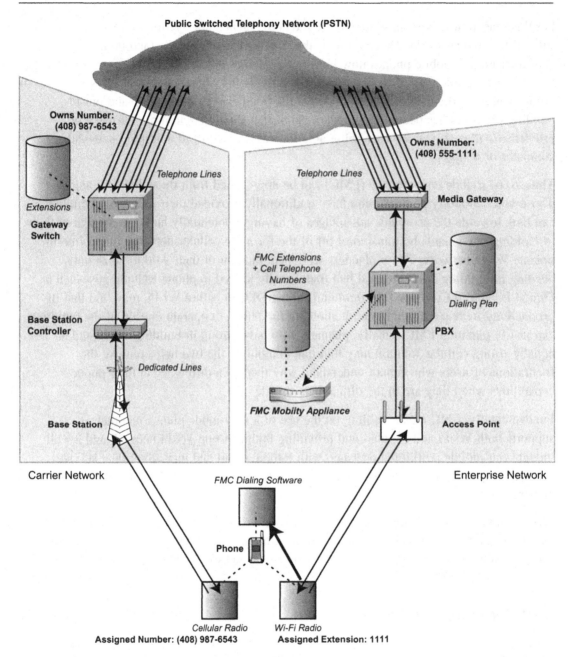

Figure 7.4: The Enterprise-Centric FMC Architecture

In this figure, the two separate networks are shown. The cellular network remains exactly as it would normally be, unmodified in any way. The enterprise network is similar to how it was before, except for the addition of an *FMC mobility appliance*, appropriately placed in the center of the picture. In the enterprise-centric FMC architecture, the user's telephone number is owned by the enterprise. This enterprise extension becomes the number that people use to reach the user, wherever she may be.

Everything centers around the FMC mobility appliance. This device integrates seamlessly into the existing SIP-centric private enterprise network. Every enterprise FMC phone requires an additional piece of FMC dialing software installed on the mobile phone. With this software, the phone registers with the mobility appliance. The mobility appliance has the database of every mobile extension that participates in the FMC operation, as well as each phone's cellular telephone number.

The main idea behind the enterprise-centric FMC architecture is that the mobility appliance becomes the phone's substitute when it is out of the office. The appliance accepts and places all calls for the roaming user, as if he or she were present. To accomplish that, the appliance bridges the calls back and forth to the mobile phone, over the cellular network.

When the telephone is in the Wi-Fi network (see Figure 7.5), the mobility appliance maintains just a management role, for the most part. The mobile phone's dialing software becomes the focus. This software provides a complete SIP-centric soft telephone stack, using the voice phone speakers and microphone, recording the audio and sending it over the Wi-Fi data network. From the user's point of view, the telephone appears to be operating as a typical telephone, for the most part. From the enterprise voice mobility network's point of view, the mobile phone and dialing software appear as a standard SIP extension.

When the telephone leaves the Wi-Fi network, however, the mobility appliance springs into action. The dialing software on the mobile phone registers back with the mobility appliance in the enterprise. To do this, the cellphone may use its mobile data service to connect, over the Internet, to the mobility appliance in the enterprise. The mobility appliance then turns on its own SIP client engine. The engine assumes the identity of the user's extension, and registers on the user's behalf with the PBX. The PBX is unaware that an FMC solution is in operation. Instead, it is aware only that the user temporarily changed physical devices. As far as the PBX is concerned, the user's phone is located, physically, in the FMC mobility appliance.

When a call comes in for the user, the PBX simply routes the telephone call to the FMC appliance, expecting it to answer. The mobility appliance gets the incoming call, and then places a second, outgoing call to the mobile phone, as shown in Figure 7.6. Once the mobile phone answers, the two calls will be bridged into one. At the same time as placing the outbound call, the appliance sends a message over the cellular network to the FMC dialing software, informing it that a call will be coming in and relaying the phone number of the caller.

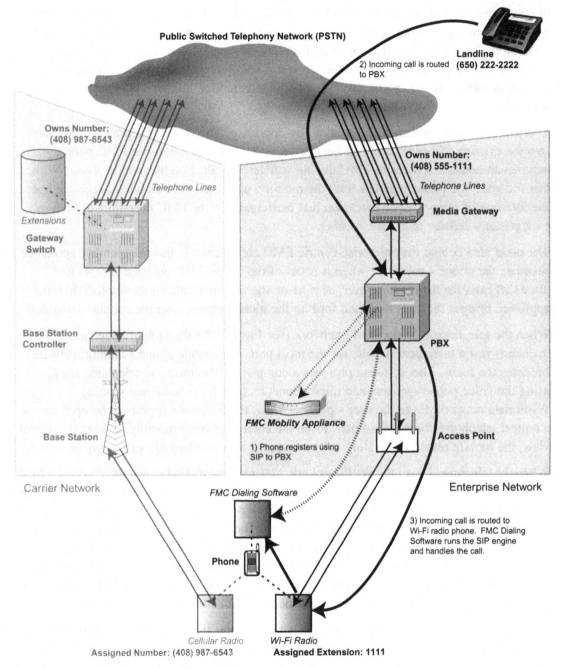

Public Switched Telephony Network (PSTN)

2) Incoming call is routed to PBX

**Landline
(650) 222-2222**

Owns Number:
(408) 987-6543

Telephone Lines

Owns Number:
(408) 555-1111

Telephone Lines

Media Gateway

Extensions

**Gateway
Switch**

**Base Station
Controller**

PBX

FMC Mobility Appliance

Base Station

1) Phone registers using SIP to PBX

Access Point

Carrier Network

FMC Dialing Software

Enterprise Network

3) Incoming call is routed to Wi-Fi radio phone. FMC Dialing Software runs the SIP engine and handles the call.

Phone

Cellular Radio
Assigned Number: (408) 987-6543

**Wi-Fi Radio
Assigned Extension: 1111**

Figure 7.5: An In-Building Call with Enterprise-Centric FMC

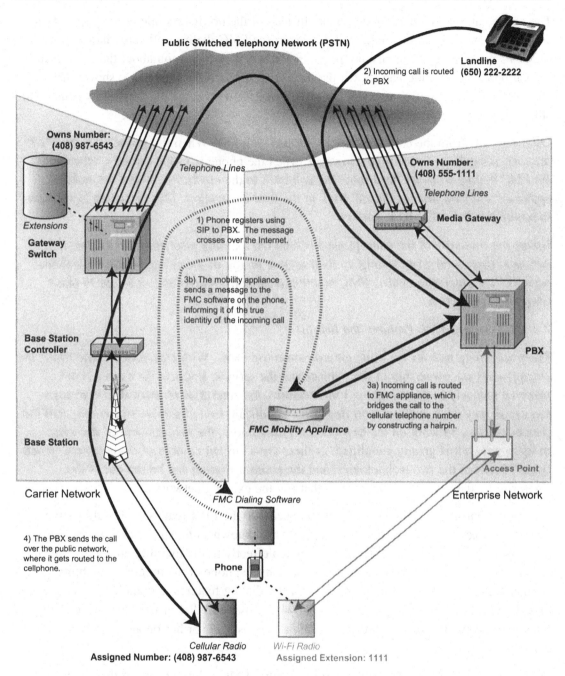

Figure 7.6: An Out-of-Building Call with Enterprise-Centric FMC

The phone call reaches the mobile phone. Instead of the normal mobile phone application and display showing an incoming call, however, the FMC dialing software takes over and shows its own display. This display provides the correct information about the call—if the normal phone application had shown the incoming call as is, it would have shown the phone number the mobility appliance used to place the outgoing call, or possibly some other incorrect enterprise number.

Outgoing calls from the cellphone are performed similarly. When the phone is in the Wi-Fi network, the call works as usual. However, when the phone is in the carrier's network, the FMC dialing software must route the call back to the enterprise, where the mobility appliance will then bridges the call back to the final destination. This extra step is necessary to ensure that the caller ID is correct.

Notice the similarity of the concept used behind the roaming mechanism to the one the mobile carriers used when setting up their service on top of the public landline telephone network. The enterprise-centric FMC solution regards the carrier as just a pipe to place telephone calls across.

7.3.1.1 Enterprise FMC Features and Benefits

There are strong reasons for using enterprise-centric FMC. With enterprise-centric FMC, the enterprise retains ownership of the number and the service. Because the carrier is not involved uniquely in providing the FMC features, the enterprise remains able to change carriers as they wish, without worrying about significant loss of service—providing that the phones continue to work on the new carrier. Furthermore, the management of the voice mobility network is greatly simplified, as there are a limited number of management "touch points" between the two technologies, and the entire operation can be run out of the enterprise by the same IT or voice staff that run the existing networks.

Enterprise-centric FMC also provides direct access to the PBX features. This allows the user to have access to *direct-extension dialing*, also known as *four-digit dialing*. Because the outbound calls from the cellphone are routed directly to the mobility appliance, which then bridges the call, the PBX does not know that the phone is roaming. The mobility appliance can then use all of the in-building PBX dialing features to complete the call, even though the call itself is coming from the outside. A user can dial just the extension of the other party, using the same abbreviated dialing plan as if his or her phone were in the enterprise.

Furthermore, by bridging the call, enterprise-centric FMC solutions ensure that the PBX remains in the call path, whether the phone is local or roaming. PBX features such as call transfer, conferencing, and autoattendant are available to the outside user. As far as the outside user is concerned, the phone is a true part of the enterprise PBX, precisely because the call is always going through the PBX.

On the other hand, this model locks the voice mobility network into the specific FMC vendor, and further constrains voice mobility planners by working on a subset of available dual-mode phones, and not being able to provide the same feature set and quality across each phone. Enterprise-centric FMC is in its early days. The FMC vendors have to interoperate with both the PBX vendor and the handset vendor, and often, because of the FMC vendors' sizes, this interoperability is not a relationship of peers. What works with today's phones may not continue to work in the next generation, and every upgrade cycle runs the risk of causing the FMC solution to suffer.

The area of greatest risk lies with the phone software. This dialing software is required to replace, or interfere with, the standard telephone dialing application. The reason is to ensure that outgoing calls are routed first to the enterprise, and that incoming calls have the requisite caller information displayed. Furthermore, the software needs to take over or use the correct set of speakers and microphones. Cellphones often have two speakers, one dedicated for phone calls and another used to play music or ringtones. For some of the older models of phones, it was difficult or impossible for the FMC software to access the voice speaker, and the audio would come out the speakerphone speaker instead. Issues such as these restrict the phones that the FMC vendors have available to produce the highest-quality experience. Problems with audio routing, layered applications, and user interfaces can lead to widely different experiences on phones from even the same vendor. Finally, there is an education process required to convince users of dual-mode phones to use the FMC dialer and not the standard telephone dialer. The FMC vendors do a good job making their application look and feel like a native dialer, but the presence of two dialers can cause severe issues with user experience. If the caller uses the wrong dialing software, the digits dialed may make no sense, the call may go through the wrong path, and the hand-in and hand-out processes may be disrupted. This is a critical problem for calls placed in-building, but on the wrong dialer, when FMC is explored as a way to improve the in-building voice coverage. Dialing on the wrong network can lead to rejected or poor-quality calls.

Enterprise-centric FMC also has an issue of draining enterprise resources. Whereas the previous problems are all shorter-term issues, and, over time, the enterprise-centric FMC players will be able to work through these engineering challenges, enterprise-centric FMC has a major long-term challenge.

Enterprise-centric FMC does occupy additional PBX and telephone line resources that would not otherwise be occupied. Every phone that is out-of-building requires *twice* the number of resources when in a call than when it is in-building. This is not a significant resource drain when the number of callers who are roaming is small, but if the FMC option grows in popularity among the users of a voice mobility network, planners and administrators may find that a large percentage of their calling resources become occupied performing just the hairpinning (Figure 7.7). The enterprise PBX is usually not provisioned

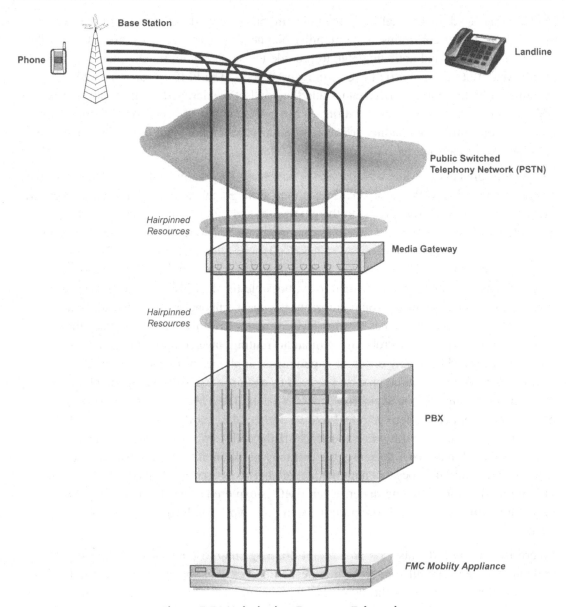

Figure 7.7: Hairpinning Resource Exhaustion

to perform the sort of bridging that a cellular gateway switch is. Classic PBX provisioning techniques rely on assuming that most calls stay in-building, and do not occupy outside lines. With every indoor extension possibly consuming one outside line, the demand on provisioning of both the PBX and the links to the PSTN can become greatly inflated. This is a distinct problem with enterprise-centric FMC that will not likely be resolved, because the FMC appliance will bridge the call, and the hairpin cannot be torn down.

7.3.2 Cellular-Centric FMC

The alternative to enterprise-centric FMC is *cellular-centric FMC*. In cellular-centric FMC (Figure 7.8), the mobile operator owns the phone number. The phone is given one and only one phone number. In this case, the enterprise PBX is not at all involved in the FMC operation.

In cellular-centric FMC, all of the changes are in the carrier. The carrier adds another platform, an FMC *mobility gateway*, to their network. This gateway fits into their existing architecture, but is given an IP address and access directly to the Internet.

When the phone is in the cellular network, it behaves as a typical mobile phone. However, when the phone roams into the enterprise, its carrier-centric software—built-in to the phone, so there is no distinction between the cellular and Wi-Fi software—registers over the Internet, back to the preprovisioned address of the cellular FMC gateway. The FMC gateway updates the cellular network's systems, informing them that any incoming calls should be directed to the FMC gateway. When an incoming call is placed, the mobile operator's switch routes the call to the FMC gateway, which converts the voice into IP packets and sends those packets across the wide-area Internet, to the enterprise. The packets are placed on the Wi-Fi network, and arrive at the cellular telephone, which decodes the packets and plays them out.

Outgoing calls are performed in the same way. The phone uses the Wi-Fi radio to signal the start of the call, and then directs the bearer traffic over the Wi-Fi network, through to the Internet, where the FMC gateway converts them back to the digital circuit and through the mobile operator's network.

Cellular-centric FMC fits directly into the existing mobility architecture for the carrier. As far as the rest of the carrier's network is concerned, the phone is using a typical base station on a typical base station controller. The mobility gateway just happens to be that controller, and tunnels the signaling and bearer traffic over IP, rather than continuing down on circuits to the remaining parts of the network. The elements of the base station itself is left out, as the air interface has been replaced with an IP interface.

7.3.3 Cellular-Centric FMC Features and Benefits

Cellular-centric FMC's advantage is that it does not use or tax the enterprise PBX. There are no voice network provisioning requirements for cellular-centric FMC. This is a great advantage to environments where either a large percentage of the workforce will be roaming outside the enterprise. It is also an advantage to organizations where their existing PBX infrastructure does not have the capacity to provide the mobility forwarding, or even the extensions for the mobility devices. This latter part is true for organizations that are just embracing voice mobility, and may have had a smaller number of desk phones but are

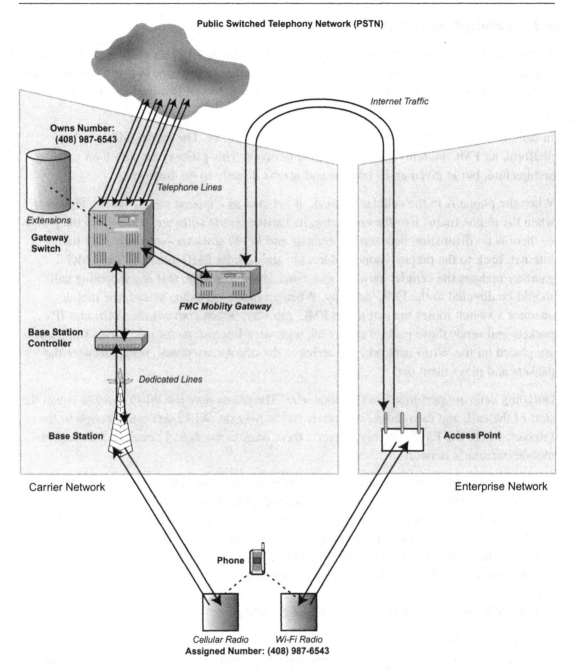

Figure 7.8: The Cellular-Centric FMC Architecture

looking for expanding the reach of the network at the same time as providing mobility, as a part of a packaged upgrade.

The economic advantages are still preserved in cellular-based FMC. When the phone is using the Wi-Fi network, the mobile operator will usually charge either less for the call or not at all. They do this in part because it frees up resources on their network, allowing them to add extensions for nearly free, and thus making a higher profit on the monthly service fees. This can provide a method for them to expand their business without expanding their expensive cellular resource allocation step-by-step.

Furthermore, the integration of the technology is far more complete than with enterprise-centric FMC. There are no issues with having software to install or manage on the phone itself, as the software is a part of the phone's operating system. There is only one dialing application, one address book, and one feature set for a user to be concerned with. This provides both less usability headaches and reduced management complexity.

There are two obvious challenges for cellular-based FMC. The first one, a short-term challenge, concerns the lack of enterprise PBX integration. Cellular-based FMC today has no integration with the enterprise network. PBX features are not available, therefore, and every extension becomes a public phone. Cellular-based FMC has primarily, up until now, been targeted mostly at consumers. As carriers enter the enterprise market, they are likely to add other methods of enterprise integration. Most certainly, given that many of the cellular-centric FMC phones on the market today are enterprise-oriented phones such as the Research in Motion BlackBerry, the enterprise data integration is already strong. Specifically, Research in Motion offers software solutions to extend the PBX dialing plan and features out to the phone, with or without FMC. The need for enterprise PBX integration is somewhat diminished, and the avenue for further enterprise services integration is available. Even in environments with well-provisioned private telephone networks, there are some where users will insist on trying a person's cellphone first, and the desk phones remain mostly idle. These environments will do well with cellular-centric FMC.

The second challenge is one of provisioning a high-quality Internet link to the carrier. Cellular-centric FMC relies on sending real-time streaming video over the Internet, a network not directly designed for quality-of-service traffic. Especially as the number of in-building users grows, the traffic on this one link to the outside can become congested. Just as enterprise-centric FMC makes additional demands on the PBX and PSTN link, cellular-centric FMC makes additional demands on the router and Internet link. Additional access bandwidth to the Internet may be required, as cellular-based FMC becomes more adopted. While the FMC bearer packets are within the enterprise, they are stamped with the correct DSCP tags to allow routers, switches, and the access points to provide the proper quality of service. Nevertheless, there is no natural quality-of-service on the path out from the enterprise to the carrier. It remains to be seen if carrier-centric FMC operators will work with Internet service providers to provide a more reliable and quality-protected network between the enterprises and the mobile operator's networks.

7.4 Handoff Between Different Networks: Handing In and Handing Out

The descriptions so far have concentrated on the two types of architectures for creating an FMC voice mobility network. Both architectures address the problem of finding a mobile phone, when it is roaming in a different network than the one that owns the phone number. However, because FMC provides the illusion of seamlessness or homogeneity between the carrier and enterprise networks, users may expect that they can carry a live phone call between the networks, without having to drop it. Just as with cellular operators alone, the combined solution must provide a way to hand off between the two networks, or at least allow the user to predict where one network ends and another begins. This is the FMC handoff problem.

7.4.1 The Handoff Problem

The problem of *call continuity* itself—transferring the call without disruption between the two networks—may not need to be solved, for many FMC voice mobility networks. For networks and organizations where users are either in-building or on the road, the lack of *call continuity* may not be a concern. Of course, it is inconvenient to lose a call as the user walks inside or outside a building, but physically entering or exiting a building is sufficient an event that the user can be trained to expect a disruption, and plan accordingly.

However, all FMC technologies work to provide call continuity, as one of the expected benefits of a unified system and as a powerful selling tool. Each of the two architectures handles the call connection or transfer in a different way, but the principles on actually deciding when or how to make the transition are similar for both.

The problem the phone needs to solve is to determine when it makes sense to put in for a transition. Unlike with straight mobile networking, where different base stations can see the phone and can possibly coordinate to decide when the best transition time is, with FMC each network can only see its own side of the solution. The cellular network cannot use pure cellular control—not without substantial mobile assistance—because only the phone knows whether it is in range of the appropriate Wi-Fi network. On the other side of the puzzle, the Wi-Fi network is good any knowing where the client is, but does not know about any other network.

From the phone's point of view, it is given two tasks that must be performed simultaneously: evaluate the Wi-Fi network for suitability for voice, and ensure that the cellular connection is good. If either side breaks, the phone must immediately register back with the central mobility platform to inform it of the loss of a link. When a call is in progress, this can be a substantial task.

Wi-Fi complicates the picture. Cellular networks are designed to keep the phone associated with the base station that will provide continuous access to voice resources. The phone's cellular engine will, under the network's control, maintain a connection to the mobile operator unless the phone is truly going out of coverage. The Wi-Fi radio does not work that way. Unfortunately, as you saw in Chapter 5, the principles of Wi-Fi require that Wi-Fi make numerous handoffs as the mobile makes its typical motions throughout the network. These Wi-Fi to Wi-Fi handoffs, also called *horizontal handoffs*, because they work within the technology that is providing service, appear to the phone as constant service disruptions. These disruptions may be brief, or they may last for long periods of time, but they are generally false, as the phone is not leaving coverage from the network itself, but just leaving coverage from elements of the network and entering coverage of other elements. Because the phone cannot know when it is leaving the network for good, versus when it is just exiting one small Wi-Fi cell and entering another, the phone's FMC engine has to weigh whether the waning Wi-Fi coverage is sufficient enough to warrant a *handout*, or a handoff outside of the Wi-Fi network. Should the phone and supporting services make the wrong decision, the phone can alternatively end up with no Wi-Fi coverage when the phone has not prepared to transition to cellular, or end up on cellular when the phone is within the building. The latter can have a direct monetary impact on the voice mobility network, because minutes spent on the carrier's network may not be free, and the point of many FMC solutions is to minimize the amount of money spent on the carrier.

There are broadly two methods that this is being handled by FMC today, both of which can be complimentary. One method works specifically for enterprise-centric FMC, and involves using Wi-Fi intelligence on the phone and FMC appliance to make a better call. The general idea is to require the voice mobility planner to enable location tracking on the Wi-Fi infrastructure, using location training (walking from point to point and marking where the point is on a digital map of the building) to let the FMC appliance learn where the edges of the buildings are. When the phone remains within the building, the FMC appliance and the phone will resist handing out to the cellular network, prioritizing Wi-Fi over cellular. This does subject the caller to whatever handoff quality the Wi-Fi network has, which in itself can be difficult to tolerate, but it does allow the enterprise to save money on the transitions. As the phone enters the edge of the building, the appliance and the phone switch the preference, favoring the cellular network, if not requiring it outright. Therefore, as a user heads to the edge of the building, the mobility appliance will detect this and start preparing the phone for a potential transition to the outside. In reverse, as a phone enters the building, the mobility appliance can detect this and allow the phone to predictably *hand in*, or hand off to the Wi-Fi network, without requiring the cellular coverage to be completely diminished.

The second method is to attempt to avoid Wi-Fi handoff altogether, or at least to mitigate the effects. The use of channel layering wireless architectures for Wi-Fi allows the Wi-Fi network to perform the same task for its inter-access-point handoffs as the cellular network provides for its inter-base-station handoffs. By centrally controlling and regularizing the handoff process, the network frees the phone to having to make only a very simple choice. Is the Wi-Fi network sufficiently strong to allow for a phone call, or is the cellular network the correct network? This can eliminate the problem of poor FMC call quality for mobile users in-building.

7.5 Cellular-Centric Technology with UMA

For voice mobility networks that may be considering a cellular-centric FMC solution, the currently available technology is known as *Unlicensed Mobile Access* (UMA). UMA integrates into the GSM cellular architecture, and uses the cellular-centric FMC architecture to encode GSM signaling and bearer traffic directly over IP.

UMA requires the phone to have UMA-compatible software as well as the dual-mode radio. The degree of integration of phone and handoff engines required for UMA is very high—UMA interacts directly with the GSM engine—and thus UMA must be installed into the cellphone by the manufacturer.

Within the carrier's network, the FMC mobility gateway is installed as a UMA base station controller. This controller uses what is known as the *A* interface to connect to the MSC, or gateway switch. As far as the mobile network is concerned, the entire Internet-connected phone population is local, not roaming, and connected to one special base station. The UMA name for the FMC mobility gateway is the *UMA Network Controller* (UNC). As the user hands out and in from the mobile operator's network, the transitions which occur appear to the GSM network as nothing more than an inter-BSC handoff. The one wrinkle to this is that the mobile phone can be in a geographically distant area from the original mobile network. In this case, because the IP traffic always goes back to the home MSC, the handoff could very well be across MSCs and the PSTN.

UMA connects the phone the carrier directly using IPsec (Chapter 8). The phone uses a preprogrammed DNS name for the UNC, after which it negotiates directly to the UNC using ISAKMP, IPsec's negotiation protocol used to set up the IPsec link. The IPsec link uses encryption to ensure that the call can be neither eavesdropped on nor interfered with, and protects the phone and carrier resources. UMA specifically uses a *network address translation* (NAT)–friendly form of IPsec, which allows the UMA phone to connect from behind the enterprise firewall. The outgoing message from the phone opens up the port in the firewall, and the IPsec-encrypted payload is placed into a special UDP header for UDP port 4500 (IPsec NAT traversal). The UNC accepts these packets and lets the encrypted tunnel operate. The signaling traffic is exchanged as needed between the phone and the UNC using this IPsec tunnel and GSM signaling within.

Once a call is placed, the bearer traffic starts. The bearer channel runs within the same IPsec-encrypted connection, but has the DSCP for Expedited Forwarding (see Chapter 4 for details on this). This ensures that the enterprise can apply quality-of-service handling to the packets, in both directions, and enables the access point to send the traffic as a high-priority packet—specifically for Wi-Fi, these packets will come out with the *video*, not the *voice* prioritization, which means that you may wish to consider reclassifying the packet, as long as WMM Power Save is not in operation. Bearer traffic is sent in 20ms intervals.

The only provisioning tasks required for enterprises are to ensure that the bearer channel has priority and to provide enough capacity for the link to the carrier.

As of the time of writing, T-Mobile offers UMA for voice mobility networks, based on a variety of handsets that encompass enterprise-grade features.

7.6 Potential Alternatives to FMC: Cellular-Only Technology

If the goal of FMC for a voice mobility network is to improve the in-building coverage, rather than to provide true convergence of services, then there are alternatives which may begin to gain in popularity. A number of U.S. mobile operators have begun to investigate the idea of installing smaller base stations directly into buildings. The strategies divide into two parts.

The more established strategy is for the mobile operator to work in cooperation with the enterprise, to add what are called *picocells* into the building. A picocell is a small base station meant to be installed within buildings to allow cellular coverage to be provided by radios located within the building. The picocell devices are owned and managed by the cellular carrier, who connects the devices back to their network using Ethernet cables or similar. Picocells interact with the carrier's network at the level of a base station or base station controller. An alternative to this strategy is one in which the mobile operator takes the burden of installing proper base stations outside the building, but uses sectorized antennas and closer base stations to attempt to boost the signal inside the building. These two approaches generally require the organization to have an outstanding contract with the carrier for significant mobile services, or for the campus to be a place that tends to attract public users.

For more private users, the carriers are now embarking on a strategy based around the idea of *femtocells*. A femtocell is a device that looks much like a Wi-Fi access point, but is owned by the carrier and operates in the licensed spectrum, connecting back to the mobile operator over the public Internet. In terms of traffic flow, much like UMA (but not requiring any changes to the handset), femtocells have the ability to place points of coverage anywhere necessary. Currently, femtocells are being targeted for the home, for customers who cannot get good access where they live or may be enticed by special discounts to install one. However, femtocells may have an upcoming application for the deployment of enterprise voice networks, and it would be wise to watch the developments to see where they may lead.

CHAPTER 8

Securing Voice

8.0 Introduction

Security is the most important aspect for voice mobility networks. Of course, quality of service must be there, or the network will not be used for long or with any sense of appreciation. And the devices must be simple, and the network must be run well enough that users do not notice when things are changed or the network is grown. However, no one would use a phone if they thought that others were listening in on their calls. Enterprises would abandon their networks left and right if strangers, criminals, or their competition could listen in on the goings on in the network.

Voice mobility means no wires, and no wires means that there are signals. These signals do not know the difference between a friendly receiver and an eavesdropper. As long as the signal can make it to the attacker, the voice call is in jeopardy.

Furthermore, if the network is not protected from intrusion, it may serve as a free ride for people looking to place anonymous, long-distance phone calls to wherever they may please. The idea of an open connection to the voice network did not have to be confronted before voice mobility, because the network ran on wires, and someone needed to gain physical access into the building. The worst possible breach that was possible was trusted employees making unauthorized calls from other workers' desks. True, access to the actual management consoles of the voice network needed to remain under lock and key, but otherwise there wasn't much that a malicious person could do with access to the phone itself or the digital line it ran in on while inside the building.

But now, with voice mobility, new signals can come in, ones that are designed to access the network from the safety of the parking lot. And if devices are stolen, there could be a free ride into the enterprise from hundreds, or even thousands, of miles away.

The purpose of this chapter is not to scare anyone interested in voice mobility, but to bring them comfort—comfort that security is an integral part of voice mobility technologies of all forms. This chapter builds upon all of the different voice mobility technologies already mentioned to show how the entire network can be secured.

8.1 Principles of Security

A secure network provides the following:

- *Confidentiality*: No device other than the intended recipient can decrypt the message.

- *Outsider rejection*: No device other than a trusted sender can send a message correctly encrypted.

- *Authenticity and forgery protection*: The recipient can prove who the original composer of the message is.

- *Integrity*: The message cannot be modified by a third party without the message being detected as having been tampered with.

- *Replay protection*: A older but valid message cannot be resent by an attacker later, thus preventing attackers from replaying old transactions.

Each factor is a crucial part in protecting the network. Confidentiality, also known as privacy, is the most obvious aspect of security. No one wants anyone unauthorized to hear the message to be able to understand it. No one wants someone who is not a legitimate party to a call to be able to listen in, take notes, and maybe record it for later.

Outsider rejection and authenticity are closely related, and are also necessary for secure networks. Outsider rejection means, in this context, that no device that is not a part of the secure network is able to send messages that can be used within the network. It is actually possible for some networks to accept both encrypted and unencrypted data. It is also possible for some encryption schemes, such as public key encryption, to allow any device to send an encrypted message. For security, the network must stop both of these from happening, for all real enterprise voice and data flows. Forgery protection and authenticity is the concept that the encryption algorithms must be such that the sender's authenticity is known if the message correctly decrypts.

Integrity is the property that messages that are otherwise proper cannot be modified en route to artificially inject data that was not there in the original message. Even if that injected data were to normally appear as garbage that is carried deeper into the network an attacker can use the weakness to affect the network. Networks, therefore, must not accept any message that has been tampered with in any way.

Even when the message is authentic and free from tampering, and the contents are unknown, an attacker can save up messages and replay them at later times. This replay attack attempts to try to trigger behavior that was already requested, again, to take advantage of whatever changes the request might make. Replay protection prevents this from happening by not letting older messages be reused.

8.2 Authentication, Authorization, and Accounting Services with RADIUS

Authentication and authorization are two different elements of networking. Authentication is the process of using cryptographic procedures to find out what user is using a device and verify that the user is actually who she says she is.

Authorization, on the other hand, is a matter of policy, and determines what a user can do on the network. Authentication and authorization tend to be provided by similar infrastructure, because authorization systems require that the users all be defined and managed, and authentication uses the user database to prove out who each user is.

Accounting is a third aspect of network access. Accounting is a side activity to see what a user has done with the resources she has available to her.

All three concepts are packaged together in the joint concept of Authentication, Authorization, and Accounting (AAA). AAA is the heart of the security operations in any network.

Simple user databases can be established independently in a number of devices. However, the IETF created a network protocol and architecture definition, called Remote Authentication Dial-In User Service (RADIUS), that has a separate server located within the network dedicated to providing AAA services.

The term RADIUS belies its origins as the dial-up login mechanism for old modem banks. When users dialed into a modem, the serving model would send text to the user, requesting a username and password. This username and password would be matched in the user database, and the user would be let on. Because modem banks were made of large number of equipment, each piece of which needed to access the same user and password database, the idea of placing the user database in one central machine was created. This became the first incarnation of the RADIUS server.

The RADIUS server maintains the username and password database, or at least proxies it from an alternative-technology central store (such as LDAP). Moreover, the RADIUS server has the job of running the advanced, multiple-step cryptographic protocols that are used to ensure tight authenticity of the user in question.

8.2.1 The Basic RADIUS Protocol

RADIUS runs on UDP, and is defined in RFC 2865. Port 1812 is set aside for RADIUS access, with port 1813 used for sending accounting messages. The RADIUS server listens on that IP address for a request for authentication, and then responds with a series of challenges, or a final yes-or-no answer.

The device that does the requesting is never the user. Instead, users communicate through some undefined means with a *Network Access Server* (NAS). The NAS itself must have a shared secret in common with the RADIUS server, in order to use its services.

The NAS server gets the request for authentication from the user, and converts it into a RADIUS message called an *Access-Request*. The format of the basic RADIUS message is shown in Table 8.1. The code is the type of the message, as shown in Table 8.2. The Length field is the length of the entire message. The Authenticator is used to verify the authenticity not of the user, but of the NAS or RADIUS client, in a manner that preserves the user's password that the original RADIUS server used. Finally, the Attributes (see Table 8.3) carry the important information about the user.

Table 8.1: RADIUS packet format

Code	Identifier	Length	Authenticator	Attributes
1 byte	1 byte	2 bytes	16 bytes	*variable*

Table 8.2: RADIUS codes

Code	Meaning
1	Access-Request
2	Access-Accept
3	Access-Reject
4	Accounting-Request
5	Accounting-Response
11	Access-Challenge

Attributes are of the format *type-length-value*, as shown in Table 8.4.

Let's look at an example authentication, using a simple password login pushed to the RADIUS server.

The first message is an *Access-Request* from the NAS, as shown in Table 8.5. For an *Access-Request*, the authenticator is a random nonce to be used just once and forgotten. (The term "once" and "nonce" are, in fact, related, though it is not quite as clear as it might seem.) Within the request comes the identifying information for the user. In this case, let us assume that the client is logging in from login terminal, called "login-bank-01", on that terminal's 23rd port. The username is "user@corp.com". This is a strict username and password login, so the user's password, "password", is padded out to be 16 bytes, then an exclusive-or is applied with it to the MD5 of the shared secret and the nonce.

That is enough information for the server to determine whether the user should come in. The RADIUS response is of the type *Access-Accept*, as in Table 8.6.

Table 8.3: Selected RADIUS attributes

Type	Meaning
1	User-Name
2	User-Password
3	CHAP-Password
4	NAS-IP-Address
5	NAS-Port
6	Service-Type
7	Framed-Protocol
8	Framed-IP-Address
9	Framed-IP-Netmask
10	Framed-Routing
11	Filter-Id
12	Framed-MTU
13	Framed-Compression
14	Login-IP-Host
15	Login-Service
16	Login-TCP-Port
18	Reply-Message
19	Callback-Number
20	Callback-Id
22	Framed-Route
23	Framed-IPX-Network
24	State
25	Clear
26	Vendor-Specific
27	Session-Timeout
28	Idle-Timeout
29	Termination-Action
30	Called-Station-Id
31	Calling-Station-Id
32	NAS-Identifier
33	Proxy-State
60	CHAP-Challenge
61	NAS-Port-Type
62	Port-Limit
64	Tunnel-Type
65	Tunnel-Medium-Type
79	EAP-Message
80	Message-Authenticator
81	Tunnel-Private-Group-ID

Table 8.4: RADIUS attribute format

Type	Length	Value
1 byte	1 byte	variable

Table 8.5: Access-Request for user/password login

Code	Authenticator	NAS-Identifier	NAS-Port	User-Name	User-Password
Access-Request	nonce	login-bank-01	23	user@corp.com	encoded

Table 8.6: Access-Accept for user/password login

Code	Authenticator
Access-Accept	signature

Short and simple. The Authenticator field has an MD5 hash over all of the fields in the response, as well as the shared secret. Of course, this authentication mechanism is not likely to be appropriate for the voice mobility network. However, it gives a flavor for the transaction.

The accept message from the RADIUS server can carry a significant amount of useful information from the user database to the NAS. For example, the Filter-Id field can carry the name of the filter policy to use for the client's traffic. This is used on networks such as 802.1X (wired or wireless) to determine what firewall policy should be applied to the client that is authenticating. The specific network the user is to connect to can be provided, as well. The Tunnel-Type, Tunnel-Medium-Type, and Tunnel-Private-Group-Id can be used to force a client onto a specific VLAN, for example. The meaning of the attributes and their use depends on the NAS that is requesting the information.

8.2.2 EAP

The problem with the RADIUS protocol as previously described is that the authentication mechanism is very limited. Authentication mechanisms need both to be cryptographically strong—avoiding being intercepted or interfered with as they traverse the network—and to support the authentication model that the network administrator wishes to use.

The cryptographic mechanisms used to provide the stronger authentication must be run from the authenticating device to the RADIUS server. RADIUS was originally created with the idea in mind that the authenticating device would pass the password to the NAS, which would then authenticate the user on her behalf. But that cannot work for secure authentication, in which the user's device does not leak out the credentials. Instead, the user's device and the RADIUS server need to have a back-and-forth, partially encrypted exchange.

Furthermore, the authentication mechanisms that exist need to be extensible. Policies change in organizations, and the administrator may want to change the authentication method required. This must work transparently, so that the entire network does not need to be retooled, just the RADIUS server and the client devices.

For these reasons, the Extensible Authentication Protocol (EAP) was created. EAP is described in RFC 3748. EAP is a change in architecture for authentication. The goal of EAP is not just to authenticate users and the network securely. It also must be able to cryptographically derive keys that can be used for encryption sessions, for as long as the authentication or login remains valid. Therefore, the EAP protocol, more than ever, needs to not involve a third party.

EAP therefore makes the NAS into nothing more than a proxy, or a carrier. EAP itself is a message-based request/response protocol and can be carried in different transports. EAP over Ethernet or Wi-Fi is called *EAPOL*, for *EAP over LAN*. The same message can be taken out of the Ethernet frame and placed into a RADIUS request or response, and is simply called *EAP over RADIUS*.

EAP introduces, into the picture, the roles of *supplicant* and *authenticator*. The *supplicant* is the software or module on the user's device that performs the cryptographic authentication protocols for EAP. The *authenticator* is an intervening device that puts the EAP messages into the format the supplicant will receive. The authenticator could have been the last step, but in real networks it proxies the EAP messages without processing them, to the RADIUS server.

The result of a successful authentication will be a success or failure. On a success, both the supplicant and RADIUS server will have negotiated a *Master Session Key* (MSK). The MSK will be used as a root key to derive whatever session keys are necessary by the protocol the client uses to access the network. In practice, the MSK is transported off of the RADIUS server, which does not provide real network services, to whatever network transport or application needs to decrypt the user's messages.

The format of a generic EAP message is shown in Table 8.7. The Code field specifies the type of the message, and is either 1 for *Request*, 2 for *Response*, 3 for *Success*, and 4 for

Table 8.7: EAP message format

Code	Identifier	Length	Type	Data
1 byte	1 byte	2 bytes	1 byte	*variable*

Failure. The Identifier serves the same purpose as with RADIUS (though it need not be equal), and keeps track of requests and responses. The Type field is for requests and responses only.

A type field of 1 is for *Identity.* EAP transactions start off by the RADIUS server sending a *Request-Identity* message, meaning that the Code is Request and the Type is Identity. The client then responds with a *Response-Identity*, with the Data field containing the text username the client has.

After this transaction, the EAP mechanism is entirely dictated by the type of authentication in use. These mechanisms are referred to as *EAP Types*, and are where the administrator has the most influence. The following sections describe the authentication type and architecture, then dig into the protocol.

8.2.3 Certificate-Based Authentication

Certificate-based authentication replaces username and password authentication. Every user (or his machine, as the case may be) is issued a *certificate.* This certificate of authenticity contains the name and information about the user, including what roles they can take with that certificate. The certificate is then *signed* cryptographically by a *signing authority*, a trusted agent who signs certificates as proof of validity.

8.2.3.1 Public-Key Cryptography

We now enter the world of *public-key cryptography.* Public-key cryptography is a set of mathematical procedures and theorems that allows one party to send an unauthenticated message that only the other party can read. This might sound shocking—why would we want to allow a party to send messages without authentication, especially for an authentication protocol?—but you will see shortly how this is put together.

The first notion of public key cryptography is that there is not one key, but two keys. One of the keys is the private, secret key. The other key, however, is a public key. Using the public key algorithms, a message encrypted with one key can be decrypted only with the other. The two possible keys gives us two possible ways of sending messages. Encrypt with the private key, decrypt with the public, or encrypt with the public key, decrypt with the private.

From this point, we can now begin to see shadows of the usefulness of public key cryptography appear. Shared key cryptography requires that both parties have the same information, and there is no *cryptographic* difference in their roles. Any party that can send messages can just as easily receive them. But that's not what we want. We want to make a distinction that allows us to draw a clear line.

The two keys—public and private—are derived at once by the entity that wishes to use them. The choice of which key is public and which is private is normally arbitrary, because of the mathematics behind the procedure, but once one is chosen as public and one is private, they must each keep to their respective roles forever. The public key is widely disseminated. The private key is always kept hidden.

Now, we can take the two keys and derive two different operations, based on applying the same public key encryption functions but using the opposite keys. Any entity whatsoever can take any arbitrary message and encrypt it using the public key. This message has no proof of which entity sent it. But now, once encrypted, it is private, and only the owner of the keys can decrypt the message. This is called the *encryption* operation, because it establishes privacy. On the other hand, the key owner can take a message and encrypt it using the private key. This message has no privacy whatsoever, because everyone has the public key, and anyone with the public key can decrypt this message and read the contents. Instead, this message has authenticity: it has now been *signed*.

The mathematics of public key cryptography are more advanced than that of the private key cryptography we have seen so far, and drift into the world of computational algebra, a much more involved subject than the algebra people learn in high school.

The most famous public key cryptography mechanism, and by far the most popular, is *Rivest, Shamir, and Adleman* (RSA). Named after the three authors of the work that laid out this mechanism, RSA uses the now well-known fact that the product of two very large prime numbers is computationally difficult to factor back into those prime numbers. The RSA algorithm has both keys created from three numbers, rather than two. Each key has an *exponent*. Furthermore, both keys share a common *modulus*. Let's see quickly how this works.

The key generator creates the modulus n as the multiple of two very large, random prime numbers p and q. The public exponent e is chosen to be relatively prime to the number $(p-1)(q-1)$. Practically, this number is often fixed to be 65,537, and the selector of p and q ensures that this works. The public key, then, is simply the number e and the more random number pq. The private exponent d is derived, by ensuring that de is equal to 1 modulo $(p-1)(q-1)$.

The operation for encryption is taking an appropriately padded message m and raising it to the power of the public exponent e (mod n). The decryption step is longer, and raises the encrypted message c to the power of the private exponent d (mod n). In the end, the original message is recovered. The opposite actions of signing and verification work the same way, but by swapping the keys.

The other algorithm that is used is the *Digital Signature Algorithm* (DSA), created by the United State government. This was popular for a while, as RSA was under patent. However, the RSA patent has expired a while ago, and DSA is not as common as it once was.

8.2.3.2 Certificates

The usefulness of public-key cryptography has been baked into the concept of the certificate. The certificate contains the public key of the party it represents. Thus, the owner of the certificate has the matching private key, which he keeps safe.

Certificates are used in web security, so most major website operators are in possession of one. Certificates are stored in the *X.509* format, an ITU standard (just as G.711 is) that defines what is stored in a certificate. The following is the textual representation of a publicly available certificate used by Google for their website (the symbol ⇒ is used to show where one line wrapped to another):

```
Certificate:
    Data:
        Version: 3 (0x2)
        Serial Number:
            3c:8d:3a:64:ee:18:dd:1b:73:0b:a1:92:ee:f8:98:1b
        Signature Algorithm: sha1WithRSAEncryption
        Issuer: C=ZA, O=Thawte Consulting (Pty) Ltd., CN=Thawte SGC CA
        Validity
            Not Before: May  2 17:02:55 2008 GMT
            Not After : May  2 17:02:55 2009 GMT
        Subject: C=US, ST=California, L=Mountain View, O=Google Inc, ⇒
          CN=www.google.com
        Subject Public Key Info:
            Public Key Algorithm: rsaEncryption
            RSA Public Key: (1024 bit)
                Modulus (1024 bit):
                    00:9b:19:ed:5d:a5:56:af:49:66:db:79:fd:c2:1c:
                    78:4e:4f:11:a5:8a:ac:e2:94:ee:e3:e2:4b:c0:03:
                    25:a7:99:cc:65:e1:ec:94:ae:ae:f0:a7:99:bc:10:
                    d7:ed:87:30:47:cd:50:f9:af:d3:d3:f4:0b:8d:47:
                    8a:2e:e2:ce:53:9b:91:99:7f:1e:5c:f9:1b:d6:e9:
                    93:67:e3:4a:f8:cf:c4:8c:0c:68:d1:97:54:47:0e:
                    0a:24:30:a7:82:94:ae:de:ae:3f:bf:ba:14:c6:f8:
                    b2:90:8e:36:ad:e1:d0:be:16:9a:b3:5e:72:38:49:
                    da:74:a1:3f:ff:d2:87:81:ed
                Exponent: 65537 (0x10001)
        X509v3 extensions:
            X509v3 Extended Key Usage:
                TLS Web Server Authentication, TLS Web Client Netscape
                    Server Gated Crypto
            X509v3 CRL Distribution Points:
                URI:http://crl.thawte.com/ThawteSGCCA.crl

            Authority Information Access:
                OCSP - URI:http://ocsp.thawte.com
```

```
        CA Issuers - URI:http://www.thawte.com/repository/
            Thawte_⇒SGC_CA.crt


        X509v3 Basic Constraints: critical
            CA:FALSE
    Signature Algorithm: sha1WithRSAEncryption
        31:0a:6c:a2:9e:e9:54:19:16:68:99:91:d6:43:cb:6b:b4:cc:
        6c:cc:b0:fb:f1:ee:81:bf:00:2b:6f:50:12:c6:af:02:2a:36:
        c1:28:de:c5:4c:56:20:6d:f5:3d:42:b9:18:81:20:b2:dd:57:
        5d:eb:be:32:84:50:45:51:6e:cd:e4:2e:2a:38:88:9f:52:ed:
        28:ff:fc:8d:57:b5:ad:64:ae:4d:0e:0e:d9:3d:ac:b8:fe:66:
        4c:15:8f:44:52:fa:7c:3c:04:ed:7f:37:61:04:fe:d5:e9:b9:
        b0:9e:fe:a5:11:69:c9:63:d6:46:81:6f:00:d8:72:2f:82:37:
        44:c1
```

Let us take this line by line and extract the relevant information. The first important line is the version, which states that this is an *X.509v3*, or version 3, certificate. The next line is the serial number for this certificate, and is unique for the authority who signed the certificate. Following that is the signature algorithm that is used by the signing authority. Specifically, the signature algorithm here is the SHA-1 hash over the entire certificate, which is a 128-bit number which is then signed using RSA. The issuer of the certificate is "Thawte Consulting," one of the main certificate authorities and a part of Verisign. The next two lines state when the certificate is good. It is not good past a certain time, to allow for the signing authority to make sure the possessor of the private key is still represented by the certificate.

The subject line is the name of the organization who owns the certificate. This one is Google's. The CN field at the end is matched directly with the web address to verify that the name of the machine matches.

The meat of the certificate is in the public key. The algorithm, as it mentions, is RSA. The key itself is given in two parts. The first part is the 1024-bit modulus, which was mentioned in the previous section. Not surprisingly, the public exponent is 65,537.

The X.509 v3 extensions provide what rights the certificate owner has with the certificate. This one is marked for use for web servers and clients. More information appears about the signing authority, including the location where certificates that have been revoked can be found.

Finally, the signature appears.

When the other party to the transaction—in this case, a web browser—looks up Google's services, the server sends this certificate. The client will then look up the certificate of the authority from a locally stored, always-trusted set of signing authority certificates. Thawte's certificate is on every computer, so the client will take the public key from it and use it to undo the encryption on the SHA-1 signature in Google's certificate, then to compare the SHA-1 hash it calculates with the one in the certificate. If they match, then the certificate checks out and the server is authenticated.

Certificates can be signed in chains. The top-level authority can sign certificates, for a fee, of a next-level authority, who can sign more certificates, and so on. This allows for a longer hierarchy, which is useful in enterprises, in which an enterprise will get a signing certificate for itself, validated with one of the well-known root authorities, and then sign the client certificates that way.

The reverse process is used for client authentication. The server will validate the client's certificate.

Of course, there is a bit more going on behind the scenes.

8.2.3.3 TLS and SSL

Certificate exchanges are useful for validating that the certificate is valid, but it does not say anything about the holder of the certificate. For that, a protocol is needed.

The *Transport Layer Security* protocol (RFC 5246) is designed to both validate that everyone who has a certificate should have it, and to establish a session key that can be used for encrypting later traffic. TLS is the replacement for SSL, the *Secure Socket Layer* that is used for HTTPS.

As a protocol, TLS has a simple structure. First, the client connects to the server. The server provides its identity with a certificate, as well as a nonce. The client then responds with its certificate—an optional step for some applications, such as HTTPS—and its own nonce. The client also picks a special key, called the *premaster secret*, which is used for any subsequent steps. This is a random number that is encrypted with the server's public key and sent, encrypted, to the server, which decrypts and installs it. From that point, the server and the client have a secure channel with which to finish the operation. For TLS, this operation is rather straightforward, and the server and the client can derive their master session keys.

8.2.3.3.1 EAP-TLS

TLS can be embedded directly into EAP to allow RADIUS servers to authenticate clients using it.

TLS itself uses its *record protocol* to provide basic messages as a part of the exchange. This is shown in Table 8.8.

Table 8.8: TLS record format

Type	Version	Length	Payload
1 byte	2 bytes	2 bytes	*variable*

The Version field is given as 0x0301, which means TLS version 1, but looks like SSL version 3. This allows for SSLv3 peers to work with TLS. The types fall into the categories listed in Table 8.9.

Table 8.9: TLS record types

Type	Subtype	Name	Meaning
20	—	Change Cipher Spec	Used to switch all further TLS communication to encryption using the premaster key.
21	—	Alert	Sends a warning or error that can abort the TLS operation, if one of the parties does not have the right credentials.
22		Handshake	Used for back-and-forth requests
	0	Hello Request	Sent by the server to start the protocol, if the client hasn't already sent a Client Hello.
	1	Client Hello	Starts the process with a nonce, and provides what cipher suites the client supports.
	2	Server Hello	Responds to the Client Hello with a nonce, the selected cipher suite, and a session ID for resuming TLS later.
	11	Certificate	Provides a certificate or chain of certificates
	12	Server Key Exchange	Used for Diffie-Hellman key exchanges.
	13	Certificate Request	Sent to request a certificate from the client.
	14	Server Hello Done	Marks the end of the Server Hello group of records.
	15	Certificate Verify	Sent by the client to prove that it has the private key to the certificate it offered.
	16	Client Key Exchange	Sends the premaster secret, encrypted to the server using the server's public key
	20	Finished	The TLS session is finished.
23	—	Application Data	For extensibility; used in derivatives such as TTLS and PEAP.

This concept of one or more record protocols layers very nicely onto the EAP message. One or more of the client TLS records are sent in one or more EAP Request-TLS messages, and the server's records come in EAP Response-TLS messages. This communication is layered on top of the protocols from the client to the NAS, and from the NAS, using RADIUS, to the RADIUS server. The only last detail is the EAP type for TLS, which is 13.

The example exchange for security in Chapter 5, on Wi-Fi, shows a PEAP-based TLS session, and gives a good idea on how the exchange works when the client does not have a certificate.

8.2.4 Secure Password-Based Authentication

Password authentication can be dicey, because sending passwords or anything predictably derived from them over the network will expose the user to having his or her credentials stolen. Passwords need to be sent in an already-encrypted channel.

HTTPS web servers already do this right, by setting up a TLS connection without a client certificate, and then using traditional web pages to request the passwords. The password is just sent over the encrypted HTTP session, so no one can eavesdrop.

For more automated applications, something else is needed. The most popular way to do password authentication across the entire network without the risks of exposure is to do Protected EAP (PEAP). PEAP works by setting up a TLS connection, normally without client certificates. Once the TLS session gets to the point where both sides have agreed to a premaster key, then instead of TLS finishing up and establishing a session key, the client and RADIUS server drop into using the encrypted Application Data protocol of TLS to embed a second, inner EAP session. This nesting of an encrypted EAP session over an unencrypted TLS session allows PEAP to protect any of the EAP methods that needlessly expose information.

As it turns out, PEAP version 0, the most common PEAP version, happens to use a particular form of a challenge-and-response password protocol called *MSCHAPv2*. This protocol was designed for integration directly with Microsoft Active Directory systems, and it uses the typical Windows NT password hashing mechanisms to verify the username and password. PEAP itself is flexible enough to conceivably have any EAP protocol run over it, but implementations are strictly limited.

PEAP uses a different EAP type than TLS: 25.

A far less common alternative to PEAP is EAP-TTLS, for *tunneled TLS*. TTLS is conceptually very close to PEAP, and was actually created a bit earlier than PEAP. It uses a few more fields between the inner and outer tunnel, and the details are not significant enough to go into here.

8.2.5 Cardkey Authentication

In cardkey authentication, each user carries on his or her person a small device. This device, which is anywhere from the size of a miniature pocket calculator to a small USB flash, carries a cryptographic random number generator and a clock that has been synchronized to a central authentication server. When the user wants to log in, the server asks the user to pull out the cardkey and read the number on the display. This number, which changes every minute or less, will serve, along with the user's name and password, that the user is at least in physical possession of the device. As these devices are very difficult to copy—they self-destruct electronically—this provides a higher grade of authentication than just a password alone.

These systems have the requirement, however, that some authentication server be able to engage the user in a two-way interactive session, with text messages and so on. To prevent those messages from being shown in the clear—especially because an attacker who might have electronic access the username, password, and cardkey display could open up all of the network if it acted quickly enough—these exchanges should be tunneled as well.

PEAPv1 is an extension to PEAP that uses an inner EAP method called *EAP-GTC*, for *Generic Token Card*. EAP-GTC is an arbitrary text-based protocol, riding over EAP, to perform challenges and responses with the user. EAP-GTC is defined in the same RFC that defines EAP itself, and is EAP type number 6.

Cardkey authentication can also be done using proprietary protocols, and are often integrated into virtual private network (VPN) access clients for large networks.

8.2.6 SIM and AKA: Cellular Authentication over IP

SIM authentication uses the GSM SIM card to authenticate the user. This can have an immediate application to voice mobility. EAP-SIM is defined in RFC 4186, and uses multiple invocations of the GSM SIM authentication mechanism to verify that the client is what is claims to be. The *Authentication and Key Agreement* (AKA) protocol is used in UMTS networks to provide the mutual authentication between the phone and the network. EAP-AKA is defined in RFC 4187, and layers the request for AKA negotiation on top of EAP.

Both of these mechanisms would allow for out-of-the-box authentication of mobile devices. Unfortunately, they do require that the RADIUS server have access to the identity information for the clients, something that only the mobile operator has. Therefore, private voice mobility networks will not have access to these mechanisms.

8.3 Protecting Your Network End-to-End

Once the user is authenticated using a strong authentication technique, the problem becomes protecting the network. *End-to-end security* means that the eventual endpoints in any conversation directly authenticate each other, establish a session key, and then enforce the encryption from there. Any device in the middle will see only integrity-protected, encrypted, and authenticated traffic, and thus can only observe that some unknown activity is taking place.

End-to-end protection does require that both ends, or every end, have access to enough information to authenticate the other party. Furthermore, with voice mobility, it is always somewhat difficult to figure out exactly who the other end is. SIP, RTP, and other protocols are often proxied by the PBXs and switches. Therefore, we will use the term "end-to-end" to refer to the application-level endpoint, in the appropriate architecture, that the traffic is being sent to.

8.3.1 Generic IP Encryption: IPsec

The first approach to encryption works for any IP network. IPsec is a protocol that works for any IP packet, assuming the two endpoints are participating. IPsec (RFC 4301 and related) defines an encrypted or authenticated transport mechanism that allows individual IP frames to be protected.

IPsec, by itself, does not assume how the two endpoints authenticate or derive keys. What it does is provide a format for the encryption or authentication cryptographic algorithm to be applied to the IP packet.

IPsec defines two types of encryption or authentication that can be done to a packet: *transport mode* or *tunnel mode*. Transport mode is designed to protect an IP packet which is being send from an originating source to a final destination. In this case, IPsec only encrypts or authenticates the payload of the packet, of course binding the header fields to the cryptographic operation to prevent static fields from being modified in flight. Tunnel mode works on top of an IP-IP tunnel, in which one IP packet is placed inside the other. In tunnel mode, the packet to be encrypted is encrypted, in its entirety, and is then placed inside another IP packet. This allows the outer IP packet to be modified without concern to the security properties of the connection.

The two flavors of IPsec cryptography—encryption and authentication—reflect that some traffic does not need to be encrypted but just signed so that the receiver can know that the packet was received without modification from the authentic source. Encryption is more important for voice mobility networks, so we will start there.

The encryption protocol is called *Encapsulating Security Payload* (ESP), and uses IP protocol type 50. The format of the ESP header is shown in Table 8.10.

Table 8.10: IPSec encapsulating security payload format

SPI	Sequence Number	Payload (encrypted)	Padding (encrypted)	Pad Length (encrypted)	Next Header (encrypted)	ICV
4 bytes	4 bytes	variable	variable	variable	1 byte	variable

The SPI, or *Security Parameters Index*, is an arbitrary four-byte value, used to identify the security association between the two endpoints. This allows multiple sessions of IPsec to be set up between two devices, or even one session to be shared among multiple devices (such as for multicast). The Sequence Number defines the replay counter for the packet. This number always increments by one for each packet that is sent by the sender in the secure connection. If a receiver gets a valid packet with one sequence number followed by a valid packet with a lesser sequence number, it can drop that lesser one as being a replay.

The payload is the data that is being encrypted. In transport mode, the data is the payload of the unencrypted IP packet. In tunnel mode, that unencrypted payload is another IP packet. At the start of the payload, the type of encryption negotiated beforehand may require an initialization vector (IV). This will come before the encrypted payload, and the decryption and encryption algorithms are responsible to know where to look for them. Following the payload may come any padding required by the encryption, followed by a byte specifying the number of pad bytes that were used. The Next Header field has the type of the packet that was encrypted. A special value (56) exists, which refers to a dummy packet. A dummy packet is inserted by the sender to mask the overall statistical properties of the encrypted flow, if the sender wishes to do that. This value is otherwise the value in the original packet, such as 6 for TCP. If this is a tunnel-mode packet, the value will be 4 for IP-in-IP. The receiver needs to drop all dummy packets. Finally, the integrity check value (ICV) is calculated, which covers the entire ESP header, the payload, and the trailer fields. This field, if present, provides integrity protection for the packet.

Packets that just have authentication information use the *Authentication Header* (AH) protocol, IP protocol 51. AH provides integrity and replay protection, on top of authentication, but does not protect the contents of the frame from being spied on. Usually, authentication is a computationally faster process, which explains why some devices may want to use it. It is admittedly hard to see how AH would be useful with voice mobility. The format for AH is shown in Table 8.11.

Table 8.11: IPSec authentication header format

Next Header	Payload Length	Reserved	SPI	Sequence Number	ICV	Payload
1 byte	1 byte	2 bytes	4 bytes	4 bytes	*variable*	*variable*

Apart from a bit of rearranging, and the elimination of any potential IV, the fields mean the same as with ESP.

For both approaches, when integrity is applied to the IP header, some of the fields are allowed to change, and others are not. Specifically, the only fields that can change in the outer IP header (or only header for transport mode) are the DSCP field, the flags, the fragment offset, the TTL, and the header checksum.

Replay protection is provided by employing a window algorithm, to allow for a limited amount of reordering. The size of the window is up to the negotiation between the two parties, although everyone must support 32 and 64 is recommended. The idea of the window is that the receiver keeps the highest sequence number that it has seen from a packet that has been successfully authenticated. (Forgeries may try to push the window around, and so must be ignored for setting the window.) Any packet received with a sequence number older than the current receive one, minus the window size, is dropped right away. That leaves the packets in the middle of the window. For those packets, a list of sequence numbers already seen is kept. If the packet with the same sequence number comes in twice, the second one is dropped. Otherwise, the packet is allowed in and its sequence number recorded.

IPsec is flexible enough to allow for a number of different encryption and authentication protocols to be negotiated. Common encryption protocols are *3DES-CBC* and *AES-CBC*. A common authentication protocols is HMAC-SHA1. Recall that an HMAC is a special type of signature that requires a private key to validate. If a message is received, the key and the packet data together produce the signature, which is then compared to the one on the packet. If they match, the sender has the right key. So the possession of the key by the sender is proof of the authenticity of the packet.

8.3.1.1 IPsec Key Negotiation

Because IPsec is only a transport, there must be a protocol to set up the tunnels. The simplest protocol allowed is to use none, and IPsec connections are allowed to be set up on both sides manually. However, it is usually far simpler for management of the connections to use some sort of user authentication and negotiation protocol.

The *Internet Security Association and Key Management Protocol* (ISAKMP) is used between devices to negotiate the type of IPsec connection and to establish the security association. The two endpoints decide on the type of tunnel, the type of encryption or authentication algorithm to use, and other parameters using this protocol. ISAKMP is defined in RFC 2408 and uses UDP port 500 for its communication. Related to ISAKMP is the key exchange protocol itself, the *Internet Key Exchange* protocol (IKE). IKE takes care of the key exchange portion of the setup, and thus piggybacks with ISAKMP as a part of the setup.

ISAKMP has similar message exchanges as the other security negotiation protocols do, including certificate requests and responses, nonce exchanges, and capabilities exchange. ISAKMP is a complex protocol; it would not be useful to go into the same level of detail for ISAKMP here as it was for TLS. However, let us take a look at the basic exchange.

The first phase for ISAKMP is for one endpoint—usually a VPN client—to reach out to the authenticating server. The first message sent contains a significant amount of information. When using *Aggressive* mode, which is common because it reduces the amount of messages that have to be exchanged, the first message contains nearly everything needed in one shot to set up the connection. The first major piece of information in this message is the proposal list for the algorithms to use for authentication or encryption, for the ISAKMP/IKE exchange itself. This list is an ordered set of the combinations of encryption and authentication payloads methods that the client wishes to request, as well as the authentication methods and the expected key lifetimes. The next important set of information kicks off the key exchange. This key exchange starts off the key negotiation, using Diffie-Hellman keys to create a session key. A nonce is included, followed by the identification of the endpoint, and a number of options.

The next phase, in aggressive mode, comes back from the server. This selects the IPsec encryption and authentication that will be used, and how the user is to authenticate. Following this is a nonce, then the server's identity. Options conclude the packet.

At this point, the ISAKMP/IKE session can be encrypted. The third phase involves the two endpoints establishing the IPsec security association proper. The two endpoints select what IPsec authentication and encryption mechanism they will use.

After this has completed, the information is pushed down to set up the IPsec connections themselves.

User (not packet) authentication can occur one of a couple of ways. Each side can have a preshared key, which is then used in the validation of the ISAKMP session. Or, each side can use certificates. The certificate-based scheme is undoubtedly more secure, but is harder to manage. This is precisely the same tradeoff that is experienced on link protection, such as using 802.1X verses pre-shared key for Wi-Fi.

8.3.2 Application-Specific Encryption: SIPS and SRTP

The difficulty of using end-to-end encryption is that all of the endpoints must support it, which may not be the case. Instead, protocol-based security can be used.

For SIP, one option is to use SIP over TCP, protected by a TLS session. This is identical to the approach HTTP uses for protection, by using TCP and requiring a TLS negotiation first. The advantage to doing this is simplicity, as TLS is a well-understood technology, and vendors do not have a difficult time implementing it. Furthermore, using TLS allows the voice mobility administrator to enable the built-in SIP authentication system, based on WWW digests, without fear of eavesdropping. Using SIP authentication greatly decreases the complexity of an authentication-based network, because SIP clients are far more likely to support it out of the box. The major issue with SIPS, or SIP with TLS, is that the processing requirements on the PBXs go up significantly, which may affect the scale that the PBX can operate at.

Protecting SIP does nothing to protect the payload. To protect the bearer channel, SRTP is an option. SRTP uses a AES (and only AES) encryption to encrypt each RTP packet. The

AES encryption is used in a stream setting, by running in *counter mode*, which ensures that AES can be restarted if intervening packets are lost. SRTP is effective, in that it protects the packets from eavesdropping and modification.

8.3.3 Consequences of End-to-End Security

There is a major consequence of using end-to-end security. Devices that may not have been built for fast cryptographic operation, such as PBXs and media gateways, will be forced to use computationally expensive protocols on the real-time voice path to ensure privacy. The protocols themselves are not ubiquitously supported. For example, IPsec is common for VPNs, and can be used from router to router or even laptop to laptop, but phones are unlikely to have a VPN client at all, let alone one that is fast enough to be appropriate for real-time voice traffic. PBXs are less likely to support IPsec. Using application specific security makes more sense in this case, but even then, the protocols are relatively new, and are not commonly deployed. For this reason, it is usually far easier to dispense with end-to-end security, and instead to focus on protecting the mobile, exposed portion of the network.

8.4 Protecting the Pipe

The pipe, in voice mobility networks, can be a number of things. When the mobility network is heavily wireline, the problem becomes authenticating over Ethernet. (Encryption for wireline networks is considered less necessary.) When voice mobility uses Wi-Fi, the problem transforms into finding the right WPA2 settings for both authentication and encryption. When traffic is coming in from the outside world, using fixed-mobile convergence solutions or remote access clients, the pipe that needs protecting crosses the Internet.

The advantage of protecting the pipe, and not the entire path, is that the part of the path most vulnerable can be addressed using specific, dedicated security infrastructure, whereas the less vulnerable parts can be placed in physically or logically secure networks. This allows "legacy" voice mobility equipment to have a high chance of operating with strong security.

For wireline networks, especially dedicated wireline voice networks that have exposed jacks, one of the major concerns is that someone might plug into the voice network rather than the data network, either by mistake or to cause mischief. To preserve the sanctity of the voice wireline network at the edge, one solution available is to use 802.1X on the wireline ports. 802.1X works on wireline in almost the same way as it does for Wi-Fi. (See Chapter 5 for details.) The major difference is that the end of the 802.1X EAP exchange does not lead to a continuation into any sort of key exchange or encrypted session. Rather, the edge switch, acting as authenticator, unlocks the port for use for more than just authentication. The issue with using 802.1X authentication for wireline networks is that the desktop phones may not support it. In that case, a practical, though not terribly secure, alternative is to implement MAC address filtering on the switch. This can be done per port, or better, switch-wide. The goal is to only let phones onto the network, and ensure that any traffic that ends up on the network that comes from an accidentally connected device is dropped before it starts consuming resources. One of the biggest concerns on that front is that a client may come in and exhaust the phones' DHCP address space. This can happen when the accidentally connected laptop is looking for an IP address in a specific range, and gets a completely different one from the DHCP server. If the client ends up rejecting that address for being in, say, a private address range that it has been configured not to use, there is a chance that the device will try again. If this happens enough times, the DHCP server will lose all of its addresses, and any phones that get plugged in or introduced to the network wirelessly will not be able to gain basic connectivity.

Wi-Fi security is a must. Chapter 5 went into significant detail on how preshared keys work, compared to usernames or certificates. The advantage of using Wi-Fi's own security, rather than an end-to-end piece, is that the phone is likely to have a high-performance security function built into the Wi-Fi chip, just for WPA2. This is because Wi-Fi certification requires that every device support WPA2, and every Wi-Fi chipset manufacturer embeds just that process into the chips that they make. Phone manufacturers need only turn on those features; there is no heavy lifting that needs to be done. Compare this to SRTP, for example, which requires that the voice coder engine, which is usually an optimized engine for producing real-time payloads, must also either know how to encrypt the traffic by itself or must pass it along to a slower software process to encrypt. This can cause significant battery drain on the phone, if such a configuration is even supported.

FMC solutions beg the use of a remote security product. Again, the physical and resource capabilities of the phone come into play here. Some phones do, in fact, have VPN clients, which can be used for access into the enterprise. These VPNs terminate long before enterprise server infrastructure is reached. Running voice protocols, using FMC soft clients, over the VPN can make sense, although the common mode of operation is for voice to remain on the mobile operator's network and for data to go through the VPN. One interesting twist, however, is that voice devices that have VPN capabilities must have the VPN logged into when the user is on the road. They can sometimes be configured to log in by themselves, but more often than not, they must be enabled manually. This is especially important for converged, dual-mode phones that also operate within the enterprise. When in the enterprise, associated to the corporate Wi-Fi network, there is no need or benefit for enabling the VPN link. In fact, the VPN server may not even be accessible from within the network, as its major interface is meant to point outside, to the Internet. Because the VPN should be on for some uses of the Wi-Fi network and not others, it can stand in the way of convincing users to access the enterprise network on the road, reducing the productivity gains that the FMC solution was looked at for in the first place. One option that many Wi-Fi infrastructure vendors offer, which takes advantage of the concept of protecting the pipe and not the end-to-end application, is for remote users to be offered remote access points. These access points are similar to the campus access point, and yet are designed for operation when on the road. The remote access point is essentially a VPN client and a normal access point combined. The access point's VPN client tunnels through the Internet to the corporate network, where it terminates at the wireless controller. Once connected to the controller, the access point pulls the same enterprise configuration down as it would if it were in the office, and provides it to the remote user. This way, the remote user can use the same cellphone as on campus, with the same WPA2 security policies, without having to be bothered with the VPN. This does provide a measure of privacy from the phone to the physically protected office.

8.5 Physically Securing the Handset

The handset itself is still a weak link. Handsets are designed to be portable, so in that sense, they are also designed to be transported away from their rightful owners. Many converged handsets are quite impressive in their capabilities, and so they make for an interesting target for thieves. Furthermore, busy, high-productivity enterprise users are unlikely to set up a complicated, strong phone lock password. It is simple to imagine a hospital on-call attending physician not wanting to be bothered with typing an eight-character, letter, number, and punctuation-mark password on a phone with itself no more than 16 buttons.

The problem becomes, then, how to prevent phones from being stolen in the first place, and how to take care of the problem once they have left the building.

8.5.1 Preventing Theft

There are many practical considerations for preventing the theft of a voice mobility handset. Unfortunately, limiting the network access is not one of them. Take Wi-Fi-only handsets as an example. It is pretty clear to the voice mobility administrators that a Wi-Fi-only handset will not be of much use at a person's home, or a hotspot, or even another office. Most phones provide tight administrative privilege requirements to change the Wi-Fi network and SSID that the phone uses. Even if someone can accomplish the feat of penetrating those restrictions, the phone will work only with the PBX it was configured for, unless someone changes it. They do not perform well over the Internet, and are not going to be useful as a personal wireless voice phone for someone's house. However, the devices do look like cellphones, and the people who might take one such phone are not likely to know or care about the difference until they have the phone securely in their possession.

Some common-sense approaches can work. Sticking labels onto phones that state just the fact that they will not work outside can have a slight deterring effect, much the same that which restaurant pagers' warnings have on their patrons. A more workable approach is to use telephones that do not look like telephones. In the limited environments where voice mobility can be conducted without outdoor support, such as warehouses or hospitals, specialized devices like two-way communicator badges or ruggedized handsets can discourage some amount of theft. For environments where devices have specialized chargers, a daily checkout policy and a central charging station can at least keep track of the phones that do exist and strongly discourage users from taking devices with them to places where they might get stolen. A good example of this is with nursing—the nurses' station makes an acceptable location to place the chargers.

But, ultimately, if someone wants to take the phone, they can. A better way to protect against theft is to detect theft. Theft detection can be performed somewhat readily on converged, dual-mode phones or Wi-Fi-only phones using *location tracking*. Location tracking is a feature of Wi-Fi networks or of overlay systems that use Wi-Fi for monitoring, where the rough positions of each device are recorded within the system. Location tracking systems can be built with automated policies, such as email alerts that are sent when devices enter or exit certain positions. There are a number of networks that use the location tracking system to monitor the exits to the buildings, to send alerts if a device passes through there when it should not. This can send an email out to the person who owns the phone, informing them of its activity, and gently reminding them that the phone is not to leave the building. Such as system seems like it may do nothing for people who intend to steal the device and already have it within their possession, as removing the battery or powering down the device will disable the location tracking. Another option is for the system to then send a message if the phone is taken off the air. In many of these environments, phones are generally kept on 24 hours a day, and so it is possible to come up with the right set of rules to make a deterrent useful.

Unfortunately, there is no foolproof way to prevent theft of voice mobility devices. Making sure that the devices do not look like high-end cellphones, unless the users need those high-end features, and a stronger educational campaign about locking phones and keeping track of them, are likely to be the most effective.

The second half of the discussion is what to do when the device is stolen. This question is either more simple or less simple than it looks, depending on whether the device has a cellular radio. If the device does have a cellular radio, then the options are wider. Cellular phones, even dual-mode phones, connect to the mobile network. Once connected, a phone that was reported stolen can be disabled remotely by the mobile operator or administrator, and even potentially tracked, if the phone is being used in the commission of a serious crime. On the other hand, if the phone does not have a cellular radio, then there is a good chance that the thief will not be able to use the phone for much of anything, so it may already be in a state that is close enough to disabled for the comfort for the administrator.

This is a reminder for voice mobility administrators to explore the encryption options for smartphones that may be used with the enterprise network. Phones set up for encryption may find that the information locked up in the stolen phone will be lost, but it is better than the information being exposed.

8.6 Physically Protecting the Network

Physically protecting a voice mobility network is a very strong way to preventing unauthorized access. People may be tempted to leave voice mobility networks more exposed than their data counterparts, because the only traffic flowing across them is real time call data, and not, say, highly sensitive corporate strategies. Nonetheless, it is important that voice mobility networks be given the necessary physical security to prevent problems with direct intrusions.

The wireline portion of the voice mobility network should be treated with the same level of concern as the service network for data usage would. Most IT organizations are good at ensuring that email servers are locked up, if not for security, then at least to prevent accidental disruption. The same goes for IP PBXs and gateways. However, there has been a historically different way of thinking about voice networks compared to data. Data networks invariably terminate in a switch that is placed in a locked switching closet. Voice lines, however, used to terminate in a series of punch blocks, which had been placed in locations convenient for the wire-pullers. If a network is subsequently upgraded to voice over IP, those same locations may be used for placing the Ethernet switches that concentrate the desk ports and send them to the voice network, towards the PBX. Clearly, those areas should be kept locked with the same scrutiny.

Another area of concern is with the accidental confusion of voice and data network services. Voice networks should always be kept physically separate and distinctly marked from data networks. Before wireline voice over IP, the phone port was never electrically like a data port. A user making an error in reconnecting the devices on his desk would have found that the phone and computer would both not work if the machines were plugged in the wrong way. But, unfortunately, with voice-over-IP services, it is possible for the user to get it wrong and still have the appearance that everything is working correctly—that is, at least, until he tries to place a call. This is a different aspect of physically protecting the network. Previously, we saw an example of how a misconnected device can exhaust network resources, such as with accidental DHCP exhaustion. Physically protecting the port of the network would also solve that problem. The simplest way to do that is not necessarily to secure the jack but to place a wall plate over it that makes it difficult to unplug the phone. This serves as a potential deterrent to accidental swapping.

On the wireless side, however, real physical security is a necessary. Again focusing on Wi-Fi deployments, the fact that the signals do pass through walls and outside the building requires that the network be well planned for security.

Depending on the nature of the environment, even strictly followed WPA2-Enterprise security with certificate exchanges using TLS can reveal information about the caller that should not be exposed. Given that a reasonable number of voice mobility devices use preshared keys, and not WPA2-Enterprise, and a physical security approach can help provide an additional layer of protection. Furthermore, there are a few environments where it is important that the very fact that a user places a phone call should be hidden. Voice traffic over Wi-Fi is designed to be distinctive, with a regular pattern of fixed-rate, fixed-length frames coming on high-priority services tipping any observer off that a call has been initiated.

Here, the concern is the exposure to the outside areas of the building of the in-building voice mobility network. Physical layer Wi-Fi firewalling solutions have recently been introduced that use RF activity to mask the presence of the network and its traffic to different physical regions. By deploying these physical layer blocking systems on the outside walls of the building, the systems can provide a curtain that separates the inside, where the network is accessible, from the outside, where the network is not even recordable. This is an inherently different solution from attempting to use specialized antennas and beam technologies to concentrate the Wi-Fi network towards the inside of the building. Doing the latter does not prevent an attacker from recording the voice traffic. It requires only that the attacker get a slightly bigger antenna, with a dB gain improvement for every dB of isolation that the network installer was able to provide. Using RF firewalling instead blocks the signal from being intelligible past the curtain of coverage, preventing eavesdroppers from having useful access to the leaked signals from within the building. Be careful that some products may be labeled "RF firewalls" if they simply use the location of the device to influence the firewall policies of the network. These are not true RF firewalls, since they do not provide any security at the RF layer itself, and thus are completely vulnerable to the passive leakage attacks mentioned here.

CHAPTER 9

The Future: Video Mobility and Beyond

9.0 Introduction

This entire book has so far been looking at the present day, with voice mobility taking center stage. But mobile devices have begun the dramatic transition from voice-only phones to multipurpose, "converged" systems that fit in the pocket but perform the work that a laptop computer would have just a couple of years ago.

The main difference is that the type of applications that these devices run has expanded. From a quality-of-service perspective, voice is no longer alone as the main application. Video is here, in the form of webcasts, corporate events, and videoconferences.

In this chapter, we will look at video mobility, building upon the concepts already covered. Then we will look towards the future, and try to see where mobility may be going, in the enterprise.

9.1 Packetized Video

Video is an interesting thing. Besides that it contains voice as a proper subset, video also is responsible for carrying quite a bit more information. Whereas voice recording and playback technology was perfected in miniature form decades ago, video is always a work in progress, needing bigger and bigger screens with higher resolutions and more sharpness. In some senses, it's hard to imagine how video and mobility go hand in hand, or simply in your hand, as video requires watching on a device that is not constantly shaking and jiggling, and requires nearly constant attention, whereas voice can be used on the run.

However, video has the ability to connect with the user in a way voice can never. If a picture is worth a thousand words, a moving picture should be worth at least a thousand times more again. More practically speaking, videoconferencing and webcasting has become increasingly attractive, as travel budgets constrain companies from hosting large fly-in gatherings, and video technology has improved in concert. In the end, voice and data have already been able to make the transition from wires, but as video becomes more prevalent as a part of networking in general, the video must naturally follow the user, and thus become wireless as well.

doi:10.1016/B978-1-85617-508-1.00001-3.

Let us look into some of the fundamental differences between voice and video. The first difference is the most obvious: video requires significantly more throughput than voice. Video has to carry the moving picture along with the same voice stream as before, and so the overall content must be quite a bit larger. In voice mobility, the throughput of voice is usually not the constraint except in very large voice aggregation centers. Instead, voice makes its presence known by its increase in the number of packets over the network. Video, on the other hand, is bandwidth constrained from the outset. The second difference is that video requires synchronizing multiple media streams. This impact is felt especially when loss rates rise or bandwidth constraints are hit, and some part of the video must be sacrificed. In the ideal case, the software on the endpoints keeps everything in synchronization, but this can often slip when poor network quality or lack of capacity begins to challenge the ability of the video client to find its way without all of the appropriate information. The third difference has more to do with how video is used today. For the most part, video is one-way. The user watches a video that is being streamed. Because of this, video can build up reasonable latency, and so video is not as latency-constrained as voice is. This may change, as mobile devices are given more sophisticated cameras and videoconferencing on a mobile device becomes possible. But, as of today, video is still fundamentally a broadcast mechanism. And for this reason, video also differs from voice at a fundamental network level, because it can be effectively multicast over the network. The final difference is that video can be significantly more sensitive to loss than voice. Without the two-way conversation, lost information may not be covered up by asking the other side to repeat, and the user may end up with a poorer opinion on the quality of the network.

9.1.1 Video Encoding Concepts

As mentioned earlier, video has many more dimensions (quite literally) of information to it than voice does. Video is nothing more than a series of still *pictures*, whereas voice is a series of nothing more than still point-in-time readings of sound pressure: just one small number at a time. Where voice has just this one sample at any given time, video has the entire picture at that time. This picture, itself, is made of two dimensions of *pixels*, or small areas that possess the same color.

Let's dig into this a bit more deeply. A picture, or still image, has hundreds of thousands of pixels, or picture elements, arranged in a rectangular grid. Each pixel has some fixed dimension, often measured in millimeters, with the most important aspect being the ratio of the width to the height of the pixel. The entire screen is made of hundreds or thousands of pixels in each of the horizontal and vertical directions, resulting in images with a given *resolution*. Common resolutions are the small 640×480, meaning 640 pixels wide by 480 pixels high; 1280×720, used in *720p* high-definition; and 1920×1080, used in *1080p* and *1080i* high-definition video. Resolution can also be a function of the actual number of pixels in the device's display, as there is not much sense in using a video stream that has a higher resolution than the display.

Each pixel holds the color information for that pixel. This makes a pixel a rectangular area of uniform color. Computer representations of human-visible color can be measured in many different ways—which is both a problem and an opportunity. For raw images, which have no compression, one common way to measure color is to use the *RGB* method. RGB relies on how the human eye can see three primary colors—red, green, and blue—and that all colors are some combination of those three. For this reason, computer and television displays are made with pixels that are themselves made of the three different colors, each lit with independent intensity. The perception of color is a rich and detailed subject, but there are some concepts to it that we should briefly address here to give a sense for what is going on. Pure light, as electromagnetic radiation, can be thought of as having one frequency. The colors of the rainbow are all colors of one pure "tone," or light of one single frequency. The eye has *cones*, the receptors that show color, and these cones come in three types, not surprisingly one for red, one for blue, and one for green. The three cones will respond differently for each pure, spectral color, giving the eye good coverage over all of the colors that can exist in the world. When light of one frequency falls upon the cones of the eye, each type cone responds differently and predictably. From this comes the perception of the color of that frequency.

Of course, not all colors are made of pure color. White is most certainly not a pure color, as it is able to be split by a prism into the entire rainbow spectrum. It must be more than just one frequency, then. In fact, it is a balanced mixture of many frequencies, and comes out white because it is able to excite the three primary color cones the eye sees with even intensity. This is no coincidence, as while happens to represent the total mixture of frequencies the sun radiates down on the planet. Because everything we see is reflected from normally white light, our most common experiences of colors have to belong to the subset of white light itself.

We can think of representing color by, then, a triplet of intensities, representing the amount of activation of the three cones in the eye. This is a good start. As it happens, however, video displays are not built around the exact excitement of the three cones. The red, blue, and green are a bit different than what the eye sees, mostly because display screens need to generate light from something that could be mass produced, and it was sufficient to use a different red, green, and blue, so long as the entire range of colors could get close to running the *gamut* of what the eye can see. (*Gamut* is the correct term for the space of colors that a particular way of representing colors can cover.) This means that RGB can represent essentially all of the colors we know of and use on a regular basis, though often as only an approximation to the real thing. There are some colors in nature that can never be seen on a screen—often those that are the most vibrant.

However, we can still get a sense of how the primary RGB colors of light add up. Red and blue readily mix to form purple, which is easy to describe as a reddish-blue hue. Blue and green mix to form a blue-green color, such as cyan. The odd one is red and green together, which produces yellow, a color most people would not describe with the term red-green, being a pair of colors which seem as opposite as can be imagined. (It is difficult to imagine that the two colors of a poinsettia can ever combine to form the color of a banana.) All three together form white, and all three absent form black. Staring at any screen up close will show exactly how this process works in video displays to blend together to form a color, as each pixel is made of *subpixels*, or regions of only one color, that can only vary by intensity. When viewed from a reasonable distance, the eye does not do a good job seeing the separate subpixels, and so it blends the three different hues together to form the final, intended, color.

In computer displays, the widely-used sampling method is to given each primary color an *intensity* of eight bits, producing a 24-bit overall color sample per pixel. This lets us step back and look at the size of video. Compared to raw voice, which must encode only one 16-bit sample at a given time, raw video must encode hundreds of thousands of 24-bit color samples. Even though the number of times a second the set of pixels, the picture known as a *frame* is far less—the size of a frame quickly dominates. On standard video, the picture (and thus pixel color intensities) changes up to 30 times a second, far less often than the 8000 times a second that the voice intensity changes. Multiplying it out, a raw voice stream requires 128,000 bits a second; a raw 640×480 video at 30 frames per second requires 221,184,000 bits per second for just the video portion, not including any associated audio streams.

9.1.2 Video Compression

The large size of video clearly begs for compression. Fortunately, much of the detail in video is wasted on the viewer, and video has a tremendous amount of room for lossy compression to be employed. Video compression is an area of active research, but the basics are easy to understand, and are used in modern compression algorithms, such as those for *MPEG* video.

9.1.2.1 Still Image Compression

The first area we can look to compress video is with the representation of color. Many of you who may read this book may also remember how color digital displays evolved, and thus would have a natural understanding of how excessive 16,777,216 possible colors per pixel once seemed. However, it does bear repeating. The eye can be challenged, by certain color transitions, or changes of color from one to the next, to need all 24 bits to capture most of the range of differences the eye can see. But, ignoring those specific, challenging situations, the eye can really see only a few thousand colors. Furthermore, if exact color reproduction is not the point—and it is not, for video—making minor approximations here or there is quite acceptable. Therefore, the first technique for video compression is to radically reduce the number of colors, following the usual media compression technique of focusing the bits to where they are most perceivable by the human observer, and then filling in the details with bits that may not be kept or can be afforded to be lost.

Although the red, green, and blue representation works quite well for representing what the video display needs to do, it is not the most obvious choice for representing human-perceived color. We can take a hint from the development of analog television. Intensity—the overall intensity of the pixel—matters the most. Thus, black and white works quite well for representing the subject of the video, so long as people are not dressed with garish colors that would go missed. Because intensity is so important, representing the intensity is the best use of the bits of information for a pixel. Once the black-and-white intensity is known, the remaining bits can be used to give the shade of pixel a tint, or hue. Three primary colors mean three additional hue values, right? Thankfully not. If we think of color as a mathematical vector of three dimensions, knowing the intensity is like knowing the length. We only need to know how far the color extends into two of the three dimensions, along with the length, to get the intended color back out. (This is embedded in the very definition of having three dimensions to color.) The two choices that television designers settled on was to record both how red and how blue the color is, leaving green to be inferred. It starts with an intensity, also known as a *luminance* in this representation, such as highest for white and lowest for black. The tinting, using the two *chrominances*, each add or subtract an amount of the primary color they represent. Together, the luminance and two chrominances create another 24-bit value, often represented by the abbreviation YC_bC_r (Y is the standard symbol for luminance, and C is for chrominance). White can be represented, in percentages, as 100% Y, 0% C_b, and 0% C_r, for (100, 0, 0). Black is (0, 0, 0), and middle gray is (50, 0, 0). From white, we can get cyan, which has full blue and green but no red. This requires subtracting off the full measure of red. Therefore, a nearly pure cyan is (100, −100, 0). Similarly, a nearly pure green is (100, −100, −100), removing both the red and blue components. The qualification of *nearly* is used only because the standard weightings for the YC_bC_r space require a bit of tweaking to get to the pure RGB versions of the color. (YP_bP_r, the name for component video cables such as used with HD televisions, refers to the same concept, but for analog signals.)

Given that the eye is most particular to the precision of the intensity, and less so to the precision of the tinting, video compression will commonly change the ratio of information rate for each. Specifically, the video compression can require that some of the information be held the same across multiple pixels. By halving the amount of information for the chrominance, the *4:2:2* encoding stores twice as much information to luminance than to the blue and the red chrominances, by requiring that the chrominances change only every other pixel horizontally. Simply halving the amount of information in one pixel dedicated to chrominance would result in saving one-third of the bits. The sacrifice is in color fidelity when the color changes, but for video—and often for still images as well—this may not matter, as the eye considered to be roughly have as sensitive to the tint as it is to brightness. There is also a *4:2:0* "ratio," which is actually more common. 4:2:0 is a special term that means making squares of two pixels by two pixels share the same chrominance.

This sort of color compression falls into the category of *quantization* compression, where the goal is to represent either a continuous or more precise value with a less precise quantized value. Quantization compression was also used in voice, for the two logarithmic encoders in G.711. That compression cut the number of bits in half, but it was smarter than just dividing the signal by 256, as the quantization steps between two encoded values do not have to be even, and logarithmic encoding concentrated the slices more towards the smaller signal values. 4:2:2 compression for video concentrates the bits more to luminance.

Even with reducing the chrominance, video compression will also quantize the luminance, based on the range and precision that it actually achieves over the area being quantized. This is a fundamental part of encoding still images and moving videos, as well. Because the range of intensities within an image varies from part to part, this sort of quantization is best done in small chunks, or regions, of the image. Once quantized, the encoding must have the particular parameters used to make the quantization happen.

As a general rule, the hard part of media compression is finding a representation that has the bits shuffled and applied to categories that matter *differently*—from there, compression can be achieved simply by chopping bits from the categories (high-intensity audio samples, chrominance) that matter the least. This can be clearly seen than with the base for modern image and video compression, as found in *JPEG* and *MPEG* formats. Cutting colors and rescaling to pack in bits where they matter most is one thing. But the designers of JPEG thought of the image differently, looking at its frequency components. Just as audio has the obvious frequency components, representing the set of pitches that are being heard at a given time, video has two dimensions (horizontal and vertical) of frequencies at any given point. The thinking behind JPEG is that the higher frequencies represent the presence of detail, and the lower frequencies represent the slight variations. In audio compression, converting the signal to frequencies is useful because some frequencies are not as important, and producing what amounts to a rank ordering of the most important frequencies at a given time allows the important parts of the signal to be preserved, while the less important parts—such as the faint but often highly detailed noises in the background of a recording—can be erased or approximated more easily. The same applies to video.

(If you are thinking, by now, that color, being comprised of multiple pure tones or frequencies of light, could benefit from being represented this way, then you are on the right path. Color itself, however, happens to not be a good example for being represented and then compressed this way, because the eye already does such a good job removing most of the information out of light by forcing it from an infinite-dimensional space of continuous functions to a three-dimensional space of primary colors, with enough tolerance for approximation.)

One method used to convert an image (or any signal) from space-defined pixels to frequencies is to take the *Fourier transform* of the image. It's easier to think of Fourier transforms first with audio. We know that a sound can be made of one or more pitches. A chord can contain, for example, the four sounds of middle D flat, E flat, F, and A flat (producing a Dbadd2 chord), and those four tones will be the four most important frequencies while the chord is being played. Of course, the instrument or instruments playing the chord each produce a number of both similar and widely different tones around the main tone of the note, and losing those would lose the character of the instrument completely. But, by seeing that some pitches are more important than others, we can begin to see how the pitches can be ranked. The Fourier transform does not do any ranking itself. It is purely mathematical—a change in representation (really, linear basis) from time to frequency and back. The notion is rather simple. Overlap the signal with the one for each pure tone or pitch. The higher the overlap, the more that pitch is present in the signal. We can think of the Fourier transform as testing the signal for overlap with each and every pure tone. Overlap for two signals, or functions, can simply be thought of as the sum over the entire signal of the product, or multiple, of the signal and the "test" tone. If the tone overlaps well with the signal, then each value in the pure test tone will multiply with that of the pure tone embedded in the real signal, and the sum of this will add up to produce a large number. However, if the pure tone is not present in the original signal, then the sums will all go out of step, and the result will be small. Figure 9.1 shows this in action.

If two signals are out of phase, but have the same frequency, then the sum of the product can go to 0, even though there is a match. But using complex numbers (see Chapter 5), the phase can be captured without ambiguity or mistake. It is this process that produces the Fourier transform. For math's sake, we can write it as

$$F(\omega) = \int f(t) e^{-i\omega t} dt$$

where F is the Fourier-transformed representation of f, based on the *angular frequency* ω, which is 2π times the frequency.

What does all this mean? It means that we have a mathematical way of converting a signal to its frequencies. A continuous signal has a continuous (infinite) number of frequencies. But, with the digital world, signals are always finite and discrete. We can shift from the continuous Fourier transform and go to the discrete variant, however. This variant uses the same math, but replaces the integral with a discrete sum. The result of a *discrete Fourier transform* of a signal of a certain number of samples is a new signal with the same number

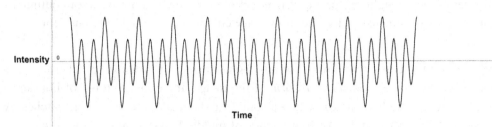

a) Original signal *2 cos(10·2πx) + cos(30·2πx)*, composed of two signals: one at a frequency of 30Hz, and another, half as strong, at a frequency of 10Hz.

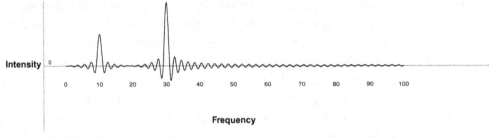

Frequency

b) Fourier Transform, or frequency plot, of the same signal. Noice how there are strong peaks at both 10Hz and 30Hz, with the 30Hz peak having twice the intensity. This is the advantage of the Fourier transform, which pulls out the frequencies present in a signal.

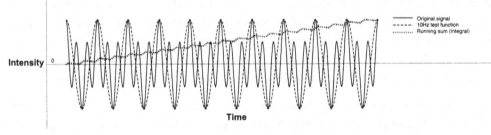

c) Overlap for 10Hz test signal in original signal, which does have a 10Hz component. Notice how the running sum, representing the amount of overlap the test signal has with the original as a running sum of the product of the two signals at each point, steadily increases. This 10Hz test signal is a match, and there will be a peak in the Fourier transform at 10Hz.

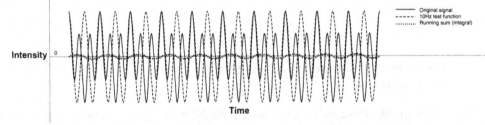

d) Overlap for 20Hz test signal in original signal, which does not have a 20Hz component. The running sum now doesn't steadily increase, from left to right, but instead vacillates around zero, as expected because the test signal now does not overlap with the original signal. This 20Hz test signal is not a match, and there will be no peak in the Fourier transform at 20Hz.

Figure 9.1:   Fourier Transform

of samples, but with the first sample representing the lowest frequency—the larger the sample, the more this frequency is present in the original signal—and so on.

The method now begins to become clear. Take the signal, then the discrete Fourier transform of it. The most important frequencies will have the largest value, and less important frequencies will have smaller values. Assigning more bits to the larger frequencies and less bits to the smaller by quantization will compress the signal. Most of the frequency components will actually get compressed to zero, or completely removed, when compression is successful.

Coming back to video, the thought is the same. The discrete Fourier transform—and its variant, the *discrete cosine transform* (DCT), which works with entirely real numbers (no imaginary numbers)—can work in both the horizontal and vertical directions, to capture the frequencies present in the image. Now, it may not seem that a video image obviously has frequencies. Over the entire image, it probably doesn't have any ones that can be seen intuitively. But the trick is to divide the image up into small rectangles—the size depends on a number of factors, such as how much or little the part of the image in the rectangle varies—and do the frequency transform and subsequent quantization for each rectangle. Now, the benefit for frequency information can begin to make sense. A rectangle whose image barely changes, such as background shading, has very little frequency information, and so can be compressed greatly. But even areas that represent real shapes can be compressed rather well, with a lot of loss but preserving the rough character of that part of the image. As long as each rectangle does not have a lot of irregularity in it, the compression will be good for each rectangle. From here, the compressors try to figure out how to make each rectangle as large as possible for the same bits, looking for areas with lots of similarity.

This adaptive sizing of rectangles, and the rectangles themselves, can be seen fairly easy on compressed images, but usually happens away from the action (intentionally, as we will see). Let's take the example in Figure 9.2.

The lefthand image is the original, and the righthand is with compression set high—as would happen away from the action. Right away, you can see the outlines of the rectangles, in this case, all of the same size, as this is a JPEG image. Most of the squares, such as those for the sky and the solid parts of the building—wherever there is not a lot of detail—got compressed down to no frequency components in any direction: a solid color. But some areas had more than one frequency component that wasn't compressed to zero. If you look at the right side of the tower, where it meets the sky, you will notice that all of the squares seem to have vertical bands, and thus are horizontally smooth. This happens because one frequency component in the horizontal direction got retained, which makes sense for a mostly vertical shape. Squares with more than one component in each direction can be seen where people's heads are.

Figure 9.2:   Compression Artifacts

9.1.2.2 Motion Compression

Once a still frame has been compressed, we can move on to compressing the motion itself. The simple way of doing this would be to just have a sequence of compressed images, one compressed image for each frame. This is actually done in a format called *Motion JPEG*. However, doing so would not take advantage of the fact that most frames are nearly identical, with similar backgrounds but different images of the active subjects as they move around.

We can think of compressing the motion itself by starting out with the first frame, a compressed still image like any other. But, for the following frame, imagine not storing another compressed image. Instead, just store which pixels in the image, corresponding to a moving subject, have moved, and in which direction they have moved. The decoder will just copy those pixels forward, moving them according to the directions the encoder gives. The only new pixels the encoder needs to send are those for the background that got revealed as the moving subject moved away from them. As most objects are pretty solid or large, the pixels in them move together, and so encoding regions of essentially uniform motion is fairly simple and highly efficient. From this comes the dual concept of the *key frame* (or *I frame* for *intra-coded*), the first frame in the sequence with the complete set of pixels, and the *intermediate frames* (or *P frame* for *predictive*), the following frames that only carry the motion and the newly revealed pixels.

This would work just fine, except that viewers like to jump around in videos, or intermediate frames might get lost somewhere. If there were only one key frame, the video would be ruined for any absence of even one bit of information of the intermediate frames. To overcome that, the encoder starts off with a new key frame every so often. This effect, too, is something you can see rather easily. DVDs and digital video recorders use similar compression algorithms, with key frames spaced fairly far apart and a few dozen intermediate frames in between. When you fast forward or rewind the video, you may see these key frames go by, one by one, as still pictures. This is quite different than the old analog VCR days, where the video would just animate faster, and happens because processing the intermediate frames takes too much time for fast forwarding.

Figure 9.3 illustrates the parts of the intermediate frame that are used for added compression.

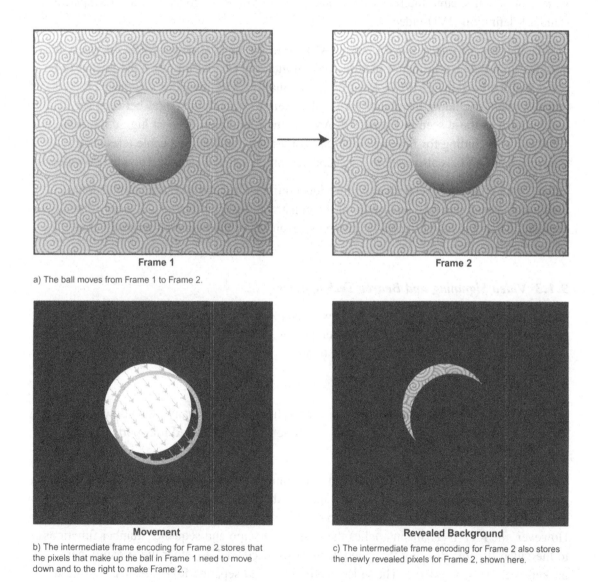

| Frame 1 | Frame 2 |

a) The ball moves from Frame 1 to Frame 2.

Movement

b) The intermediate frame encoding for Frame 2 stores that the pixels that make up the ball in Frame 1 need to move down and to the right to make Frame 2.

Revealed Background

c) The intermediate frame encoding for Frame 2 also stores the newly revealed pixels for Frame 2, shown here.

Figure 9.3: Motion Compression

Video compression leaves much of the intelligence to the compressor: the decompressor is required only to execute the instructions. More intelligent compressors can focus on the subject matter of importance, and the decompressor itself does not need to understand what matters more in the subject in order to do its job.

There are a few types of video codecs, all of them similar to each other, but with some differences in the degree of intelligence and the type of formatting. ITU H.262, the codec used in MPEG-2 video, is the most common codec, used in DVDs, as well as a wide variety of downloadable Internet video content. The bitrate can go as high as 10Mbps for standard-definition DVD video.

ITU H.264, for *Advanced Video Coding* (AVC) and the foundation for MPEG-4 video, was designed to produce a far better picture at a significantly smaller bit rate—the goal is to be about half the bit rate of MPEG-2, for the same quality. AVC includes a number of improvements, including the ability for the decoders to smooth the edges between the blocks, such as those seen in Figure 9.2. AVC is the foundation of most high-definition (HD) video, including for *Blu-ray* video discs and many satellite and cable television transmissions. AVC is also used in YouTube and other Adobe Flash–based video downloads.

Other, often proprietary, codecs exist for videoconferencing and webinar (web seminar) broadcasts, which can take advantage of the constrained subject matter—a series of heads or presentation slides, for example—to compress even better than general-purpose video compressors.

9.1.3 Video Signaling and Bearer Technologies

Video must be carried in much the same way as voice. The video flow or call may need to be set up—this is especially true for conferencing—and then the video stream itself must be transported, along with the related audio streams.

9.1.3.1 Video Bearer

Let's start with the video transport, as the bearer, first. Because many video downloads are streaming, rather than conferencing, both real-time and stream-based transports can be considered.

Real-time video transport is often based on the same RTP mechanism that is used for voice. When transported this way, each of the frames in the video may span multiple RTP packets. The opposite also will happen, and multiple frames may meet in any given RTP packet. However, the RTP mechanism applies the same timestamp and sequence number functions to the video stream, allowing the video decoder to piece back together the stream when packets are lost or reordered. The video sender can send separate RTP streams, sharing the same timestamp clock, for each of the media streams that make up the video. This can be used, for example, to synchronize the one or more audio channels with the one video channel. On the other hand, encoded video itself usually comes with its own frame encapsulation mechanism. The reason for this is that it allows all of the media streams for a video to be kept together. These mechanism allow the embedding of the audio streams into the video stream, even on a real-time basis, and thus provides a higher-layer representation of the order and flow of the video, on top of RTP. Whether the audio streams are combined into the video is as much a function of the device that is sending the video and the signaling application as it is the video encoding.

Because all of the media can be kept together in one stream of bytes, TCP-based streaming for video also works. The disadvantage of using TCP, of course, is that the real-time nature of the stream is broken, as TCP will delay the stream if not every byte can make it through at the moment. On the other hand, TCP-based video streaming is the popular method of transporting video over the Internet, especially when having real-time video is not the requirement, and approximately real-time video is acceptable. This is commonly used for news streaming, where the content needs to be almost live, but it is acceptable for different viewers to see versions of the video with different delays, based on whatever the transport required. Nevertheless, it is important to maintain a distinction between serving video clips and streaming live video. Both may use TCP, but the latter will be rate-limited to the speed of the real-time video, and thus special quality-of-service considerations can be applied that might not be appropriate for the former.

9.1.3.2 Video Signaling

Video signaling depends on the application being used to send the video. However, the requirements for effective video conferencing share a great deal with that of voice calling, and so the concepts learned in previous chapters can be applied. Video conferencing is often performed with H.323, which was designed to simply having multiple users enter a conference. In much the same way, SIP can be used by some applications for point-to-point video calls, or for where there is a video conferencing server that combines multiple videos and provides one flow for each client. Nevertheless, standardization for video conferencing is not as ubiquitous, as many video conferencing platforms have advanced features and codecs that only work with their own products—again, the fact that video is still an area of active research and development makes putting together a video system more complicated than doing the same for voice.

TCP-based video streams usually come from a streaming HTTP-based video server. The signaling protocol, in this case, is the trivial one of the client requesting a URL corresponding to a particular video stream—even a live one—and the server responds with the bits of the video, encoded as if it were a standalone, already-generated video file (such as MPEG-4) that could conceivably even be captured, saved to a file, and played later.

9.1.4 Video Networking

Now that you have seen what goes into a video stream, we can look at how video is transported over real networks—especially mobility networks. As with voice, Wi-Fi will play a prominent role in the delivery of video, especially now that users have become more mobile.

One of the hardest challenges for video networks is identifying the video stream and applying quality-of-service differentiation or protection to it. Unlike voice, which is often marked on a packet-by-packet basis with the correct DSCP tags to designate it and differentiate it from the best effort traffic on the network, video often lurks in best effort traffic streams. For example, TCP-based video is designed to look like best effort, and detecting whether the stream is video may require actively inspecting the traffic, looking for markers in the flow (such as HTTP MIME types designating video media), and then trying to apply the appropriate tagging. TCP applications are notoriously difficult to apply quality-of-service to. For starters, the networking stack may not have the facility for applying DSCP tags to the packets of the TCP stream. Whereas UDP applications can often write whatever DSCP tag they like, because they deal with writing individual packets or datagrams, TCP packets are produced by the underlying operating system. Applications only see a *socket*, or a place to write bytes in the order they should go over the network, and the underlying system produces packets based on the requirements of TCP itself. But even more, applying quality-of-service prioritization to one TCP stream can lead that TCP stream to crowd out the other traffic on the network. Most TCP server applications are not bandwidth-limited, and therefore will provide traffic at whatever rate the client can process. Applying *prioritization* to TCP traffic can then cause that traffic to dominate. When TCP is being prioritized, it is best to apply band shaping or tight throughput bounds, to prevent the TCP connection from racing ahead at a bandwidth too much higher than that the underlying video codec expects. Unfortunately, this is not automatically provided by most routers and network systems—there is no "limit to codec" switch or configuration option—and so setting the bounds may have to be by hand. Some Wi-Fi networks and wireline bandwidth shapers can at least apply the upper limits on a per-flow basis, rather than lumping all flows together into the same bandwidth constraint.

Multicast is another area where video stands out. As mentioned before, both video conferencing and video broadcasting can take advantage of IP-level multicast to greatly reduce the amount of packets that are necessary for real-time video streaming. IP multicast works by the video encoder sending traffic to a particular IP multicast *group address*, which begin with the first octet of 224 to 239. These IP packets are not destined to any one device, so no technology such as ARP is used to determine the next hop Ethernet address. Instead, the multicast traffic is simply sent out on the link. All devices on that layer-2 subnet can receive the traffic. Multicast can cross from one subnet to another when a multicast router sits across the two subnets. Multicast routing with IP requires that the clients that wish to receive the multicast traffic advertise their desires using the *Internet Group Management Protocol* (IGMP), for IP version 4. (The similar *Multicast Listener Discovery* [MLP] protocol exists for IP version 6.) The client sends an IGMP packet to a specific subnet-local multicast address. The IGMP packet contains the real multicast addresses that the client wants to subscribe to. The router listens in on these IGMP messages and collects the list of multicast group addresses on that subnet, joining or leaving the group from its own upstream router as necessary. When a router receives a multicast packet for a group on the upstream link, it repeats it onto every downstream link that has at least one device that is a member of this group. In that way, a tree is built back to the multicast source, covering all of the links that lead to multicast group members. There are specific routing protocols that manage this between routers as needed. Setting up multicast networks requires a fair amount of work, and it would not be appropriate to enter into a discussion on all of the details here. However, it should be clear that multicast does allow the sender to send only one packet, which is then copied efficiently—one and only one per subnet, no matter how many clients are listening in that subnet—across the entire network of interested devices.

Multicast for Wi-Fi, however, has a snag. In wireline Ethernet, a multicast packet takes up the same amount of per-port networking resources as a unicast packet. long as the switches are aware of multicast, and are doing *IGMP snooping*—listening into the IGMP messages and placing multicast packets only on ports that have listening devices for that multicast group on them—multicast is more efficient that unicast. On the other hand, Wi-Fi multicast may be significantly less efficient than unicast. There are a few reasons for this. The first reason is that multicast on Wi-Fi cannot require the clients to acknowledge the receipt, because there are multiple listeners. The multicast packet is sent only once, and devices that do not receive it cannot benefit from the retransmissions that unicast packets offer when they are not received the first time. Furthermore, the multicast packet must go at a very low Wi-Fi physical data rate. Unicast packets on an 802.11n network can go as fast as 300Mbps, but multicast packets from access points must go at the lowest data rate that all of the clients associated to that access point can hear. This can be as low as 1Mbps. No such restriction exists in wireline. Finally, multicast quality-of-service prioritization may not necessarily be used in Wi-Fi networks. The rule is, unfortunately, that if any one client on the access point uses non-WMM for traffic, then none of the multicast traffic can use WMM. This weakest-link restriction makes multicast over wireless a challenging proposition, compared to wireline.

Look to advances in wireless and wireline video technology to occur in the next couple of years, as video takes hold in the enterprise. Many of the issues just mentioned may potentially be solved by then, and video mobility may become a reality.

9.2 Beyond Voice and Video

It is always tough to predict what lies ahead. Mobility technology is not embraced by the majority of organizations simply because it is exciting, or can do amazing things, but only because it solves core problems of productivity that were not addressed before and ended up costing, in terms of time or dollars. Video mobility will likely become useful for the enterprise as applications embrace video. This is especially true in industries such as health care and education, which each have obvious needs for moving video media and can immediately benefit from the network supporting video mobility. But no matter what the application, mobility itself will be the main driver. Wi-Fi started the trend, and other technologies may pick up if Wi-Fi fails to remain in the lead, but the removal of wires as an edge networking technology, and their replacement with wireless radios, seems likely to only intensify in pace and degree, as budgets get tight and IT organizations become forced to justify why they should spend a lot of money running copper to each user, when a wireless signal works better for less money. In terms of applications, these too will be driven by mobility. It is clear that the user's device is shrinking. Mainframes and terminals gave way to desktops, which gave way to portable laptops, and which are themselves giving way to a brand new generation of handhelds. Devices such as the Apple iPhone show that ease of use and a rich feature set—including three-dimensional graphics and a strong audio offering—can entice users to focus their efforts onto just one device. The transition to enterprise-class devices and applications is tough and is still in its infancy, but it would not be unreasonable to expect that the future lies in being able to provide the entire enterprise experience on one device. This is not to say that desktops or laptops are dead, as there are so many things that you cannot do on a small screen: the sheer evidence that televisions started out small and got larger only over time is proof that people do want to see things on a large scale. However, users will demand that their information and services are available, in some familiar, if modified, form, everywhere they go.

I hope you have enjoyed this book and have found it to be useful in explaining the concepts behind voice mobility and has helped point the way for those who have decided to implement such a network.

References

Chapter 2

Internet Engineering Task Force. (2002) RFC 3261. *SIP: Session Initiation Protocol.* http://www.ietf. org/rfc/rfc3261.txt

Internet Engineering Task Force. (2003) RFC 3550. *RTP: A Transport Protocol for Real-Time Applications.* http://www.ietf.org/rfc/rfc3550.txt

Internet Engineering Task Force. (2006) RFC 4566. *SDP: Session Description Protocol.* http://www. ietf.org/rfc/rfc4566.txt

Internet Engineering Task Force. (2006) RFC 4733. *RTP Payload for DTMF Digits, Telephony Tones, and Telephony Signals.* http://www.ietf.org/rfc/rfc4733.txt

International Telecommunication Union. (2006) Recommendation H.323. *Packet-based multimedia communications systems.* http://www.itu.int/rec/T-REC-H.323

International Telecommunication Union. (1998) Recommendation Q.931. *ISDN user-network interface layer 3 specification for basic call control.* http://www.itu.int/rec/T-REC-Q.931

International Telecommunication Union. (1988) Recommendation G.711. *Pulse code modulation (PCM) of voice frequencies.* http://www.itu.int/rec/T-REC-G.711

International Telecommunication Union. (2007) Recommendation G.729. *Coding of speech at 8 kbit/s using conjugate-structure algebraic-code-excited linear prediction (CS-ACELP).* http://www.itu. int/rec/T-REC-G.729

Chapter 3

International Telecommunication Union. (1996) Recommendation P.800. *Methods for subjective determination of transmission quality.* http://www.itu.int/rec/T-REC-P.800

International Telecommunication Union. (2001) Recommendation P.862. *Perceptual evaluation of speech quality (PESQ): An objective method for end-to-end speech quality assessment of narrow-band telephone networks and speech codecs.* http://www.itu.int/rec/T-REC-P.862

International Telecommunication Union. (2008) Recommendation G.107. *The E-model: a computational model for use in transmission planning.* http://www.itu.int/rec/T-REC-G.107

International Telecommunication Union. (2007) Recommendation G.113. *Transmission impairments due to speech processing.* http://www.itu.int/rec/T-REC-G.113

International Telecommunication Union. Recommendation G.711

International Telecommunication Union. Recommendation G.729

Chapter 4

Internet Engineering Task Force. (1981) RFC 791. *Internet Protocol.* http://www.ietf.org/rfc/rfc791. txt

Internet Engineering Task Force. (1998) RFC 2460. *Internet Protocol, Version 6 (IPv6) Specification.* http://www.ietf.org/rfc/rfc2460.txt

Internet Engineering Task Force. (1980) RFC 768. *User Datagram Protocol.* http://www.ietf.org/rfc/rfc768.txt

Internet Engineering Task Force. (1997) RFC 2205. *Resource ReSerVation [sic] Protocol (RSVP)—Version 1 Functional Specification.* http://www.ietf.org/rfc/rfc2205.txt

Internet Engineering Task Force. (1999) RFC 2597. *Assured Forwarding PHB Group.* http://www.ietf.org/rfc/rfc2597.txt

Internet Engineering Task Force. (1999) RFC 2598. *An Expedited Forwarding PHB.* http://www.ietf.org/rfc/rfc2598.txt

Chapter 5

Institute of Electrical and Electronics Engineers (IEEE) Computer Society. (2007) IEEE Std 802.11-2007. IEEE Standard for Information technology—Telecommunications and information exchange between systems—Local and metropolitan area networks—Specific requirements Part 11: Wireless LAN Medium Access control (MAC) and Physical Layer (PHY) Specifications

Institute of Electrical and Electronics Engineers (IEEE) Computer Society. (2007) IEEE P802.11n/D2.00. Draft STANDARD for Information technology—Telecommunications and information exchange between systems—Local and metropolitan area networks—Specific requirements Part 11: Wireless LAN Medium Access control (MAC) and Physical Layer (PHY) Specifications Amendment: Enhancements for Higher Throughput

Paulraj, Arogyaswami, Rohit Nabar, Dhananjay Gore. Introduction to Space-Time Wireless Communications. Cambridge University Press, 2009

Chapter 6

Institute of Electrical and Electronics Engineers (IEEE) Computer Society. (2008) IEEE Std 802.11k-2008. IEEE Standard for Information technology—Telecommunications and information exchange between systems—Local and metropolitan area networks—Specific requirements Part 11: Wireless LAN Medium Access control (MAC) and Physical Layer (PHY) Specifications Amendment 1: Radio resource Measurements of Wireless LANs

Institute of Electrical and Electronics Engineers (IEEE) Computer Society. (2008) IEEE Std 802.11r-2008. IEEE Standard for Information technology—Telecommunications and information exchange between systems—Local and metropolitan area networks—Specific requirements Part 11: Wireless LAN Medium Access control (MAC) and Physical Layer (PHY) Specifications Amendment 2: Radio Fast Basic Service Set (BSS) Transition

Voice over Wireless LAN 4.1 Design Guide, Cisco Systems, 2009. http://www.cisco.com/en/US/docs/solutions/Enterprise/Mobility/vowlan/41dg/vowlan41dg-book.html

Chapter 7

Institute of Electrical and Electronics Engineers (IEEE) Computer Society. (2006) IEEE Std 802.16e-2005. IEEE Standard for Local and metropolitan area networks Part 16: Air Interface for Fixed and Mobile Broadband Wireless Access Systems Amendment 2: Physical and Medium Access Control Layers for Combined Fixed and Mobile Operation in Licensed Bands

Chapter 8

Internet Engineering Task Force. (2000) RFC 2865. *Remote Authentication Dial In User Service (RADIUS).* http://www.ietf.org/rfc/rfc2865.txt

Internet Engineering Task Force. (2004) RFC 3748. *Extensible Authentication Protocol (EAP).* http://www.ietf.org/rfc/rfc3748.txt

Internet Engineering Task Force. (2008) RFC 5246. *The Transport Layer Security (TLS) Protocol Version 1.2.* http://www.ietf.org/rfc/rfc5246.txt

Internet Engineering Task Force. (2006) RFC 4186. *Extensible Authentication Protocol Method for Global System for Mobile Communications (GSM) Subscriber Identity Modules (EAP-SIM).* http://www.ietf.org/rfc/rfc4186.txt

Internet Engineering Task Force. (2005) RFC 4301. *Security Architecture for the Internet Protocol.* http://www.ietf.org/rfc/rfc4301.txt

Internet Engineering Task Force. (1998) RFC 2408. *Internet Security Association and Key Management Protocol (ISAKMP).* http://www.ietf.org/rfc/rfc2408.txt

International Telecommunication Union. (2005) Recommendation X.509. *Information technology—Open Sytems Interconnection—The Directory: Public-key and attribute certificate frameworks.* http://www.itu.int/rec/T-REC-X.509

Chapter 9

International Telecommunication Union. (2000) Recommendation H.262. *Information technology—Generic coding of moving pictures and associated audio information - Video.* http://www.itu.int/rec/T-REC-H.262

International Telecommunication Union. (2007) Recommendation H.264. *Advanced video coding for generic audiovisual services.* http://www.itu.int/rec/T-REC-H.264

Index

Printed in the United States
By Bookmasters